Program
Evaluation
&
Performance
Measurement
An Introduction
to Practice

James C. McDavid
University of Victoria
Laura R. L. Hawthorn

 SAGE Publications
Thousand Oaks ▪ London ▪ New Delhi

For information:

Sage Publications, Inc.
2455 Teller Road
Thousand Oaks, California 91320
E-mail: order@sagepub.com

Sage Publications Ltd.
1 Oliver's Yard
55 City Road
London EC1Y 1SP
United Kingdom

Sage Publications India Pvt. Ltd.
B-42, Panchsheel Enclave
Post Box 4109
New Delhi 110 017 India

Printed in the United States of America

Library of Congress Cataloging-in-Publication Data

McDavid, James C.
Program evaluation and performance measurement:
An introduction to practice / James C. McDavid and Laura R. L. Hawthorn.
        p. cm.
Includes bibliographical references and index.
ISBN 978-1-4129-0668-5
    1. Organizational effectiveness—Measurement. 2. Performance—Measurement.
3. Project management—Evaluation. I. Hawthorn, Laura R. L. II. Title.
HD58.9.M42 2006
658.4'013—dc22   2005008583

This book is printed on acid-free paper.

        07   08   09   8   7   6   5   4   3

Acquisitions Editor:	Lisa Cuevas Shaw
Editorial Assistant:	Karen Wong
Production Editor:	Diane S. Foster
Copy Editor:	Joanne Yang
Typesetter:	C&M Digitals (P) Ltd.
Proofreader:	Kevin Gleason
Indexer:	Teri Greenberg
Cover Designer:	Ravi Balasuriya

Program Evaluation & Performance Measurement

This book is dedicated to the memory of Ronald C. Corbeil, public servant, consummate evaluation professional and contributor to the field.

FIGURES AND TABLES

ACKNOWLEDGMENTS

*P**rogram Evaluation and Performance Measurement* owes much of its development to students in the School of Public Administration at the University of Victoria, British Columbia, Canada. This textbook began as course notes for a graduate seminar in program evaluation, and successive groups of students made key contributions by challenging the ideas of the course instructor, demanding relevance to their experiences as mid-career practitioners and as co-operative education students, and inspiring ways of presenting complex ideas in an understandable fashion.

An opportunity to develop and teach a distance education course in program evaluation and performance measurement offered a way to build first drafts of many of the chapters in this book. Successive offerings of that course to both undergraduate students and mid-career practitioners from all over the world prompted revisions that took advantage of the comments and suggestions by those students.

The Social Sciences and Humanities Research Council in Canada provided a grant that enabled the senior author to assess how legislators, as the end users of public performance information, use it in their decision making. Findings from that project have been included in this book.

Anonymous reviewers, commissioned by Sage Publications at several points in the process, have provided invaluable comments and suggestions. This textbook has benefited enormously from their reading of drafts of the chapters, and we will always be indebted to their willingness to provide this service as evaluation professionals.

Members of the evaluation community in Canada and the United States have heard key parts of this book presented as papers and commentary on evaluation theory and practice. We consider ourselves fortunate to be members of this community of scholars and practitioners.

We thank the editorial staff at Sage for encouraging us at a point when we thought that our idea for this textbook was not feasible. In particular, Lisa

Cuevas Shaw, Katja Fried, and Karen Wong have all been supportive, understanding about the demands on our time that delayed completing chapters of the book, and unfailingly responsive to our questions and suggestions.

And finally, we would like to thank Irene Huse for her contributions to this book. In addition to being a co-author of two chapters, she provided research and editorial support as we prepared the book for publication. She is a principal reason that we succeeded in getting this book done.

PREFACE

*P*rogram *Evaluation and Performance Measurement: An Introduction to Practice* is a textbook that offers readers both a conceptual and practical introduction to program evaluation and performance measurement for public and nonprofit organizations. The conceptual framework for the book is the performance management cycle in organizations, which includes strategic planning, program and policy design, implementation, evaluation, reporting and utilization of results to adjust strategic objectives.

Program evaluation and performance measurement are presented as integrated and complementary activities in public and nonprofit organizations. Skills that are essential for building program evaluation expertise are also important foundations for performance measurement.

From the outset of the textbook, the key role played by professional judgment in the work that we do as evaluators is highlighted. Ways that professional judgment is involved in different phases of program evaluation work are made explicit in key chapters. The final chapter in the book is devoted to discussing the nature and practice of sound professional judgment, as well as ways that evaluation professionals can develop their professional judgment.

With its conceptual framework rooted in the performance management cycle, the book is suitable for managers in public and nonprofit organizations who either have responsibilities for conducting program evaluations and developing and implementing performance measurement systems or are supervising those who are responsible for these functions.

Program Evaluation and Performance Measurement discusses the realities of organizational politics and incentives for managers and evaluators. Although exemplars of good evaluation practice are highlighted, the book emphasizes the effects that resource and other organizational constraints have on both the design and the conduct of evaluations.

As a textbook, it will be useful for senior undergraduate or introductory graduate courses in program evaluation, performance measurement, and performance management. The book does not assume a thorough understanding of research methods and design, instead guiding the reader through a systematic introduction to these topics. Nor does the book assume knowledge of statistics, although there are some sections that do outline the role that statistics play in evaluations. These characteristics make the book well suited for students in professional fields such as public administration and management, where research methods may not be a central focus, as well as for practicing public and nonprofit managers who are not research specialists.

CHAPTER 1

KEY CONCEPTS AND ISSUES IN PROGRAM EVALUATION AND PERFORMANCE MEASUREMENT

INTRODUCTION ●

Program evaluation is a rich and varied combination of theory and practice. It is widely used in public, nonprofit, and private sector organizations to create information for planning, designing, implementing, and assessing the results of our efforts to address and solve problems using policies and programs. **Evaluation** can be viewed as a structured process that creates and synthesizes information intended to reduce the level of uncertainty for stakeholders about a given program or policy. It is intended to answer questions or test hypotheses, the results of which are then incorporated into the information bases used by those who have a stake in the program or policy.

This book will introduce a broad range of evaluation approaches and practices, reflecting the richness of the field. An important, but not exclusive, theme of this textbook is evaluating the effectiveness of programs and policies, that is, constructing ways of providing defensible information to stakeholders as they assess whether and how a program accomplished its intended outcomes.

As you read this chapter, you will notice words and phrases in bold. These bolded terms are defined in a glossary at the end of the book. These terms are intended to be your reference guides as you learn or review the language of evaluation. Because this chapter is introductory, it is also appropriate to define a number of terms in the text that will help you get some sense of the "lay of the land" in the field of evaluation.

The richness of the evaluation field is reflected in the diversity of its methods. At one end of the spectrum, students of evaluation will encounter **randomized experiments** where people have been randomly assigned to a group that receives a program that is being evaluated, and others have been randomly assigned to a control group that does not get the program. Comparisons of the two groups are usually intended to estimate the incremental effects of programs. Although these are relatively rare in the practice of program evaluation and there is some controversy around making them the **benchmark** for sound evaluations, they are still often considered as exemplars of "good" evaluations (Scriven, in progress, unpublished).

More frequently, program evaluators do not have the resources, time, or control over program design or implementation situations to conduct experiments. In some cases, an experimental design may not even be the most appropriate for the evaluation at hand. A typical scenario is to be asked to evaluate a program that has already been implemented, with no real ways to create **control groups** and usually no baseline (preprogram) data to construct before-after comparisons. Often, measurement of program outcomes is challenging—there may be no data available and resources available to collect information are scarce.

Alternatively, data may exist (program records would be a typical situation) but closer scrutiny of these data indicate that they measure program characteristics that only partly overlap with the key questions that need to be addressed in the evaluation. Using these data can raise substantial questions about their **validity**.

Integrating Program Evaluation and Performance Measurement

Evaluation as a field has been transformed in the last 15 years by the broad-based movement in public and nonprofit organizations to construct and implement systems that measure program and organizational performance. Often, governments or boards of directors have embraced the idea that increased accountability is a good thing and have mandated performance measurement. Measuring performance is often accompanied by requirements to report performance results for programs.

Performance measurement is controversial among evaluation experts—some advocate that the profession embrace performance measurement (Bernstein, 1999) while others are skeptical (Perrin, 1998). A skeptic's view of the performance measurement enterprise might characterize performance measurement this way:

Performance measurement is not really a part of the evaluation field. It is a tool that managers (not evaluators) use. Unlike program evaluation, which can call upon a substantial methodological repertoire and requires the expertise of professional evaluators, performance measurement is straightforward: program **objectives** and corresponding outcomes are identified; measures are found to track outcomes, and data are gathered which permit managers to monitor program performance. Because managers are usually expected to play a key role in measuring and reporting performance, performance measurement is really just an aspect of organizational management.

This skeptic's view has been exaggerated to make the point that some evaluators would not see a place for performance measurement in a textbook on program evaluation. But this textbook shows how sound performance measurement, regardless of who does it, depends on an understanding of program evaluation principles and practices. Evaluators who are involved in developing and implementing performance measurement systems for programs or organizations typically encounter similar problems to program evaluators. A scarcity of resources often means that key program outcomes that require specific data collection efforts are either not measured or are measured with data that may or may not be intended for that purpose. Questions of the validity of **performance measures** are important, as are the limitations to the uses of performance data.

Rather than seeing performance measurement as a quasi-independent enterprise, we *integrate* performance measurement into evaluation by grounding it in the same tools and methods that are essential to assess program processes and effectiveness. Thus, program **logic models** (Chapter 2), **research design** (Chapter 3), and **measurement** (Chapter 4) are important for both **program evaluation** and performance measurement. After laying the foundations for program evaluation, we turn to performance measurement as an outgrowth of our understanding of program evaluation (Chapters 8, 9, and 10).

We see performance measurement approaches as complementary to program evaluation, and not a replacement for evaluations. Analysts in the evaluation field (Newcomer, 1997; Mayne, 2001) have generally recognized this complementarity, but in some jurisdictions, efforts to embrace performance measurement have eclipsed program evaluation (McDavid, 2001). We see an important need to balance these two approaches, and our approach in this textbook is to show how they can be combined.

Connecting Evaluation and Performance Management

Both program evaluation and performance measurement are increasingly seen as ways of contributing information that informs **performance management** decisions. Performance management, which is sometimes called **results-based management**, has emerged as an organizational management approach that depends on performance measurement.

Increasingly, there is an expectation that managers will be able to participate in evaluating their own programs, and will also be involved in developing, implementing, and reporting the results of performance measurement. Information from program evaluations and performance measurement systems is expected to play an important role in the way managers manage their programs. Changes to improve program operations and effectiveness are expected to be driven by evidence of how well programs are doing in relation to stated objectives.

Canadian and American governments at the federal, provincial (or state), and local levels are increasingly emphasizing the importance of accountability for **program outcomes**. Central agencies (including the U.S. Federal Office of Management and Budget [OMB] and the General Accounting Office [GAO] and the Canadian Federal Treasury Board of Canada) as well as state and provincial finance departments and auditors are developing policies and articulating expectations that shape the ways program managers are expected to be able to inform their administrative superiors and other stakeholders outside the organization about what they are doing and how well they are doing it.

In the United States, the OMB publishes reports that serve as policies/guidelines for evaluations across federal departments and agencies. An example is the "Program Evaluation" report that outlines criteria for assessing program effectiveness (Office of Management and Budget, 2004). It is worth noting that the OMB's view is that randomized control treatment designs are the best for assessing whether and how programs achieved their intended outcomes. In Canada, publications such as "Evaluation Policy" (Treasury Board of Canada Secretariat, 2001) emphasize the importance of creating, implementing, and using an integrated approach for program evaluation and performance measurement and reporting.

The **performance management cycle** includes an iterative planning–implementation–evaluation program adjustments process in which program evaluation and performance measurement play important roles as ways of providing information to decision makers who are engaged in managing organizations to achieve results.

In this book, we will use the performance management cycle as a framework within which evaluation activities can be situated for managers in public sector and nonprofit organizations. Figure 1.1 shows how organizations integrate strategic planning, program and policy design, implementation and evaluation into a cycle. Although this example is taken from a Canadian jurisdiction (Auditor General of British Columbia and Deputy Ministers' Council, 1996), the terminology and the look of the framework are similar to others that have been adopted by many North American, European, and Australasian jurisdictions.

The five stages in the performance management cycle begin and end with formulating clear objectives for organizations and, hence, programs and policies. "Effective strategies" includes program design, and "aligned management systems" incorporates implementation of programs.

"Performance measurement and reporting" includes both program evaluation and performance measurement, and is expected to contribute to "real consequences" for programs. Among these consequences are a range of possibilities from program adjustments to elections. All can be thought of as parts of the accountability phase of the performance management cycle.

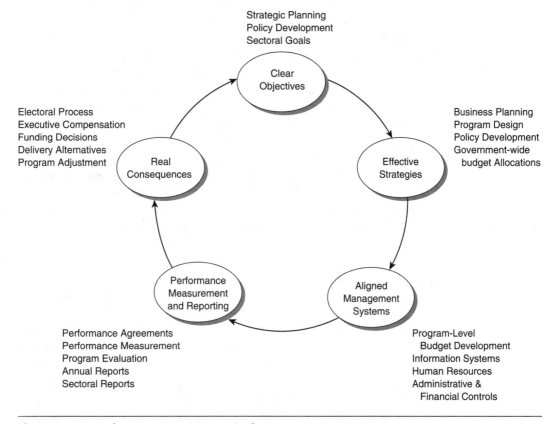

Figure 1.1 Performance Management Cycle

SOURCE: Adapted from Auditor General of British Columbia and Deputy Ministers' Council. April, 1996. *Enhancing Accountability for performance: A framework and an implementation plan.*

Finally, objectives are revisited, and the evidence from earlier phases in the cycle is among the inputs that result in "new" or, at least, revised objectives— usually through a strategic planning process.

In this book the performance management cycle illustrated in Figure 1.1 is used as a framework for organizing different evaluation topics and showing how the analytical approaches covered in key chapters map onto the performance management cycle. Performance measurement and reporting are the main focus of this textbook and include both program evaluation and performance measurement. Chapters 1, 2 (**logic modelling**), 3 (**research designs**), 4 (**measurement),** and 5 (**qualitative methods**) serve as foundations for both program evaluation and performance measurement.

Chapters 8 (introduction to performance measurement), 9 (designing and implementing performance measurement systems), and 10 (using performance measures) elaborate performance measurement.

Needs assessments (Chapter 6) build on topics covered in Chapter 4 (measurement) and can occur in several of the phases in the cycle: setting clear objectives, designing effective strategies, and measuring and reporting performance. As well, **cost-benefit analysis** and **cost-effectiveness analysis** (Chapter 7) build on topics in Chapter 3 (research designs) and can be conducted as we design programs (the effective strategies phase) or as we evaluate their outcomes (the performance measurement and reporting phase).

Finally, the relationships between organizational management and evaluation activities (Chapter 11) are key to understanding how performance management and evaluation are linked. Chapter 12 (the nature and practice of **professional judgment**) emphasizes that the roles of managers and evaluators depend on developing and exercising sound professional judgment.

The Practice of Program Evaluation: The Art and Craft of Fitting Round Pegs Into Square Holes

One of the principles underlying this book, which is referred to repeatedly, is the importance of exercising professional judgment as program evaluations are designed, executed, and acted on. The methodological tools we learn, and the pluses and minuses of applying each in our practice, are often intended for applications that are less constrained in time, money, and other resources than are typical of evaluations. One way to look at the fit between the methods we learn and the situations in which they are applied is to think of trying to fit round pegs into square holes. Even if our pegs fit, they often do not fully meet the criteria specified for their application. As evaluators, we need to learn to adapt the tools we know to the uniqueness of each evaluation setting. In some situations, we find that no approach we know fits the circumstances, so we must improvise.

Our tools are indispensable—they help us to construct useful and defensible evaluations, but like craftspersons or artisans, we ultimately create a structure that combines what our tools can shape with what our own experience, beliefs, values, and expectations furnish and display.

The mix of technique and professional judgment will vary with each evaluation. In some, where causality is a key issue and we have the resources and the control needed to construct an experimental or perhaps **quasi-experimental** research design, we will be able to rely on well-understood methods, which the field of program evaluation shares with social science disciplines. Even here, evaluators exercise professional judgment. There are no program evaluations that can be done without the evaluator's own experiences, values, beliefs, and expectations playing an important role.

In many situations, program evaluators are expected to "make do." We might be asked to conduct an evaluation after the program has been in place for some time, in circumstances where control groups are not feasible, and resource constraints limit the kinds of data we can gather. Or, we are confronted by a situation where the evaluation design that we had developed in consultation with stakeholders is undermined by the implementation process. Fitzgerald and Rasheed (1998) describe an evaluation of a program intended to increase paternal involvement in inner-city families where the father does not share custody of the children. The evaluation design started out as a randomized control and treatment experiment, but quickly evolved in ways that made that design unfeasible.

As we shall see, this kind of situation is not intractable. But it demands from us the exercise of professional judgment, and a self-conscious recognition that whatever conclusions and recommendations we produce, they are flavored by what we, as evaluators, bring to the project. Fitzgerald and Rasheed (1998) salvaged the evaluation by including qualitative data collection methods to develop an understanding of how the program actually worked for the participants at the three implementation sites. Although their approach did not meet the standards that they had in mind when they began, they were able to adjust their expectations, take advantage of a **mix of methods** available to them, and produce credible recommendations.

It is tempting, particularly in this latter kind of situation, to conclude that we are not really doing program evaluations, but some other form of "review." Some would argue that real program evaluations are more "pure," and that the absence of some minimum level of methodological sophistication disqualifies what we do from even being considered program evaluation.

But such a stance, although it has some appeal for those who chiefly value methodological sophistication and elegance, is difficult to defend. Drawing some line between "real" and "pseudo" program evaluations is arbitrary. Historically in our profession, there was a time when experimental methods were considered to be the *sine qua non* of evaluations. During the latter part of the 1960s and the first part of the 1970s, experimental methods were applied to evaluating social programs—often with ambiguous conclusions while still being costly (Basilevsky & Hum, 1984).

Now, there is no one dominant view of "correct" evaluation methods. Indeed, qualitative evaluation methods were born out of a strong reaction to the insular and sometimes remote evaluations produced by social experimenters. Qualitative evaluators like Michael Patton (Patton, 1997) eschew much of the methodological armamentarium of their predecessors and point out that if we want our work to be used, we need to conduct evaluations in ways that encourage the users to take ownership of the conclusions and recommendations. The upshot of this diversity in how we define good evaluations is that drawing a line between real and pseudo evaluations presupposes we agree on one continuum of methods—and we simply do not.

The stance taken in this book, and reflected in the contents of the chapters, is that our program evaluation practice is rich and very diverse. Key to understanding *all* evaluation practice is accepting that no matter how sophisticated your designs, measures, and other methods are, you *will* exercise professional judgment in your work. It is true that the nature and the consequences of such judgments will differ from one evaluation to the next, but fundamentally, we are *all* in the same boat. In this book, we will see where professional judgment is exercised in the evaluation process and begin to learn how to make defensible judgments.

Some readers might have concluded by now that we are condoning an "anything goes" attitude. Readers will discover, instead, that in preparing this book, we have taken a "structured" approach to evaluations that relies on understanding the tools that have been developed in the profession and applying them in ways that maximize (within the constraints that exist) the defensibility of what we produce.

Program evaluation clients often expect evaluators to come up with ways of telling whether the program achieved its objectives, despite the difficulties of constructing an evaluation design that meets conventional standards to assess the cause and effect relationships between the program and its outcomes. The following case summary illustrates one way that one program evaluator responded to the challenge of conducting an evaluation with limited resources, while addressing questions that we might assume would require more sophisticated research designs. It also illustrates some of the features of the practice of program evaluation.

● A TYPICAL PROGRAM EVALUATION: ASSESSING THE NEIGHBOURHOOD INTEGRATED SERVICE TEAM PROGRAM

In the summer of 1995, a west coast city implemented a Neighbourhood Integrated Service Team (NIST) **program**. NIST was intended as a way to get major city departments involved in neighborhood-level communications and problem solving. A key objective of the program was to improve cross-department service delivery by making services more responsive to community needs. Related to this objective was a second one: to strengthen partnerships with the community and increase community involvement in problem solving.

The program was a response to concerns that city departments were not doing a good job of coordinating their work, particularly for problems that crossed department responsibilities. The existing "stovepipe" model of service delivery did not work for problems like the "Carolina Street House."

Citizens in the Mount Pleasant area of Vancouver had spent several frustrating years trying to get the city to control a problem house on Carolina Street. Within a 1-year period alone, neighbors noted that the police had attended the house 157 times, while the fire department had been called 43 times. Property use inspectors had attended the house on a regular basis, as had environmental health officers. In total, over a 3-year period, it was estimated that the city had spent more than CAD$300,000 responding to citizen complaints related to this property (Talarico, 1999).

The City Manager's Office reviewed this problem in 1994 and determined that each city department involved had responded appropriately within the scope of its mandate. Where the system had broken down was its failure to facilitate effective communications and collaboration among departments. NIST was intended to address this problem, and deal with situations like Carolina Street before they became expensive and politically embarrassing.

The Neighbourhood Integrated Service Teams were committees of representatives from all eight of the major city departments. The city was divided into 16 neighborhoods, based on historical and city planning criteria, and a NIST committee was formed for each neighborhood.

The committees met on a monthly basis to share information and identify possible problems, and between meetings, members were encouraged to contact their counterparts in other departments as the need arose. With the City Manager's Office initially pushing NIST, the program was implemented within a year of its start date.

Implementation Concerns

Although the program seemed to be the right solution to the problem that had prompted its creation, concern surfaced around how well it was actually working. Existing city departments continued to operate as separate hierarchies, in spite of the NIST committees that had been formed.

In some areas of the city, the committees did not appear to be very active, and committee members expressed frustration at the lack of continued leadership from the City Manager's Office. Although a coordinator had been hired, the position did not carry the authority of a senior manager.

A key concern was whether the program was making a difference: Had service delivery improved and was the community more involved in problem solving? Although the city was receiving inquiries from communities elsewhere about NIST, it could not point to any systematic evidence that the program was achieving its intended objectives.

The Evaluation

In early 1998, the Deputy City Manager commissioned an evaluation of NIST. Since she had been principally responsible for designing and implementing NIST, she wanted an independent view of the program—she would be the client for the evaluation, but the study would be conducted by an independent contractor.

The terms of reference for the evaluation focused in part on whether the program was, in fact, fully implemented: How well were the 16 NIST committees actually working? A related evaluation issue was learning from the experiences of the committees that were working well, so that their practices could be transferred to other committees that needed help.

Although the evaluation did not focus primarily on whether the objectives of the program had been achieved, the Deputy City Manager wanted the contractor to look at this question, as information was being gathered on the strengths and weaknesses of the NIST committees and the work that they did.

The contractor selected to do this evaluation had a limited budget: her time, access to city employees, use of a city vehicle, and an office in city hall. She opted for a qualitative approach to do the study. She would sample and interview persons from four key stakeholder groups: NIST committee members, department staffs, city council, and the community.

She used a combination of individual interviews and focus groups to contact 48 NIST team members, 24 departmental staff (three from each of the eight departments involved in NIST), four members of the city council, and 24 representatives from community groups that were active in city neighborhoods.

Using interview questions that were intended to get at the principal evaluation issues, she recorded (using written notes and, in some cases, tape recordings for focus groups) responses, observations, and her own reflections on the information she was gathering.

Data analysis involved content-analyzing interview notes, identifying common ideas in the material she had recorded, and organizing all the information into themes. Four main categories of themes emerged: areas where the program was reported to be functioning well, areas where there was room for improvement, stakeholder recommendations, and "other" themes. Each of these four areas was subdivided further to assist in the analysis.

Because the evaluation had solicited the views of four key stakeholder groups, the similarities and differences among their views were important. As it turned out, there was a high level of agreement across stakeholders—most identified similar strengths and weaknesses of NIST and offered similar suggestions for making the program work better.

A total of six recommendations came from the evaluation, the key ones focused on ways of better integrating NIST into the city departments. Stakeholders generally felt that although NIST was a good thing and was making a difference, it was not clear how team members were expected to balance their accountability to NIST and to their home departments.

NIST needed to be reflected in department business plans, acknowledging its continued role in city service delivery, and NIST needed stronger leadership to advocate the program within city departments.

Since this evaluation was completed, the Deputy City Manager has been appointed as the City Manager, and her commitment to NIST is reflected in her policies and initiatives. The NIST program has since been recognized for its innovative approach to community service delivery, winning a United Nations Award for Innovation in the Public Service.[1]

In addition, the city is leading a partnership with other levels of government to implement a multi-organizational **strategy** using NIST-like mechanisms to tackle the homelessness, crime, and drug problems in one neighborhood that has been the single most difficult challenge for social service agencies, the police department, and other criminal justice agencies (Bakvis & Juillet, 2004).

Connecting the NIST Evaluation to This Book

The development of this program and its evaluation are typical of many in public and nonprofit organizations. In fact, NIST came into being in response to a politically visible problem in this city—a fairly typical situation when we look at the **program rationale**. When NIST was put into place, the main concern was dealing with the problem of the Carolina Street house and others like it. Little attention was paid to how the program would be evaluated. The evaluation was grounded in specific concerns of a senior manager who wanted answers to questions about NIST that were being raised by key stakeholders. She had a general idea of what the problems were, but wanted an independent evaluation to either confirm them or indicate where and what the real problems were.

The NIST evaluation is typical of many that evaluation practitioners will encounter. It was expected to answer several important questions for the Deputy City Manager, and do so within the guidelines established by the client. The evaluation was constrained by both time and money; it was not possible, for example, to conduct community surveys to complement other lines of data collected. Nor was it possible to compare NIST to other, non-NIST communities. Other noteworthy points are as follows:

- The evaluation relied on **triangulating** evidence from different points of view with respect to the program, and using these perspectives to help answer the questions that motivated the study.
- Data collection and analysis relied on methods that are generally well-understood and are widely used by other program evaluators. In this case, the evaluator relied on qualitative data collection and analysis methods—principally because they were the most appropriate ways to gather credible information that addressed the evaluation questions.
- The recommendations were based on the analysis and conclusions, and were intended to be used to improve the program. There was no "threat" that the evaluation results might be used to cancel the program. In fact, as mentioned, the program has since been recognized internationally for its innovative approach to community problem solving.
- The evaluation, and the circumstances prompting it, are typical. The evaluator operated in a setting where her options were constrained. She developed a methodology that was defensible, given the situation, and produced a report and recommendations that were seen to be credible and useful.
- The evaluator used her own professional judgment throughout the evaluation process. Methods decisions, data collection, interpretation of findings, conclusions, and recommendations were all informed by her judgment. There was no template or formula to design and conduct this evaluation. Instead, there were methodological tools that could be applied by an evaluator who had learned her craft and was prepared to creatively tackle this project.

Each of these (and other) points will be discussed and elaborated in the other chapters of this textbook. Fundamentally, program evaluation is about gathering information that is intended to answer questions that program managers and other stakeholders have about a program. Program evaluations are always affected by organizational and political factors and are a balance between methods and professional judgment.

The NIST evaluation illustrates one example of how evaluations are actually done. Your own experience and practice will offer many additional examples (both positive and otherwise) of how evaluations get done. In this book, we will blend together important methodological concerns—ways of designing and conducting defensible and credible evaluations—with the practical concerns facing evaluators, managers, and other stakeholders as they balance evaluation requirements and organizational realities.

WHAT IS A PROGRAM? ●

A program can be thought of as a group of related activities that is intended to achieve one or several related objectives. Programs are means-ends relationships that are designed and implemented purposively. They can vary a great deal in scale. For example, a nonprofit agency serving seniors in the community might have a volunteer program to make periodic calls to persons who are disabled or otherwise frail and living alone. Alternatively, a department of social services might have an income assistance program serving clients across an entire province or state. Likewise, programs can be structured simply: a training program might just have classroom sessions for its clients; or be complex: an addiction treatment program might have a broad range of activities from public advertising, through intake and treatment, to referral, and follow-up. In Chapter 2, we look at the structure of programs in depth and apply an **open systems approach** to describe and model programs.

KEY CONCEPTS IN PROGRAM EVALUATION ●

One of the key questions that many program evaluations are expected to address can be worded as follows

To what extent, if any, did the program achieve its intended objectives?

Usually, we assume that the program in question is "aimed" at some intended objective(s). Figure 1.2 offers a picture of this expectation.

The program has been depicted in a "box," which serves as a conceptual boundary between the program and the **program environment**. The

Figure 1.2   Linking Programs and Intended Objectives

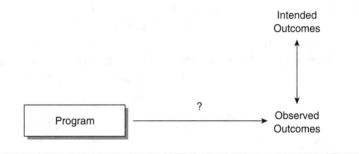

Figure 1.3   The Two Program Effectiveness Questions Involved in Most Evaluations

intended objectives, which we can think of as statements of the **program's intended outcomes**, are shown as occurring *outside* the program itself; that is, the intended outcomes are results intended to make a difference outside of the program itself.

The arrow connecting the program and its intended outcomes is a key part of most program evaluations. It shows that the program is intended to *cause* the outcomes. We can restate the "objectives achievement" question in words that are a central part of most program evaluations.

Was the program effective (in achieving its intended outcomes)?

Assessing **program effectiveness** is the most common reason we conduct program evaluations. We want to know whether, and to what extent, the program's actual results are consistent with the outcomes we expected. In fact, there are *two* evaluation issues related to program effectiveness. Figure 1.3 separates these two issues, so it is clear what each means.

The causal link between the program and its outcomes has been modified in two ways: intended outcomes has been replaced by the **program's observed outcomes** (what we actually observe when we do the evaluation), and a question mark (?) has been placed over the causal arrow.

We need to restate our original question about achieving *intended* objectives to say instead

To what extent, if at all, did the program achieve the observed outcomes?

Notice that we have focused the question on what we *actually observe* in conducting the evaluation, and that the "?" above the causal arrow now raises the key question of whether the program (or possibly something else) caused the outcomes we observe. In other words, we have introduced the **attribution** question, that is, the extent to which *our program* was the *cause* of the outcomes we observed in doing the evaluation. Alternatively, were there factors in the *environment* of the program that caused the observed outcomes?

In Chapter 3 we examine the attribution question in some depth and refer to it repeatedly throughout this book. As we will see, it is often challenging to address this question convincingly, given the constraints within which program evaluators work.

Figure 1.3 also raises a second evaluation question.

To what extent, if at all, are the observed outcomes consistent with the intended outcomes?

Here, we are comparing what we find to what the program was expected to accomplish. Notice that answering that question *does not* tell us whether *the program* was responsible for the *observed* or *intended* outcomes.

Sometimes, evaluators or persons in organizations doing performance measurement do not distinguish the attribution question from the "objectives achievement" question. In implementing performance measures, for example, managers or analysts spend a lot of effort developing measures of intended outcomes. When performance data are analyzed, the key issue is often whether the measures are consistent with intended outcomes. In other words, do the patterns in observed outcomes correspond to the trends or levels that were predicted for that program for that year (or whatever timeframe is specified)? If benchmarks were specified, did the observed outcomes meet the benchmarks? The attribution question (Were the observed outcomes caused by the program?) is not usually addressed, although some analysts with experience in both program evaluation and performance measurement are beginning to look at this issue (Mayne, 2001).

KEY EVALUATION QUESTIONS ●

The previous discussion focused on one of the key questions that program evaluations are expected to answer, namely, whether the program was successful in achieving its intended outcomes. Aside from the question of program effectiveness, there are a number of other questions that evaluations can address. They are listed below, together with a short explanation of each question.

Was the program efficient? This question needs to be unpacked—it is really two questions expressed in the same words, since it relates to both **technical efficiency** and **economic efficiency**. The question of whether the program was technically efficient refers to whether the ratios of program inputs (usually money) to program outputs (work done) were acceptable in comparison to specified **benchmarks**. Typically, evaluations that focus on **technical efficiency** or **productivity** are concerned with costs per unit of work done or service delivered (an example might be the cost per client served in a meals-on-wheels program). Sometimes, externally developed benchmarks are available to compare observed efficiency ratios to established "standards." Alternatively, technical efficiency ratios might be compared over time for one program, creating a program baseline against which future ratios could be compared.

In research conducted by the Local Government Institute in the School of Public Administration at the University of Victoria, the technical efficiency of local government services is measured and compared using national surveys. An example of this research is comparing the cost per household served for residential solid waste collectors in Canadian local governments (McDavid, 2001).

Economic efficiency, discussed in Chapter 7, focuses on the benefits of a program compared to its economic costs. Costs and benefits are expressed in monetary terms, and a number of issues having to do with attribution, measurement, and valuing benefits and costs need to be addressed if cost-benefit analysis is used.

Was the program cost-effective? We examine this question, as well, in Chapter 7. Briefly, **cost-effectiveness** is determined by comparing program costs to program outcomes. Ratios of cost per unit of outcome are among the features of this approach to evaluating programs. An example of a cost-effectiveness ratio in a burglary reduction program would be the cost per percent of burglaries reduced in a neighborhood or a community.

Was the program appropriate? The question addresses the technologies and structures incorporated into a program. Was the program constructed logically, that is, did the structure "make sense," given the intended outcomes? Given the circumstances, did the program designers (often, the program managers) do the best job possible in selecting and organizing the "knowledge base" available to achieve the intended outcomes?

What is the rationale for the program? Given the intended objectives of the program, is the program still relevant to the mission, goals, and objectives of the government or agency in which it is embedded? How well does the program "fit" current and emerging priorities and policies? Questions about program rationale can sometimes relate to needs assessments. Asking whether a program is still relevant can be answered in part by conducting a needs assessment.

Was the program adequate? Given the scale of the problem or condition that the program was expected to address, is the program large enough to do the job? Program adequacy is a question of the "fit" between program resources (and, hence, outcomes) and the scale of the problem embedded in or implicit in the **program objectives**.

Effective and efficient programs may or may not be adequate. An example might be a healthy babies program operating out of a neighborhood family center that does a good job of improving the birth weights and post-natal health of newborns in that neighborhood, but does not visibly affect the overall rate of underweight births in the whole city.

Anticipating the adequacy of a program is also connected with assessing the *need* for a program: Is there a (continuing/growing/diminishing) need for a program? *Needs assessments* are an important part of the program management cycle, and although they present methodological challenges, can be very useful in planning or revising programs. We discuss needs assessments in Chapter 6.

Figure 1.4 offers a visual model and summary of key evaluation concepts (Nagarajan & Vanheukelen, 1997). The model relies on an open systems model of a program: Program inputs are converted to program activities/processes that in turn produce outcomes. The program operates in an environment into which outcomes are delivered, and in turn, offers opportunities or constraints with which the program must work. In this figure we show how objectives drive the program process: they are connected with actual outcomes via *Effectiveness (2):* Are the observed outcomes consistent with the intended objectives? *Effectiveness (1):* Is our attribution question: Did the program cause the observed outcomes? *Cost-effectiveness* connects inputs and actual outcomes, and *cost-benefit analysis* connects the economic value of the benefits to the economic value of the costs. *Adequacy* connects actual outcome to the needs: Were the program outcomes sufficient to meet the need? *Relevance* connects need with objectives: Are the program objectives addressing the need that motivated the program? Although *appropriateness* is not included in the model, it focuses on the logic of the program—the connections among inputs, activities, outputs, and outcomes.

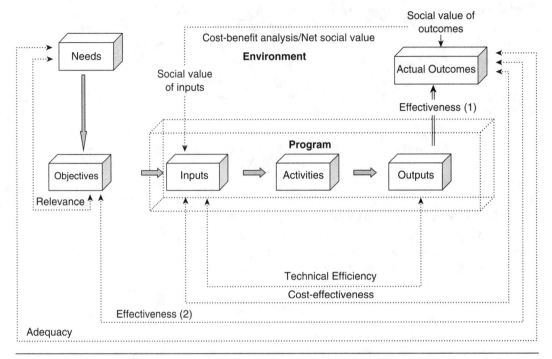

Figure 1.4   An Open Systems Model of Programs and Key Evaluation Issues

SOURCE: Adapted from Nagarajan, N. & Vanheukelen, M. (1997). *Evaluating EU expenditure programmes: A guide* (p. 25).

Each program evaluation will also generate questions that reflect the interests and concerns of the stakeholders involved in particular evaluation processes. These questions may be unique, prompted by circumstances that need to be addressed for that particular evaluation.

Another evaluation question is implicit in most program evaluations: *How well (if at all) was the program implemented?* Program implementation is often assumed as other evaluation questions are addressed, but it is obvious that unless there is evidence of the program having been implemented, it is not meaningful to ask most other evaluation questions (Weiss, 1998).

Assessing program implementation is sometimes done in the first stages of an evaluation process, when considering evaluation questions, clarifying the program objectives, understanding the program structure, and putting together a history of the program. Where programs are "new" (say, 2 years old or less), it is quite possible that gaps will emerge between descriptions of intended program activities and what is *actually* getting done. Indeed, if the gaps are substantial, a program evaluator may elect to recommend an analysis that focuses on implementation issues, setting aside other results-focused questions for a future time.

In Chapter 2, we look at program logic models that include implementation objectives. By distinguishing between program objectives (what the program is intended to accomplish) and implementation objectives (what has to happen to get the program "on the ground" so that it can produce outputs), we can examine how well the program has been implemented.

FORMATIVE AND SUMMATIVE PROGRAM EVALUATIONS ●

Michael Scriven in 1967 introduced the distinction between formative and summative evaluations (Weiss, 1998). Scriven's definitions reflected his distinction between implementation issues and program effectiveness. Scriven associated **formative evaluations** primarily with analysis of program implementation, with a view to providing program managers and other stakeholders with advice intended to improve the program "on the ground." For Scriven, **summative evaluations** dealt with whether the program had achieved intended objectives.

Although Scriven's distinction between formative and summative evaluations has become a part of any evaluator's vocabulary, it does not generally reflect the way program evaluation is practiced. In program evaluation practice, it is common to see terms of reference that include questions about how well the program was implemented; how (technically) efficient the program was; and how effective the program was. A focus on **program processes** is combined with concerns about whether the program was achieving its intended objectives.

In this book, formative and summative evaluations will be defined in terms of their *intended uses.* This is similar to the distinction offered in Weiss (1998). Formative evaluations are intended to provide feedback and advice with the intention of *improving* the program. Formative evaluations in this book *include* those that examine program effectiveness, but are intended to offer advice aimed at improving the effectiveness of the program. One can think of formative evaluations as manager-focused evaluations, wherein the existence of the program is not questioned.

Summative evaluations are intended to ask "tough questions": Should we be spending less money on this program; should we be reallocating the money to other uses; or should the program continue to operate? Summative evaluations focus on the "bottom line" with issues of value for money (costs in relation to observed outcomes) as alternative analytical approaches.

Eleanor Chelimsky (1997) makes a similar distinction between the two primary types of evaluation, which she calls evaluation for development (i.e., the provision of evaluative help to strengthen institutions and to improve organizational performance) and evaluation for accountability (i.e., the measurement of results or efficiency to provide information to decision makers). She adds to the discussion a third general purpose for doing evaluations: evaluation for knowledge (i.e., the acquisition of a more profound understanding about the factors underlying public problems and about the "fit" between these factors and the programs designed to address them).

Program evaluations can, of course, strive to meet all these objectives and be both formative (advice/feedback) and summative (future of the program). But program evaluators will discover that wearing a formative "hat" is a very different experience from being expected to conduct a summative evaluation. Understandably, program manager reactions to these two types of evaluation are quite different. We examine these differences in Chapter 11 (management and evaluation).

● *EX ANTE* AND *EX POST* PROGRAM EVALUATIONS

Typically, program evaluators are expected to conduct evaluations of ongoing programs. Usually, the program has been in place for some time, and the evaluator's tasks include assessing the program up to the present and offering advice for the future. These ***ex post* evaluations** are challenging: they necessitate relying on information sources that may or may not be ideal for the evaluation questions at hand. Rarely are baselines or comparison groups available, and if they are, they are only roughly appropriate. In Chapter 3 and Chapter 5 we will learn about the research design options and qualitative evaluation alternatives that are available for such situations.

Ex ante (before implementation) program evaluations are less frequent. Cost-benefit analyses can be conducted *ex ante,* to prospectively assess whether a program at the design stage (or one option from among several alternatives) is cost-beneficial. Assumptions about implementation and the existence and timing of outcomes are required to permit such analyses.

In some situations, it may be possible to implement a program in stages, beginning with a pilot project. The pilot can then be evaluated (and compared to the existing "no program" status quo), and the evaluation results used as a kind of *ex ante* evaluation of a broader implementation.

One other possibility is to plan a program so that before it is implemented, **baseline measures** are constructed and appropriate data are gathered. The "before" situation can be documented and included in any future program evaluation or performance measurement system. In Chapter 3, we discuss the strengths and limitations of before-and-after research designs. They clearly offer us an opportunity to assess the incremental impacts of the program, but in environments where there are other factors that could also plausibly account for the observed outcomes, this design, by itself, may not be adequate.

ANALYZING CAUSE AND EFFECT ●
LINKAGES IN PROGRAM EVALUATIONS

The attribution question—*To what extent, if any, did the program cause the observed outcomes?*—is at the heart of most program evaluations. Figure 1.3 depicts the program being "connected" to the observed outcomes, with a question mark above the causal arrow to suggest that this connection is a key subject of most evaluations.

To say that the program caused the observed outcomes is really to assert three different things.

- The program occurred before the observed outcomes.
- The program co-varied with the observed outcomes, that is, when (or where) the program occurred, the outcomes tended to occur.
- There were no plausible rival explanations for the observed outcomes, other than the program.

These are "conditions of causality," which we examine in Chapter 3, and they are a key challenge for program evaluators. Of the three, the last one is most problematic in that program evaluations are often conducted in circumstances where it is not possible to rigorously rule out **rival hypotheses**.

This model of cause and effect relationships is intended to work in situations where we typically do not have a "full" explanation for what we observe in an evaluation. For example, if we were evaluating the effectiveness of a neighborhood watch program in reducing the number of burglaries in a community, we would ideally design an evaluation that would permit us to see what effect the program had on burglaries, controlling for other factors that could also affect burglary levels. We would not expect the program to eliminate burglaries; instead, we would be looking for a reduction that we judged to be significant.

In program evaluations, we almost never expect the implementation of a program to account for all the changes we observe in the outcome in question. The relationship between the program and the observed outcome is probabilistic; there is almost always variance in the outcome variable(s) that is not accounted for by the presence or absence of the program.

Some program evaluations are conducted under conditions where data appropriate for ascertaining or even systematically addressing the attribution question are scarce. In these situations, the evaluator or members of the evaluation team may end up relying, to some extent, on their professional judgment. Indeed, such judgment calls are familiar to program managers, who rely on their own observations, experiences, and interactions to detect patterns and make choices on a daily basis. Scriven (in progress, unpublished) suggests that our capacity to observe and detect **causal relationships** is built into us. We are hard-wired to be able to organize our observations into patterns and detect causal relationships therein.

For evaluators, it may seem "second best" to have to rely on their own judgment, but realistically *all* program evaluations entail a substantial number of judgment calls, even when valid and reliable data and appropriate comparisons are available. As Daniel Krause (1996) has pointed out, "a program evaluation involves human beings and human interactions. This means that explanations will rarely be simple, and interpretations cannot often be conclusive" (p. xviii). Clearly then, systematically gathered evidence is a key part of any good program evaluation, but evaluators need to be prepared for, and accept the responsibility of, exercising professional judgment as they do their work.

● THE PROCESS OF CONDUCTING A PROGRAM EVALUATION

A key assumption in this book is that designing a sound performance measurement system benefits from an understanding of the principles and practices of program evaluation. Although some practitioners appear to take the view that performance measurement can replace program evaluations in organizations, that view overstates the benefits of performance measurement and understates the benefits of program evaluations. The two approaches share a common **goal** of illuminating program processes and outcomes to improve the quality of decisions made about programs and policies. But, as we shall see, they are not interchangeable. In Chapter 8, we compare program evaluation and performance measurement—pointing out the differences between these two evaluation approaches.

Because the textbook is structured so that program evaluation methods are introduced and explained first, it is useful, in this introductory chapter, to outline the sequence of activities in a typical program evaluation. In Chapter 9, we do the same thing for designing and implementing a performance measurement system.

The 15 steps in conducting a program evaluation (10 steps for the evaluability assessment phase and 5 steps for the evaluation study itself) are offered in the spirit of the Checklists Project led by Daniel Stufflebeam and Michael Scriven (The Evaluation Center, 2001).

Scriven (2000) advocates the use of checklists.

> There are many different types of checklist, although they have at least one nondefinitional function in common—that of being a mnemonic device. This function alone makes them useful in evaluation, since the nature of evaluation calls for a systematic approach to determining the merit, worth, etc., of what are often complex entities. Hence, a list of the many components or dimensions of performance of such entities is frequently valuable. (p. 1)

It is important to remember that each program evaluation situation is different and that the "steps" outlined below may not reflect the actual train of events as one designs and conducts a given evaluation. It is possible, for example, that clarifying the purposes of the evaluation may need to be revisited as the evaluator has discussions with program managers and other stakeholders about the structure and the objectives of the program.

General Steps in Conducting a Program Evaluation

Rutman (1984) distinguished between planning for an evaluation and actually conducting the evaluation. The **evaluation assessment** process can be separated from the **evaluation study** itself, so that managers and other stakeholders can see whether the results of the evaluation assessment support a decision to proceed with the evaluation.

Table 1.1 summarizes 10 questions that are important to answer as part of most evaluation assessments. Also are five additional steps that are included in most evaluation studies. Each of the questions and steps is elaborated in the discussion that follows Table 1.1.

The Evaluation Assessment

1. Who are the client(s) for the evaluation?

The general perspective taken in this outline of the evaluation assessment process is that program evaluations are substantially *user-driven.* Like

---

Table 1.1   Summary of Key Questions and Steps in Conducting
                Evaluation Assessments and Evaluation Studies

---

Questions to answer as part of an evaluation assessment:

1. Who are the client(s) for the evaluation?

2. What are the questions and issues driving the evaluation?

3. What resources are available to do the evaluation?

4. What has been done previously?

5. What is the program all about?

6. What kind of environment does the program operate in and how does that affect the comparisons available to an evaluator?

7. Which research design alternatives are desirable and appropriate?

8. What information sources are available/appropriate, given the evaluation issues, the program structure and the environment in which the program operates?

9. Given all the issues raised in points 1–8, which evaluation strategy is least problematical?

10. Should the program evaluation be undertaken?

Steps in conducting an evaluation study:

1. Develop the measures and collect the data.

2. Analyze the data.

3. Write the report.

4. Disseminate the report.

5. Make changes, based on the evaluation.

Michael Patton (1997), who makes **utilization** a key criterion in the design and execution of program evaluations, the view taken here is that the intended users *must* be identified early in the process and *must* be involved in the evaluation assessment. That does not mean that intended users should determine the way the evaluation should be conducted, but their information needs are a key part of the assessment process.

Possible clients of the evaluation include, but are not limited to

- Program managers
- Agency/department executives
- External agencies (including funding agencies)

- Clients of the program
- Political decision makers/members of governing bodies (including boards of directors)
- Community leaders

All program evaluations are political in that selecting what to evaluate, who to report the results to, how to collect the information, even how to interpret the data, are affected by the interests and values of key stakeholders. The evaluation's client(s) will also likely affect how the goals, objectives, activities, and intended outcomes of the program are defined for the purpose of the evaluation (Boulmetis & Dutwin, 2000). Generally, the more diverse the clients for the evaluation results, the more complex the political process that surrounds the evaluation itself. Indeed, as Ian Shaw (2000) comments, "many of the issues in evaluation research are influenced as much, if not more, by political as they are by methodological considerations" (p. 3).

Because of the political nature of program evaluations, an evaluation plan (outlining such items as the purpose of the evaluation, the key evaluation questions, and the intended audience), worked out and agreed to by both the evaluator and the client prior to the start of the evaluation, can be very useful. Owen and Rogers (1999) discuss the development of evaluation plans in some detail. In the absence of such a written plan, they argue, "there is a high likelihood that the remainder of the evaluation effort is likely to be unsatisfactory to all parties" (p. 71), and they suggest the process should take up to 15% of the total evaluation budget.

2. What are the questions and issues driving the evaluation?

Program evaluators, particularly as they are learning their craft, are well advised to seek explicit answers to the following questions:

- Who wants the evaluation done?
- Why do they want it done?
- Are there hidden agendas or covert reasons for wanting the program evaluated?
- What are the main evaluation issues that they want addressed (effectiveness, efficiency, adequacy, appropriateness, rationale, need)?
- Is the evaluation intended to be formative or summative, or both?

Answering these sorts of questions prior to agreeing to conduct an evaluation is essential because, as Owen and Rogers (1999) point out, "there is often a diversity of views among program stakeholders about the purpose of an evaluation. Different interest groups associated with a given program often have different agendas, and it is essential for the evaluator to be aware of these groups and know about their agendas in the negotiation stage" (p. 66).

Given time and resource constraints, an evaluator cannot hope to address all the interests of all program stakeholders within the context of one evaluation. For this reason, the evaluator must reach a firm agreement with the evaluation client(s) about the questions to be answered by the evaluation. This process will often involve working with clients to help them narrow the list of questions they are interested in, a procedure which may necessitate "educating them about the realities of working within a budget, challenging them as to the relative importance of each issue, and identifying those questions which are not amenable to answers through evaluation" (Owen & Rogers, 1999, p. 69).

3. What resources are available to do the evaluation?

A general issue for almost all program evaluations is the scarcity of resources available to design and complete the work. Greater sophistication in evaluation research designs almost always entails larger expenditures of resources. For example, achieving the necessary control over the program and its implementation environment to conduct experimental or quasi-experimental evaluations generally entails modifying existing administrative procedures and perhaps even temporarily changing or suspending policies (to create no-program comparison groups, for example).

Although most resources can be converted to money, it is useful to distinguish among several kinds of resources needed for program evaluations:

- Time
- Human resources, including persons with necessary skills and experience
- Organizational support, including written authorizations for other resources needed to conduct the evaluation
- Money

Agreements reached about all resource requirements should form part of the evaluation plan.

4. What has been done previously?

Program evaluators should take advantage of work that has already been done. In a given situation, there may be previous evaluations or evaluations of similar programs in other settings.

In the field of program evaluation, a sub-discipline has arisen that focuses on the **meta-evaluation** of programs. Meta-evaluations are syntheses of existing studies in a given area, and are intended to summarize what we know about, for example, head start programs.

Although we will not be covering meta-evaluation as a separate topic in this book, readers will find references to this area in other program evaluation texts (see Cook, 1994; Patton, 1997; Rossi, Freeman, & Lipsey, 1999; Shadish, Cook, & Campbell, 2002).

Questions to keep in mind as you review previous work are as follows:

- If there are other evaluations, how comparable are the program(s) to the one you are proposing to evaluate?
- Is there published research that is relevant?
- Are there unpublished assessments or opinions about the program or related programs?
- Who did the previous work, and how credible is it?

5. What is the program all about?

In Chapter 2 we will learn ways of constructing program logic models. Briefly, **program logics** are models of programs that depict the key activities in the program and the flow/conversion of resources to outcomes.

Related questions include:

- What are the program objectives?
- What is the history of the program?
- Is the program growing, remaining stable, or declining?
- What is the structure of the program (the key causal linkages among the main parts of the program)?

Although central agencies, executive managers, and even program managers all claim a commitment to stating clear objectives for programs, the nature of the political processes that create programs often means that objectives are stated in very general terms. This presents a challenge for an evaluator who wants to interpret language, which is often intended for stakeholders whose interests in a program may compete with each other, into words that lend themselves to measurement. In the absence of clearly and consistently stated goals, evaluating a program's effectiveness at achieving its goals becomes extremely difficult, if not impossible (Berk & Rossi, 1999).

Another important issue to consider is whether the program has been implemented and is being administered as the program logic intends. As Berk and Rossi (1999) point out, "questions about how a program is functioning logically precede questions about program impact. An impact assessment is a waste of time unless the intervention is known and understood"

(p. 71). In other words, a program's success or failure at accomplishing its stated objectives may have nothing to do with the design of the program or the theory behind it. If the program has not been implemented the way it was intended, its impact (or lack thereof) may be the result of its administration rather than of its design (Weiss, 1998).

6. What kind of environment does the program operate in and how does that affect the comparisons available to an evaluator?

In this book, we conceptualize programs as open systems. Although we look at what that means in Chapter 2, one implication of an open systems approach is that programs interact with the environments in which they are embedded.

For program evaluators, there are issues that affect the design choices available:

- Have any baseline data been kept?
- How large is the client base for the program?
- Is the organization in which the program is embedded stable or in a period of change?

7. Which research design alternatives are desirable and appropriate?

We discuss research designs in Chapter 3, and a key point underlying that chapter is the fact that *all* program evaluations involve making comparisons. The kinds of comparisons we make depend on the evaluation questions we are addressing *and* on the resources available.

Typically, program evaluations involve multiple research designs—it is unusual for an evaluator to construct a design that relies on, for instance, a single time series alone. An important consideration for practitioners is to know the strengths and weaknesses of different designs so that combinations of designs can be chosen that complement each other.

8. What information sources are available/appropriate, given the evaluation issues, the program structure, and the environment in which the program operates?

In most program evaluations, resources to collect data are quite limited, and many research design options that would be desirable are simply not feasible. Given that, it is important to ask what data are available, and how key constructs in the program logic would be measured, in conjunction with decisions about research designs. Research design considerations can be used as a rationale for prioritizing additional data collection.

Specific questions include the following:

- What data are currently available?
- Are currently available data reliable and complete?
- How can currently available data be used to validly measure constructs in the program logic?
- Are data available that allow us to measure key environmental factors that will affect the program and its outcomes?
- Will it be necessary for the evaluator to collect additional information to measure key constructs in the program logic?
- Given research design considerations, what are the highest priorities for collecting additional data?

In Chapter 4 we discuss the process of measuring program outputs, outcomes, and environmental factors, taking into account validity and reliability issues for different measurement processes.

9. Given all the issues raised in questions 1 to 8, which evaluation strategy is least problematical?

No evaluation design is unassailable. The important thing for evaluators is to be able to understand the underlying logic of causality and the ways we can work with that logic to *anticipate* key criticisms that could be made, and have at least a well thought out verbal response to those criticisms.

Much of the work that we do as evaluators is not going to involve randomized controlled experiments, although some consider them to be the "gold standard" of rigorous social scientific research (see, for example, Lipsey, 2000). Although there is far more diversity in views of what is sound evaluation practice, it can become an issue for a particular evaluation, given the background or interests of persons or organizations who might mount criticisms of your work. The U.S. Office of Management and Budget (2004) guidelines on assessing program effectiveness, for example, emphasize the importance of randomized experimental designs, a view also held by the Council for Excellence in Government (Coalition for Evidence-Based Policy, n.d.). This standard harks back to a time in the 1960s when the evaluation profession was dominated by the belief that social experimentation was key to understanding whether programs worked. Whether randomized experiments should be the "gold standard" for the evaluation profession is still sometimes the subject of vigorous discussion in professional forums (Scriven, in progress, unpublished).

There is value in understanding the canons of rigorous research to be able to proactively acknowledge weaknesses in an evaluation strategy. But ultimately, evaluators must make some hard choices and be prepared to accept the fact that their work can and probably will be criticized.

10. Should the program evaluation be undertaken?

The final question in the evaluation assessment process is whether to proceed with the actual evaluation study. It is possible that after having looked at the mix of evaluation issues, resource constraints, organizational and political issues, research design, and measurement constraints, the program evaluator preparing the assessment recommends that no evaluation be done at this time. Although a rare outcome of the evaluation assessment phase, it does happen, and can save an organization considerable time and effort that probably would not have yielded a credible product.

Choosing the combination of methodologies, and ultimately making a decision whether to go ahead with an evaluation, involves substantial amounts of professional judgment. Evaluator experience is key in being able to negotiate a path that permits designing a credible evaluation project. Evaluator judgment is an essential part of putting together the requirements for a defensible study and making a recommendation to either proceed or not.

The Evaluation Study

Up to this point, we have outlined the planning process for conducting program evaluations. If a decision is made to go ahead with the evaluation, there are several steps common to most program evaluations.

1. Develop the measures and collect the data.

Many program evaluations rely on a mix of existing and study-specific data. Data-collection instruments may need to be designed, pretested, and then implemented. Surveys are a common means of collecting data, and we review key issues involved in doing surveys in Chapter 4.

Relevant questions include:

- Are existing data valid measures of the constructs they are intended to measure?
- Are all the evaluation questions and sub-questions addressed by at least one data-collection effort?
- Are there ways of building in triangulation of data sources, that is, two or more independent measures of a given construct?

2. Analyze the data.

Data analysis can be simple or complex, depending on the evaluation questions, the types of data that address those questions, and the comparisons needed to sort out and reduce the threats from rival hypotheses.

Data analysis can be quantitative (involve working with variables that are represented numerically) or qualitative (involve analysis of words, documents, text, narratives, and other nonnumerical representations of data).

A general rule that should govern all data analysis is to employ the *least* complex method that will fit the situation. One of the features of early evaluations based on models of social experimentation was the reliance on sophisticated, multivariate, statistical tools to analyze program evaluation data. Although that strategy addressed possible criticisms by scholars, it often produced reports that were inaccessible or untrustworthy from a user's perspective. More recently, program evaluators have adopted mixed strategies for analyzing data, which rely on statistical tools where they are necessary, but also incorporate visual/graphic representations of findings.

In this book we will not cover data-analysis methods in great detail. Reference to statistical methods is in Chapter 3 (research designs) and in Chapter 4 (measurement). In Chapter 3, key findings from examples of actual program evaluations are displayed and interpreted. In the appendix to Chapter 3, we summarize basic statistical tools and the conditions under which they are normally used. In Chapter 5 (qualitative evaluation methods), we cover the fundamentals of qualitative data analysis; and in Chapter 6, in connection with needs assessments, we introduce some basics of sampling, and generalizing from sample findings to populations.

3. Write the report.

In preparing an evaluation report, the key part is usually the recommendations that are made. Here, sound professional judgment plays a key role in that recommendations must not only be backed up by evidence, but must also be appropriate, given the organizational and political context. Making recommendations that reflect key evaluation conclusions *and* are feasible is a skill that is among the most valuable that an evaluator can possess.

Although each program evaluation report will have unique requirements, there are some general guidelines that assist in making reports readable, understandable, and useful:

- Rely on visual representations of findings where possible
- Use clear, simple language in the report
- Use more headings and subheadings rather than fewer to generate a table of contents that is complete and explicit
- Prepare a clear, concise executive summary
- Solicit feedback on drafts of the report before finalizing it

4. Disseminate the report.

Program evaluators generally have an obligation to produce a report *and* make one or more presentations of the findings, conclusions, and recommendations to key stakeholders, including the clients of the study.

Program evaluators differ in their views of how much interaction there should be between evaluators and clients, at all stages in the process. One view, articulated by Scriven (1997), is that program evaluators should be very careful about getting involved with their clients—interaction at *any* stage in an evaluation, including postreporting, can compromise their **objectivity**.

Patton (1997) argues that unless program evaluators get involved with their clients, the evaluation is unlikely to be used. We look at this issue in Chapter 11, when we consider whether program evaluations can be objective.

5. Make changes, based on the evaluation.

Program evaluations are one means by which stakeholders acquire information that becomes part of the rationale for changes in the program or the organization in which it operates. Evaluations tend to result in *incremental* changes, if any changes can be attributed to the evaluation. It is quite rare for an evaluation to result in the elimination of a program, even though summative program evaluations are often intended to raise this question (Weiss, 1998).

The following are possible changes based on program evaluations:

- Improving the existing program
- Increasing the size of the program
- Increasing the scope of the program
- Downsizing the program
- Replacing or eliminating the program

● SUMMARY

This book is intended for persons who want to learn the principles and the essentials of the practice of program evaluation and performance measurement. Given the diversity of the field, it is not practical to cover all the approaches and issues that have been raised by scholars and practitioners in the last 30-plus years. Instead, this book adopts a stance with respect to several key issues that continue to be debated in the field.

First, we approach program evaluation and performance measurement as two complementary ways of creating information that is intended to reduce uncertainties for stakeholders who are involved in making decisions about programs or policies. We have structured the textbook so that methods and practices of program evaluation are introduced first and then are adapted to performance measurement—we believe that sound performance measurement practice depends on an understanding of program evaluation essentials.

Second, a key emphasis in this textbook is on assessing the effectiveness of programs, that is, the extent to which a program has accomplished its intended outcomes. Understanding the logic of causes and effects as it is applied to evaluating the effectiveness of programs is assisted by learning key features of experimental and quasi-experimental research designs; we discuss this in Chapter 3.

Third, the nature of evaluation practice is such that all of us who have participated in program evaluations understand the importance of judgment calls. The evaluation process, from the initial step of deciding to proceed with an evaluation assessment to framing and reporting the recommendations, is informed by our own experiences, beliefs, values, and expectations. Methodological tools provide us with ways of disciplining our judgment and rendering key steps in ways that are transparent to others, but many of these tools are designed for social science applications. In many program evaluations, resource constraints usually mean that the tools we apply are not ideal for the situation at hand. Learning some of the ways in which we can cultivate good professional judgment is a principal topic in Chapter 12 (the nature and practice of professional judgment).

Fourth, the importance of program evaluation and performance measurement in contemporary public and nonprofit organizations is related to a broad movement in North America, Europe, and Australasia to manage for results. Performance management depends on having high-quality information about how well program and policies have been implemented and how effectively and efficiently they have performed. Understanding how program evaluation and performance measurement fit into the performance management cycle and how evaluation and management work together in organizations is a theme that runs through this textbook.

DISCUSSION QUESTIONS ●

1. As you are reading Chapter 1, what five ideas about the practice of program evaluation were most important for you? Summarize each idea in a couple of sentences, and keep them so that you can check on your initial impressions of the textbook, as you cover other chapters in the book.

2. Read the entire Table of Contents for this textbook, and based on your own background and experience, what do you anticipate will be the easiest parts of this book for you to understand? Why?

3. Again, having read the Table of Contents, which parts of the book do you think will be most challenging for you to learn? Why?

4. Do you consider yourself to be a "words" person, that is, you are most comfortable with written and spoken language; a "numbers" person, that is, you are most comfortable with numerical ways of understanding and presenting information; or "both," that is, you are equally comfortable with words and numbers?

5. Find a classmate who is willing to discuss Question 4 with you. Find out from each other whether you share a "words," "numbers," or a "both" preference. Ask each other why you seem to have the preferences you do. What is it about your background and experiences that may have influenced you?

6. What do you expect to get out of this textbook for yourself? List four or five goals or objectives for yourself as you work with the contents of this textbook. An example might be, "I want to learn how to conduct evaluations that will get used by decision makers." Keep them so that you can refer to them as you read and work with the contents of the book. If you are using this textbook as part of a course, take your list of goals out at about the half-way point in the course and review them. Are they still relevant, or do they need to be revised? If so, revise them so that you can review them once more as the course ends. For each of your own objectives, how well do you think you have accomplished that objective?

● REFERENCES

Auditor General of British Columbia and Deputy Ministers' Council. (1996). *Enhancing accountability for performance: A framework and an implementation plan—Second joint report*. Victoria: Queen's Printer for British Columbia.

Bakvis, H., & Juillet, L. (2004). *The horizontal challenge: Line departments, central agencies and leadership*. Ottawa: Canada School of Public Service.

Basilevsky, A., & Hum, D. (1984). *Experimental social programs and analytic methods: An evaluation of the U.S. income maintenance projects*. Orlando, FL: Academic Press.

Berk, R. A., & Rossi, P. H. (1999). *Thinking about program evaluation* (2nd ed.). Thousand Oaks, CA: Sage.

Bernstein, D. (1999). Comments on Perrin's effective use and misuse of performance measurement. *American Journal of Evaluation, 20*(1), 85–93.

Boulmetis, J., & Dutwin, P. (2000). *The ABC's of evaluation: Timeless techniques for program and project managers.* San Francisco: Jossey-Bass.

Chelimsky, E. (1997). The coming transformations in evaluation. In E. Chelimsky & W. R. Shadish (Eds.), *Evaluation for the 21st century: a handbook* (pp. 1-26). Thousand Oaks, CA: Sage.

Coalition for Evidence-Based Policy. (n.d.). Home page. Retrieved July 13, 2004, from http://www.excelgov.org/displayContent.asp?Keyword=prppcHomePage

Cook, T. D. (Ed.). (1994). *Meta-analysis for explanation: A casebook* (paper ed.). New York: Russell Sage Foundation.

Fitzgerald, J., & Rasheed, J. M. (1998). Salvaging an evaluation from the swampy lowland. *Evaluation and Program Planning, 21*(2), 199–209.

Krause, D. R. (1996). *Effective program evaluation: An introduction.* Chicago: Nelson-Hall.

Lipsey, M. W. (2000). Method and rationality are not social diseases. *The American Journal of Evaluation, 21*(2), 221–223.

Mayne, J. (2001). Addressing attribution through contribution analysis: Using performance measures sensibly. *Canadian Journal of Program Evaluation, 16*(1), 1–24.

McDavid, J. C. (2001). Solid-waste contracting-out, competition, and bidding practices among Canadian local governments. *Canadian Public Administration, 44*(1), 1–25.

Nagarajan, N., & Vanheukelen, M. (1997). *Evaluating EU expenditure programmes: A guide.* Luxembourg: Office for Official Publications of the European Communities.

Newcomer, K. E. (1997). Using performance measurement to improve public and nonprofit programs. In K. E. Newcomer (Ed.), *New directions for evaluation, 75,* 5–14.

Office of Management and Budget. (2004). *What constitutes strong evidence of a program's effectiveness?* Retrieved July 17, 2004, from http://www.whitehouse.gov/omb/part/2004_program_eval.pdf

Owen, J. M., & Rogers, P. J. (1999). *Program evaluation: Forms and approaches* (international ed.). London, Thousand Oaks: Sage.

Patton, M. Q. (1997). *Utilization-focused evaluation: The new century text* (3rd ed.). Thousand Oaks, CA: Sage.

Perrin, B. (1998). Effective use and misuse of performance measurement. *American Journal of Evaluation, 19*(3), 367–379.

Rossi, P. H., Freeman, H. E., & Lipsey, M. W. (1999). *Evaluation: A systematic approach* (6th ed.). Thousand Oaks, CA: Sage.

Rutman, L. (1984). Introduction. In L. Rutman (Ed.), *Evaluation research methods: A basic guide: Vol. 3. Sage focus editions* (2nd ed., p. 239). Beverly Hills, CA: Sage.

Scriven, M. (1997). Truth and objectivity in evaluation. In E. Chelimsky & W. R. Shadish (Eds.), *Evaluation for the 21st century: A handbook* (pp. 477–500). Thousand Oaks, CA: Sage.

Scriven, M. (2000). *The logic and methodology of checklists.* Retrieved July 22, 2004, from http://www.wmich.edu/evalctr/checklists/papers/logic_methodology.pdf

Scriven, M. (In progress, unpublished). *Causation.* New Zealand: University of Auckland.

Shadish, W. R., Cook, T. D., & Campbell, D. T. (2002). *Experimental and quasi-experimental designs for generalized causal inference.* Boston: Houghton Mifflin.

Shaw, I. (2000). *Evaluating public programmes: Contexts and issues.* Aldershot, UK, Burlington, VT: Ashgate.

Talarico, T. (1999). *An evaluation of the Neighbourhood Integrated Service Team program.* Unpublished Master's thesis, University of Victoria, British Columbia, Canada.

The Evaluation Center. (2001). *The evaluation checklist project.* Retrieved July 13, 2004, from http://www.wmich.edu/evalctr/checklists/

Treasury Board of Canada Secretariat. (2001). *Evaluation Policy.* Retrieved March 24, 2004 from http://www.tbs-sct.gc.ca/pubs_pol/dcgpubs/TBM_161/ep-pe_e.asp

Weiss, C. H. (1998). *Evaluation: Methods for studying programs and policies* (2nd ed.). Upper Saddle River, NJ: Prentice Hall.

● NOTE

1. (http://www.city.vancouver.bc.ca/parks/board/2004/040322/accomplishments_2003.PDF)

CHAPTER 2

UNDERSTANDING AND APPLYING PROGRAM LOGIC MODELS

Evaluating a program or constructing a set of performance measures entails understanding how the program operates and what it is intended to accomplish. This chapter is intended to build an understanding of what **program logic models** are, how to apply them in describing programs, and how constructing logic models facilitates program evaluation and performance measurement. Program logic models are visual representations of programs that show how a program is intended to work, that is, how resources that are available to deliver the program are converted into program activities, and how those activities in turn produce intended results.

Logic models are also featured in our discussions of research designs (Chapter 3), measurement procedures (Chapter 4) and designing and implementing performance measurement systems (Chapter 9). Chapter 2 also features an appendix which includes a written description of a program that is intended to reduce client dependency on social assistance, together with the logic model that depicts the program. Using the program description, readers can construct their own logic model of the program and then check their model against the solution included in Appendix 1.

In Chapter 1, programs were defined as clusters of activities intended to achieve an objective or a related set of objectives. Programs can be thought of as **means-ends relationships**, wherein resources are "consumed" or "converted" into activities that in turn are intended as the means by which objectives are accomplished.

To this point in the textbook, we have represented programs as "boxes" that interact with their environments. **Program activities** in the box produce outputs that "emerge" from the box as outcomes and ideally affect whatever aspect(s) of its environs the program is aimed at, in ways that are intended. This representation of programs is in some ways metaphorical—we are asserting that programs are "box-like" and engage in activities that are directed at issues, social or physical problems, or other conditions we wish to change. The use of metaphors to describe programs is now very common among program evaluators, policy analysts, and managers in the public sector.

Ernest House, in a provocative article, points out that in conceptualizing public sector programs we tend to adopt metaphors that then become the reality we expect to see when we describe and evaluate them. At the time he wrote the article over 20 years ago, one metaphor that had influenced evaluators' view of programs was the idea that programs were forms of industrial production (House, 1984).

An industrial assembly line is highly structured. The work is repetitive, and it is important that operations on the line be repeated systematically.

Resources such as raw materials, components, human and machine energy, and knowledge are combined in specific ways to produce **program outputs**. These outputs become products that can be offered for sale to consumers, or as inputs for other manufacturing processes. Companies that manufacture auto parts, for example, can sell these to consumers or to automobile manufacturers which produce cars and trucks for sale.

More recently, Gareth Morgan (1997) in *Images of Organizations,* has elaborated an **open-systems metaphor** for describing organizations. For Morgan and others, the concept of an open system has its roots in biology and, like a biological system, interacts with its environment. That is, it both gives and receives in relation to its environment. Inputs are converted to activities within the organization (structures perform functions) and the outputs are a key form of interaction between the organization and its environment.

Unlike the earlier industrial-production metaphor with its assembly-line approach, the open-systems metaphor encourages conceptualizing organizations, or programs within them, in dynamic terms—programs "take" from their environments and "give" to their environments, and these linkages can change over time. A key part of the dynamics is feedback: Positive feedback can indicate that the program is doing well, or perhaps should do more of what it is currently doing, and negative feedback indicates the program needs modification. In fact, program evaluations and performance results can be thought of as a part of the feedback that affects programs and organizations. In a well-functioning performance management system, the entire cycle, from developing strategic objectives to using performance information to modify those objectives, can be seen as a system with feedback loops.

When House wrote his article in 1984, the field of program evaluation had reached a dividing point. Proponents of a more qualitative, anthropologically-inspired view of the field were reacting to the applied social science view of the profession, which had dominated up to that time. In the 1960s and into the 1970s, good program evaluations had been expected to exhibit the characteristics of experiments, and data were analyzed quantitatively. House pointed out that the (then) dominant view limited the ways we view and evaluate programs. Evaluations based on experimental designs tended to treat programs as "black boxes." That meant that very little effort was made to understand or model the program process. Instead, the key issue was the linkage between the program as a whole and the intended outcomes.

Twenty years later, the open systems approach has not only prospered, but now dominates our view of public and non-profit programs. We have come to a point where we generally assert that programs *are* open systems—they are designed, operated, and evaluated that way. Indeed, public and nonprofit sector managers are encouraged to see performance management in open-systems terms: strategic planning (which includes **environmental scanning**) develops objectives; program design attaches resources (and activities) to those objectives; **program implementation** emphasizes aligning management systems so that objectives can be achieved; and performance measurement and program evaluation are intended as means of providing feedback—to the managers and to other stakeholders in a network of accountability relationships. The cycle is completed when accountability-related reporting feeds into the next cycle of objective-setting.

Describing programs as open systems has a number of implications.

1. Most programs have virtual boundaries; that is, there is a conceptual or an organizational demarcation between the program and its environment, but that boundary cannot be observed directly. This affects how we "see" programs and how we model them.

2. Programs are purposeful, that is, are human constructions with which we intend to accomplish objectives we value.

3. Programs have structures, structures produce activities, and activities in turn produce results (outputs and outcomes), all of which need to be described and linked.

4. Programs exist in environments, which both afford opportunities and offer constraints to programs and to evaluators.

There are alternative ways of graphically representing programs as open systems. Collectively, these representations are often referred to as *program logic models.* They have in common the fact that all are intended as ways of categorizing program activities and **program outcomes**. They differ in the ways categories are labeled, and in the level of detail in the modeling process itself, specifically, the extent to which the logic models are also intended to be **causal models** that make explicit connections between activities, outputs, and outcomes.

In this chapter, two program logic modelling approaches are introduced and illustrated. We begin with a basic example of a program that has some of the essential features of logic models. The example of the Laurel House program will illustrate how logic models can be used to categorize program processes and intended results. Then we present a more general logic modeling framework that builds on the features of our basic example, and we define its main features. After we describe the framework, we illustrate it with two examples of logic models based on actual programs. All logic models have in common the function of communicating to stakeholders the key parts of programs and the intended relationships among them (Coffman, 1999). In addition, logic models play an important role in performance management. They can be used as a part of the strategic planning process to clarify intended objectives and the program designs that are intended to achieve outcomes (Coffman, 1999). In a review article that compares international applications of logic modeling, Montague (2000) points out that in Canada, Australia, and the United States, logic modeling has become central to performance measurement and reporting systems. Governments that have embraced performance measurement systems generally also invest in logic modeling as a way of presenting their programs and linking them to intended results. In Canada, for example, the federal agency that supports government-wide evaluation activities offers a preparation guide for federal department performance reports which includes a logic model template. The report states:

> In discussing strategic outcomes, departments need to better demonstrate how their program results and resources contribute to those outcomes. To facilitate the provision of this information, we are asking departments to use the [logic model] template provided below. (Treasury Board of Canada Secretariat, 2003, p. 25)

Likewise, in the European Union, logic modeling has been advocated as an important part of sound program evaluation practice (Nagarajan & Vanheukelen, 1997).

● A BASIC LOGIC MODELING APPROACH

Public sector organizations and nonprofit agencies are being encouraged or even exhorted to develop performance measurement systems. A key step in the development of performance measures is to describe the programs in an organization in a way that facilitates developing measures of program activities and outcomes. Program logic models are widely used for this purpose.

One illustration of such a model is the program logic for Laurel House, a nonprofit social service agency in a small west coast city in Canada. The description of the program in Laurel House is as follows:

> Laurel House provides social, recreational, and pre-vocational activities and skills development for over 200 adults with serious mental illness. Some individuals have attended for years, and for them Laurel House is

their main support in continuing community living; others may attend for a much shorter period, as needed, before moving on to more mainstream "activities." About 60 people attend each day, many of whom attend 2–3 times a week and some every day. Laurel House itself is a grand old building ideally located one mile from downtown and one mile from the hospital, with five bus routes less than a block away. The House activities are also carried out in the ground floor of an office building next door, which shares with the House a quiet shady garden. Some activities take place in the general community with Laurel House staff and members participating together (Dinsdale, Cutt, & Murray, 1998, p. 106).

The logic model for Laurel House is shown in Table 2.1. Its major features include categories for inputs, activities, outputs, and three kinds of outcomes: initial, intermediate, and long term. We will briefly discuss what these categories mean for the logic model of Laurel House and more systematically define the categories when we introduce our general framework for constructing logic models.

The activities for Laurel House are stated in general terms, and the model is intended to be a way of translating a verbal description of the program into a model that succinctly depicts the program. The outputs indicate the *work done* and are the immediate results of activities. Outcomes are intended to follow from the outputs and represent the objectives of the program. Logic models generally are displayed so that a time-related sequence is implied in the model. Often, logic models are displayed so that resources occur first, then activities, then outputs. The sequence of outcomes is intended to indicate intended causality. Intermediate outcomes follow from the initial outcomes, and the long-term outcomes are intended as results of intermediate outcomes. Distinguishing between short, intermediate, and long-term outcomes recognizes that not all effects of program activities are discernable immediately on completion of the program. For example, welfare recipients who participate in a program designed to make them long-term members of the workforce may not find employment immediately upon finishing the program. The program, however, may have increased their self-confidence and job-hunting, interviewing, and resume writing skills. These short-term outcomes may, within a year of program completion, lead to the long-term outcome of many participants finding full-time employment. Such situations remind us that some program outcomes need to be measured at one or more follow-up points in time, perhaps 6 months or a year (or more, depending on the intended logic) after program completion (Martin & Kettner, 1996).

Table 2.1   Program Logic Model of Laurel House

Inputs	Activities	Outputs	Outcomes		
			Initial	*Intermediate*	*Long-term*
• Funding: $10 lunch charge Capital Health Region and United Way funding • 179 staff hrs/wk • 25 volunteer hrs/wk	• Various self or group determined activities (in and out of the house) • Learning about their illness • Links to needed services and supports • Goal setting • Activities to enhance peer support networks • Prevocational skills sessions	• Number of clients served • Number of life skill (e.g., cooking and medicine) related sessions • Number of referrals to supports • Duration and frequency of attendance • Number of meals	Members indicate an increased mastery of life They gain a greater understanding and awareness of their illness and how to cope with it. For example: • Improved life skills (cooking and hygiene) • Are more aware of available services and supports Members gain a greater appreciation for the importance of developing a wider social network	Members develop stronger social networks and friendships, and employ coping and life skills	Members indicate an increased quality of life

A logic model like the one in Table 2.1 might be used to develop performance measures—the words and phrases in the model could become the basis for more clearly defined **program constructs**, which, in turn, could be used to develop measures. But the model has a limitation. It does not specify how its various activities are linked to specific outputs, or how particular outputs are connected to initial outcomes. In other words, the model offers us a way to *categorize and describe* program processes and outcomes, but it is of limited use as a *causal model* of the intended program structure.

LOGIC MODELS THAT CATEGORIZE ●
AND SPECIFY INTENDED CAUSAL LINKAGES

Table 2.2 presents a framework for modeling program logics that builds upon the basic logic modeling approach we have introduced so far. We will discuss the framework and then introduce several examples of logic models that have been constructed using the framework. The framework in Table 2.2 does two things. First, it *classifies* the main parts of a typical logic model into inputs, components, implementation objectives, linking constructs, and outcomes. We will define each of these, and provide examples where it is appropriate.

Program inputs are the resources that are required to operate the program—they typically include money, people, equipment, facilities, and knowledge. Program inputs are an important part of logic models. It is possible to monetize all inputs, that is, convert them to equivalent dollar values. Evaluations that compare program costs to outputs (technical efficiency) or program costs to outcomes (cost-effectiveness) or compare costs to monetized value of outcomes (cost-benefit analysis) all require estimates of inputs expressed in dollars. Increasingly, performance measurement systems that are being implemented by governments are connecting program costs to results. Examples of this trend are the PART process in the Office of Management and Budget in the U.S. Federal Government (U.S. General Accounting Office, 2004) and the commitment of the Texas State Government to include performance measurement results with program budgetary requests (Tucker, 2002).

Program components are clusters of activities in the program. They can be administrative units within an organization that is delivering a program. For example, a job training program with three components (intake, skills development, and job placement) might be organized so that there are work groups for each of these three components. Alternatively, it might be that one work group does these three clusters of activities—smaller organizations in the nonprofit sector typically expect employees to become generalists.

Implementation objectives are included for each component of a program logic model. This innovation in modeling program logics was introduced by Rush and Ogborne (1991) and is intended to focus attention on the activities in the program that are required to produce program outputs. Implementation objectives are about getting the program running, that is, getting things done in the program itself that are necessary in order to have an *opportunity* to achieve the intended outcomes. Implementation objectives simply state the work that the program managers and participants need

Table 2.2   A Framework for Modeling Program Logics

		Implementation			Intended Outcomes		
Inputs	*Components*	*Implementation Objectives*	*Outputs*	*Linking Constructs*	*Short-Term*	*Medium-Term*	*Longer-Term*
● Money ● People ● Equipment ● Technology ● Facilities	Major clusters of program activities	● To provide . . . ● To give . . . ● To do . . . ● To make . . .	● Work done ● Program activities completed	Transition factors that connect outputs to outcomes	● Intended by the design of the program ● Outcomes (or impacts) relate to program objectives		
	Program			*Environment*			

to do, *without anticipating intended outcomes*. Typical ways of stating implementation objectives begin with words like: "to provide," "to give," "to do," or "to make." An example of an implementation objective for the intake component of a job training program might be: "To assess the training needs of clients." Another example from the skills development component of the same program might be: "To provide work skills training for clients."

It is important to remember that implementation objectives are *not* the same thing as program objectives. Program objectives are what the program *outcomes* are about. They are statements of the changes that the program is expected to make in its environment, which ameliorate or solve the problem or problems that motivated the creation of the program in the first place.

Successful implementation *does not* assure us that the intended outcomes will be achieved, so it is inappropriate to include outcomes-focused language in our statements of implementation objectives. In our example of an implementation objective for the skills development of the job training program, if we had said, "To provide work skills training for clients that will enhance their employability," we would be stating an intended outcome of the program and mixing together the training activities and the intended results of the training.

If the implementation objectives are not achieved, there is no real point in trying to determine whether the program was efficient (technically efficient) or effective. Implementation objectives express program activities in a way that reminds us that successful program implementation is a prerequisite for achieving outcomes.

It is possible when programs are developed that implementation becomes a major issue. If the staff in an agency or department is already fully committed to existing programs, then a new program initiative may well be slowed down or even held up at the point where implementation should occur. Furthermore, if a new program conflicts with the experience and values of those who will be responsible for implementing it, it may be that ways will be found to informally "stall" the process, perhaps in hopes that a change in organizational direction would effectively cancel the program. In turbulent political environments where elected decision makers are under considerable pressure to resolve problems quickly, there may be an organizational expectation that those who are expected to implement will outlast those who are creating the programs.

For successful implementation of performance measurement systems, resistance to change is a key issue that must be managed. In Chapter 10 we discuss the challenge of sustaining organizational support for performance measurement, particularly as program managers and others who have a stake in the uses of the information adapt to expectations that they will participate in performance measurement and be willing to use performance results to guide program decision making.

Edward Suchman (1967) has distinguished between two kinds of negative evaluation results: program failure and theory failure. The former takes place where implementation does not occur or is inadequate. The latter can only be fully ascertained if implementation has occurred. Testing the theory of a program is really about testing whether the program outputs produce the intended outcomes. Weiss (1998) has developed a theory of change that focuses on what is required to achieve the objectives of a program. A key part of her approach is the implementation theory of the program—the program activities that are necessary to drive the program theory. She includes an example of a job training program wherein the first implementation activity is publicizing the program. This activity, once implemented, makes it possible for youth to hear about the program and indicate an interest in signing up. Another way of looking at this interplay between implementation theory and program theory is that publicizing the program (an implementation objective) leads to youths contacting the program (an output) (Weiss, 1998).

Program outputs are typically ways of representing the amounts of work that is done as the program is implemented. For example, in our job training program, "the number of clients trained" would be an output. Typically, outputs are countable and are the first and often the most tangible results of a program. In our example, we could refine the way we count the work done by including "the total number of hours of training delivered" as an output. If we knew the total cost of delivering the program, we could then estimate the costs per client trained or the costs per training hour delivered. We could also estimate the average number of hours of training per client.

Linking constructs can be thought of as *transitions* from the work done by the program to the intended outcomes. Suppose we have a fire codes inspection program that is intended to reduce the number of fires in multi-family housing units in a community. The program has been designed so that fire codes inspectors go to each apartment building and check for fire codes violations. Building owners are given six months to fix the problems. The buildings are then re-inspected to see how many have been brought into compliance with fire codes. The success of the program depends on build-ings having been brought into compliance, that is, the linking construct in this program logic. If they are not brought into compliance, then the logic "breaks down" at that point, and any actual reduction we might observe in the number of fires would be due to factors other than the program.

Linking constructs were originally introduced by Carol Weiss (1972) as *bridging variables* in her efforts to conceptualize programs as processes which convert inputs into outcomes. She saw bridging variables in much the same way that we see linking constructs:

> The theory of the program posits a sequence of events from input to outcome; in order to reach the desired end, certain sub-goals have to be achieved. In a rehabilitation program for prison inmates, it is assumed that learning a skilled trade will reduce . . . chances of resorting to crime after release. Accordingly, we can measure the extent to which inmates have mastered the skill being taught in the program (as the bridging variable), and then relate skill level to the outcome measure. (Weiss, 1972, p. 48–49)

Not all program logics will have linking constructs. Whether it is appro-priate to include them depends on whether the intended cause and effect linkages in the program logically include one or more steps between the out-puts (the work done) and the intended outcomes. The key to thinking about linking constructs is to know what the program objectives are, know what the outputs are, and then ask whether there are steps in the logic that *need to occur* in order to translate outputs into outcomes.

Program outcomes are the intended results that are linked to program objectives. Typically programs will have several outcomes and it is common for these outcomes to be differentiated by when they are expected to occur. In a program logic of a housing rehabilitation program that is intended to sta-bilize the population (current residents would stay in the neighborhood) by upgrading the physical appearance of dwellings in an inner city neighborhood, we might have a program process that involves owners of houses being offered property tax breaks if they upgrade their buildings. A short-term outcome would be the number of dwellings that have been rehabilitated. That, in turn, would hopefully lead to reduced turnover in the residents in the neighborhood.

The second thing that the framework in Table 2.2 does is distinguish the program inputs, components, implementation objectives, and outputs from the environment in which the program is embedded. One feature of program logics as open systems is that program environments are an integral part of every program logic model. Programs produce outputs that are intended to make a difference in the environment of the program. Clients of programs are a part of the environment, and many social service programs are intended to change/improve/impact clients who have been served by the program.

Environmental factors can enhance the likelihood that a program will succeed—an employment training program that is implemented when a regional call center is opened in a community may make it easier for trainees to find work. Environmental factors can also impede the success of a program—a police crackdown on the distribution network for crack cocaine in a large city may make it more difficult for the police to reduce the numbers of burglaries committed in the community: as cocaine becomes scarce, the price goes up and persons who are addicted might be more likely to resort to theft to supply their habit.

Specifying environmental factors that could affect the outcomes of programs is a step toward anticipating how they actually operate as we are evaluating a program. In Chapter 3 we examine categories of rival hypotheses that can complicate our efforts to examine the intended connections between programs and outcomes.

Figure 2.1 illustrates a logic model for a program that was implemented as an experiment in two Canadian provinces. The Income Self-Sufficiency Project (Michalopoulos, Tattrie, Miller, et al., 2002) was intended to test the hypothesis that welfare recipients (nearly all program and control group participants were single mothers) who are given an incentive to work will choose to do so. The incentive that was offered to program recipients made it possible for them to work full time and still receive some welfare support. The program was implemented in 1993 in British Columbia (on Canada's west coast) and New Brunswick (on Canada's east coast). Welfare recipients from the two provinces were pooled and a random sample of approximately 6000 families was drawn. Each family was approached by the evaluation team and asked if they would be willing to participate in a 3-year trial to see whether an opportunity to earn income without forgoing all welfare benefits would increase labor force participation rates.

Most families agreed to participate in the experiment. They were then randomly assigned to either a program group or a control group, and those in the program group were offered the incentive to work. Each family had up to 12 months to decide to participate—the welfare recipient needed to find full time employment within the first 12 months after they were included in the program group in order to qualify for an income supplement. The income supplement made it possible for recipients to work full time while retaining part of their welfare benefits. Since welfare benefits were determined on a monthly basis, for each month he or she worked full time, a participant would receive a check that boosted take home earnings by approximately 50% on average. If full-time employment ceased at any point, persons would continue to be eligible for welfare income at the same levels as before the experiment began.

Persons in the control group in this study were offered the same employment placement and training services that were available to all income assistance recipients, including those in the treatment group. The only difference between the two groups, then, was the financial incentive to find full-time employment.

The logic model in Figure 2.1 shows a straightforward program. There is one **program component** and one **implementation objective**. In Figure 2.1, there is no separate column for inputs. In this program, the principal input was the financial incentive, and this is reflected in the one program component.

The **program outputs** all focus on ways of counting participation in the program. Families who have participated in the program—that is, have made the transition from depending solely on welfare for their income, to working full time—are expected to experience a series of changes. These changes are outlined in the linking constructs and intended outcomes. The three **linking constructs** identify key results that follow immediately from the program outputs. Once a welfare recipient has decided to opt into the program, he or she searches for full-time paid work (30 hours a week or more), and if successful, gives up his or her monthly welfare payments. Participants are eligible for an income supplement that is linked to continued participation in the program—if they drop out of the program, they give up the income supplement and become eligible for welfare payments again. The three linking constructs summarize these changes: increased full-time employment for program participants; increased cash transfer payments connected with the income supplement incentive; and reduced short-term use of income assistance/welfare.

There are three columns of **program intended outcomes** in the logic model, corresponding to *short-*, *medium-*, and *longer-term* program results. In the first column, each linking construct is connected with an outcome and the short term outcomes are connected with each other. The one-way arrows are intended to convey intended causal linkages among the constructs in the model. Increased stable employment is clearly a key construct—it follows

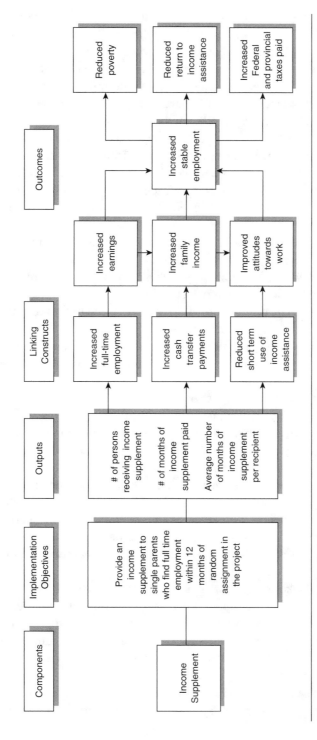

Figure 2.1 Income Self-Sufficiency Program: Logic Model

from the short-term outcomes and it causes the longer-term outcomes in the model. The overall longer-term objectives for the program are depicted as the three longer-term outcomes: reduced poverty, reduced return to income assistance, and increased tax revenues for governments.

If the program actually operates the way it is represented in the logic model in Figure 2.1, it will have achieved its intended outcomes. The logic model shows what is expected. The challenge for the evaluator is to see whether what was expected actually occurred.

The framework we introduced in Table 2.2 is a template that can guide constructing logic models for many different programs. The way the logic model for the Income Self-Sufficiency Program looks is specific to that program. Logic models will generally have the following features:

- Some will have one component, many will have multiple components, depending on the complexity of the program.
- Where a program has multiple components, there will be at least one implementation objective for each component.
- Each component will have at least one output, but there can be several outputs for particular components.
- Whether a logic model has linking constructs depends on how the program operates—linking constructs are what connect the outputs to outcomes.
- Each linking construct needs to be connected with one or more outputs which, in turn, need to be connected with one or more outcomes.
- Each logic model will have its own configuration of outcomes, with some having short-, medium-, and longer-term outcomes, and others having outcomes that all are expected to occur at the same time.
- Each short-term outcome needs to be connected to one or more subsequent outcomes.
- Although there is no requirement that the causal arrows depicted in logic models all have to be one-way arrows as they are in Figure 2.1, using two-way arrows complicates program evaluations considerably.

The key thing to keep in mind is that logic models offer a succinct visual image of how a program is expected to achieve its intended outcomes. This causal model of a program is often of great benefit to program managers and other stakeholders, who may have a good understanding of the program process but have never had the opportunity to construct a visual image of their program.

Figure 2.2 introduces a program logic for an Alcohol and Drug Services Program (Arthur, 1996). The model is more complex than the Income

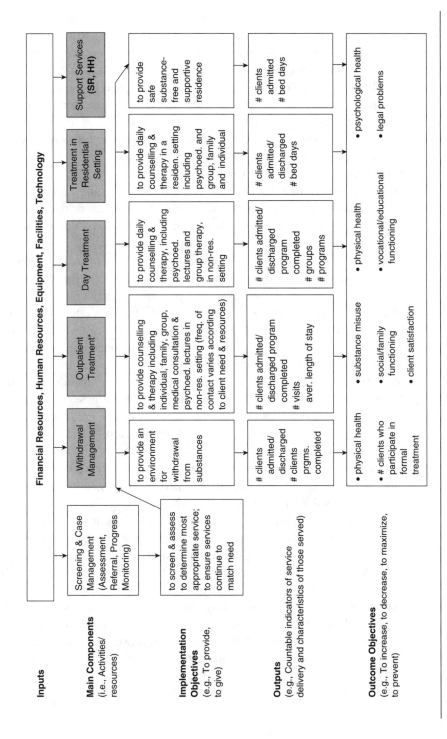

Figure 2.2   Logic Model for the Alcohol and Drug Services Program

* Not necessarily at outpatient "clinic."

Self-Sufficiency Program logic. There are six components included in the model—addictions treatment programs are often multifaceted to reflect a diversity of client circumstances—and to demonstrate success, the program evaluators would need to mount an evaluation that takes into account the complexity of the program or focuses on selected parts of the program.

The logic model in Figure 2.2 suggests that there are several outputs for each program component and that these in turn are linked to a set of short-term outcomes, and hence, to a set of longer-term outcomes. Given the complexity of the program, it is not practical to diagram each causal linkage. Instead, the logic model summarizes sets of linkages and offers a general picture of the program. Any evaluation of the program would need to unpack this logic model further to get at more specific linkages among the constructs.

If one looks at the outputs and the outcome objectives, one can see that these objectives are all client-based. Thus, any evaluation strategy would need to include methods for including clients and for measuring how well clients are doing with respect to the objectives included in Figure 2.2.

We need to keep in mind that when we are constructing logic models, we are employing templates that do not fit all situations equally well. Logic models based on an open systems approach will work best in environments where the program has clearly stated objectives, has been implemented, has a track record, and where administrative responsibility for the program is housed within the organization in which the program is embedded. Programs that are jointly designed and implemented by different government organizations or by partnerships between public and private sector organizations will be more likely to incorporate inconsistent or even contradictory expectations. This kind of environment is a challenge for logic modelers. Logic modeling is a part of the craft of program evaluation. The categories that we apply as we fit logic models to particular programs are general, and in some situations, we will find that it is unclear whether a given part of the program fits into one or another component, or just what the intended cause and effect linkages are. Developing a logic model of a program is an iterative process that relies on feedback from stakeholders to arrive at a model that is workable. It is essentially a qualitative research process that relies substantially on professional judgment, wherein documents, interviews, and other data sources are combined, weighted by the evaluator(s), rendered as a visual model, and validated with those who are able to comment on the model's accuracy. Keep in mind as well that program objectives are often constructed so that they are deliberately vague. This can mean that the programs to which they are attached may also be hard to specify. Later in this chapter, we will expand on the process of developing logic models, including specifying program objectives.

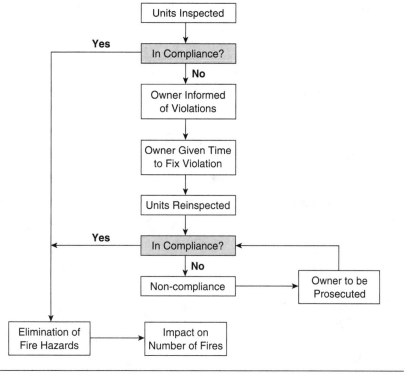

Figure 2.3   Flow Chart for Fire Codes Inspection Program

FLOW CHARTS ●

Although program logic models are frequently used to describe programs, it is possible, for some kinds of programs, to represent the activities in some detail using **flow charts**. Programs that serve clients usually have a process for intake/screening, with a set of steps that determine how the client is served. Depending on client characteristics and the evident issues/problems for that person, different paths within the program may be appropriate.

Figure 2.3 illustrates a flowchart that depicts the program process for the fire codes inspection program that was described earlier in this chapter. By reviewing the flow chart, we can see how the program is intended to operate. Where the building does not pass the first time, initial inspections are followed by re-inspections. Owners are given time to bring their property into compliance; if they do not, they are cited for fire codes violations and must deal with the legal consequences. For some buildings (apartment buildings, for example), the consequences might range from fines to condemning the building as unsafe for occupancy.

Flow charts are generally complementary to program logic models. Their advantage is that they make the program activities explicit, and if adequacy of implementation is an evaluation issue, a flow chart can reveal locations in the process where bottlenecks or breakdowns are occurring.

● CONSTRUCTING PROGRAM LOGICS IN PROGRAM EVALUATIONS

The process of representing a program as a logic model is iterative, relying on a combination of activities:

- Reviewing any documentation that describes the program and its objectives (policy documents, legislative mandates, working papers, memoranda, etc.)
- Meeting with the program managers to learn how they see the purposes and the activities of the program
- Meeting with other stakeholders in situations where the program is funded or mandated intergovernmentally or interorganizationally
- Drafting a logic model
- Discussing it with program managers/other stakeholders
- Revising it so that it is seen as a workable model of the intended processes and outcomes of the program
- Affirming that the logic model is adequate for the evaluation that is being undertaken

The evaluator plays a key role in this process. As John Owen and Patricia Rogers (1999) point out, many program planners and operators are unfamiliar with the concept of logic models and have difficulty developing them. It may be necessary and appropriate for the evaluator to explain what a logic model is and how it clarifies the structure of the program and its objectives. Then, as the model is developed, the evaluator synthesizes different views of the program structure, including the documentary view(s), and offers a visual interpretation of the program. The evaluator's background and familiarity with the type of program being evaluated play a part in how efficiently the program logic is developed.

Ultimately, the evaluator is striving for a workable logic model—there will be "rough edges" to most logic models, and there may not be complete agreement on how the model should look. It is essential to keep in mind that a workable logic model will be complete and detailed enough to represent the key parts of the program and the main causal linkages, but it cannot hope to model all the details. Ron Corbeil has pointed out that evaluators can succumb to "boxitis," that is, a desire to get it *all* down on paper (Corbeil, 1985).

Specifying Program Objectives

A key part of the process of developing and validating a program logic model is identifying the objectives of the program. The performance management cycle introduced in Chapter 1 emphasizes the importance of clear objectives; organizations should begin and end each cycle with a commitment to constructing/adjusting clear strategic objectives. These will lend themselves to clear program mandates, which in turn will facilitate implementation, then evaluation. The circle is closed when the evaluation/performance results are reported and used for the next round of refining or redefining the strategic objectives.

From both a program evaluation and a performance measurement standpoint, ideal program objectives should have at least five characteristics:

1. They should *specify the target population/domain* over which expected program outcomes should occur.

2. They should *specify the direction of the intended effects,* that is, positive or negative change.

3. They should *specify the magnitude* of the expected change.

4. They should *specify the time frame* over which expected changes will occur.

5. The outcomes embedded in program objectives should be *measurable.*

These five criteria, if all realized, will greatly facilitate the work of evaluators. It would be relatively easy to work with such objectives, and the evidence for program outcomes could easily be compared to the conditions that were specified.

An example of a well-stated program objective might be as follows:

The Neighborhood Watch Program that has been implemented in the Cherry Hill area of Boulder, Colorado will reduce reported burglaries in that part of the city by 20% in the next 2 years.

In most situations, program objectives are far less precisely stated. Programs and policies are put together in a political context. In contrast to models of performance management which rely on a view of organizations that is essentially devoid of impacts of power and **politics**, most of us have experienced the political "give and take" that is intrinsic in putting together the resources and support needed to mount a new initiative. Power dynamics are an intrinsic part of organizations, and getting things done means that individuals and groups need to work within the formal and informal structures in the organization (de Lancer Julnes & Holzer, 2001).

Morgan (1997), in his examination of various metaphors of organizations, includes a case for considering organizations as political systems. The key to understanding the necessity for politics is that persons and groups who participate in organizations do not necessarily share the same values. Resolving differences among values through bargaining or even conflict is the essence of politics. David Easton (1967) has defined politics as the authoritative allocation of values.

The implication for constructing program objectives is that competing and perhaps even conflicting views will often need to be reflected in the objectives. That will usually mean that the words chosen will reflect these different values, and can easily result in objectives that are general, even vague, and seem to commit the program to outcomes that will be difficult to measure. The objectives are, in fact, political statements and carry the freight of political discourse—they promise something to all stakeholders.

Richard Nathan (2000) provides a good example of how the political nature of defining program objectives can create a challenge for program evaluation. In his discussion of President Nixon's general revenue sharing program, enacted in 1972 to provide general purpose grants-in-aid to state, city, county, and township governments and Native American Tribes, Nathan comments:

> In textbooks on research methods, readers are told that step one in an evaluation study is to clearly define a program's objectives. Yet in the case of the revenue sharing program, as for a great many other public programs, there was not agreement . . . on the **goals** of the program.
>
> In addition to Nixon's goal of decentralization, supporters of the revenue sharing program stressed its role in achieving other goals. Among those purposes were: (1) providing fiscal relief to local governments; (2) equalizing fiscal capacity among states and localities; (3) providing new funds to deal with important public sector needs; (4) serving as a stimulus to innovation on the part of recipient state and local governments; (5) stabilizing and reducing state and local taxes; . . . and (6) altering the nation's overall tax system . . .

Which of these goals should be featured in an evaluation? How should these goals (seven in all) be defined and weighted? Views of the various players in the policy process differed. In fact, the same players often emphasized different objectives at different times. (p. 184)

Unclear language can create later challenges when the time comes to define how to measure vague constructs. This can be especially challenging with social programs where there are multiple agencies involved. For example, the Social Development Partnership Program (SDPP) "provides grants and contributions to social and non-profit organizations" (Human Resources Development Canada, 2002, p. 1). The objectives for this Human Resources Development Canada (HRDC) program, for which an evaluation framework has been developed, are stated as follows (p. 3).

The principal objectives of the SDPP are to:

- Advance nationally significant best practices and models of service delivery
- Strengthen community capacity to meet social development needs
- Strengthen the capacity of the social non-profit sector to contribute information to government and others on new and emerging social issues

These objectives are broad and actually suggest a logic for the program but lack many of the desirable characteristics of program objectives we identified above. The evaluators of the program used **case studies**, interviews, focus groups, and reviews of program files and documents to evaluate the program. They related the following among the limitations:

In a number of instances the information or data needed to confirm a particular outcome was simply not available or obtainable. . . . It should be noted that the unavailability of certain administrative/operational data is a serious problem frequently encountered in program evaluations, and is not unique to the SDPP. The need to improve this situation is a very important issue for programs with results-driven accountability frameworks. (Human Resources Development Canada, 2002, p. 11)

The evaluation team also stated:

The renewed SDPP, now built with greater clarity of purpose, and better tools to measure the incremental impacts of program investments, is in good shape to identify and nurture the innovative ideas and positive actions generated in the social non-profit sector. (p. x)

Given the amount of interpretation that is required to construct objectives that can be included in a logic model, and can eventually be measured, it is important to secure agreement on what the program is *actually intended* to accomplish, *before* the evaluation begins. Not all objectives or even parts of objectives will be equally important to all stakeholders. Depending on the purposes of the evaluation and the stakeholders involved, it may be possible to simplify and, hence, clarify objectives. This strategy relies on identifying a primary group of stakeholders and being able to work with them to translate program objectives into language that is amenable to evaluation.

Program logic models are constructed using a variety of methods and information sources, but overall, the process is essentially qualitative and involves the exercise of considerable judgment. In Chapter 5 we will learn how to apply qualitative evaluation methods and will see that evaluators have different views, some of which are grounded in philosophy and divergent assumptions about the nature of human culture and interactions.

● PROGRAM TECHNOLOGIES

James D. Thompson (1967), in a book that has made a major contribution to the study of complex organizations, introduced and elaborated the idea that organizations have purposes that are collectively specified and use **core technologies** to achieve their intended objectives. For Thompson, core technologies are means-ends relationships that define the competencies that an organization has, which are believed by organizational participants to be workable in accomplishing program/organizational objectives. Core technologies are combinations of knowledge, technique, and experience that represent the means available to organizations which are designing and implementing programs and policies. Our current efforts to identify best practices in public and nonprofit programming are in part motivated by a desire to employ state-of-the-art technologies as we deploy resources to ameliorate or solve the problems that motivate programs (Patton, 2001).

Instrumentally perfect technologies work every time—they are based on knowledge we can count on. When we have a problem or condition we wish to ameliorate or solve, applying such technologies will result in program success (Thompson, 1967). However, in the public and nonprofit sectors many of the means available to solve problems or ameliorate social conditions are imperfect. Realistically, even if we have the resources and can fully implement programs that are intended to achieve particular outcomes, we cannot be sure that the program will succeed. In fact, in some situations, we can be

Table 2.3   Program Technologies and the Probability That Outcomes Will Be Achieved

Type of Programs	Level of Certainty in the Program Technology	Workable Evaluation Approaches
Physical infrastructure programs	**High-probability technologies** include: • Highways maintenance programs • Flood control programs • Rural water supply programs	*Ex ante* cost-benefit analyses, performance measurement
	Medium-probability technologies include: • Reforestation programs • Agricultural land reclamation programs • Fire prevention programs	
Social programs (people programs)	**Low-probability technologies** include: • Workforce training for chronically unemployed • Child abuse prevention programs • Prison-based rehabilitation programs	*Ex post* outcomes—focused program evaluations Performance measures are problematic for these programs

reasonably sure that it will not succeed—we simply do not have the knowledge to solve every problem that society deems worthy of public expenditures.

Table 2.3 displays a way of looking at the relationship between the certainty of program technologies and the kinds of programs we typically evaluate. Displayed as well is an indication of the type of evaluation approaches that work best for different technologies.

The probability that a given program technology will succeed is a reflection of *how much we know* about solving different kinds of problems. Table 2.3 suggests that in the area of social programming, we know relatively less than we do for most other program areas. That does not mean that social programs are doomed to failure. Instead, it means that expectations of success, even if resources are adequate and full implementation occurs, need to be tempered by the knowledge that we do not generally have the kinds of "engineering principles" available, for example, to colleagues in a transportation and highways agency.

Several implications can be drawn from Table 2.3. First, when we are developing programs, particularly in areas where we are trying to solve problems or ameliorate conditions for people, we need to be aware of what has been tried elsewhere. We are progressively more able to learn from other people's experiences with programs aimed at solving or ameliorating a problem we want to address. There is no substitute for taking the time to find out what has been done elsewhere—and whether it has been evaluated.

An emerging subfield in program evaluation is the **meta-analysis** of existing evaluations (Cook, 1994; Rossi, Freeman, & Lipsey, 1999). Although this book will not cover meta-analysis as a separate topic (a number of technical issues involved are beyond our scope), program evaluators can search for existing meta-analyses and use them to assist in program development.

A second implication is that program managers who are responsible for "people programs" should be realistic about the likelihood of success, although it is important that realism not be mistaken for cynicism. The latter frame of mind suggests that "nothing will work." Program managers know that that is not so, but need to cultivate a capacity to learn from the mistakes that program processes make. Recall the intention behind formative evaluations—to obtain feedback to improve the program. Program managers conduct their own informal formative evaluations on a daily basis. This book can be seen as a way to sharpen and focus those evaluations, with a view to improving our knowledge of what works and why. This view is similar to the one stated in Mark, Henry, and Julnes (2000).

The third implication is perhaps the most controversial of the three. In situations where managers are working with low-probability program technologies, it is quite likely that evaluations of outcomes will offer a mixed picture. Indeed, it is possible that for most clients, the program in question will not clearly make the intended difference.

The question for program evaluators and for program managers is: Who should be held accountable if the program does not achieve its objectives? In governmental systems that have made hierarchical accountability a central feature of their bureaucratic structures, those who head these organizations are generally held accountable for what happens in their departments. The risks of being publicly connected with some kind of breach of policy, or being shown to have violated procedures, or having a program result that suggests an ineffective program, means that there is considerable emphasis on departments and agencies being seen to not make mistakes. Department heads and the elected officials to whom they report generally do not want to be publicly embarrassed by failures in their domain. Avoiding mistakes tends to create a conservative bias in decision making. "Getting it right" can mean not getting it visibly wrong.

Program managers are in a position where they are expected to achieve success—programs are usually announced as solutions to politically visible problems, but the "state of the art" in terms of program technologies is simply not going to yield a set of engineering principles that will always work or even work most of the time.

Under such conditions, is it fair to hold managers accountable for program outcomes? Where do we draw the line between managerial responsibility (and, hence, accountability) and a more diffused collective responsibility for solving problems with our programs?

One way to answer this question is to distinguish outputs from program outcomes. In fact, in Figure 2.1, outputs are categorized as part of the program, whereas linking constructs and intended outcomes are in the environment. Program managers are generally willing to be held accountable for outputs (the work done in a program). Being held accountable for outcomes can be problematic, *because outcomes occur in the program's environment, and other factors in the environment besides the program that managers cannot control can impact the outcomes.*

This problem of other causes of outcomes (rival hypotheses) is, in fact, central to the issue of attribution, which will be discussed in Chapter 3. For high-probability program technologies, outputs cause the observed outcomes, and we generally do not have to concern ourselves with rival hypotheses. For low-probability program technologies, rival hypotheses are a major concern—the logics of social programs tend to be more "open" to environmental influences.

One final comment on managerial accountability is in order. Current efforts in government departments, agencies, and nonprofit organizations to develop and use performance measures raise the same question of who should be accountable for program outcomes (or their absence). Managers are currently expected to participate in developing performance measures of their own programs and to account for program results. As we will see when we get to Chapter 9, a key thrust of current efforts to develop performance measurement systems is to focus on outcomes and hold managers accountable for achieving program outcomes. One issue we raise in that chapter is whether and under what conditions accountability for program performance is a sustainable goal in public organizations.

PROGRAM OBJECTIVES, PROGRAM ENVIRONMENTS, AND ORGANIZATIONAL OBJECTIVES •

Figure 2.1 and our earlier discussion of open systems approaches suggest that programs operate in environments, and that the outcomes that are

Table 2.4 Examples of Factors in the Environments of Programs That Can Offer Opportunities and Constraints to the Success of Programs

Factors in the Public Sector	Factors in Society
• Other programs	• Clients[a]
• Senior executives	• Interest/advocacy organizations[b]
• Other departments/agencies	• Media
• Funding agencies	• Exogenous events
• Elected officials	
• Regulatory agencies	
• Courts	

a. Some programs have outcomes that are focused within a department or across departments of a government, in which case the clients would be in the public sector.

b. Public sector labor unions could become an interested party if a program was focused on the public sector.

intended generally impact outside the program from which they originate. What are some of the factors in the environments of programs that can offer opportunities and constraints as programs work to achieve their objectives?

Table 2.4 summarizes some of the factors that exist in the environments of programs, which condition their success or failure. The factors listed have been divided into those that are internal to the public sector and those that are not. The list is not intended to be exhaustive, but instead, to alert evaluators to the fact that programs and their evaluations occur in a rich and dynamic environment, which must be taken into account in the work that they do.

Programs can be thought of as embedded open systems within organizations, which are themselves open systems. This image of nested open systems suggests that outcomes from a program can contribute to the objectives of an organization. The objectives of the U.S. Department of Agriculture, for example, are intended to contribute to the strategic objectives of the U.S. government, together with the objectives of other federal departments. We usually assume that organizational objectives arise out of, or are consistent with, program objectives. In fact, strategic plans, as a part of the performance management cycle, are intended to create organizational objectives (the vision statement, the **strategic goals**, and the mission statement) that provide a framework for program objectives. Some governments take this nesting of objectives further by constructing systems that synchronize performance objectives for individual employees with objectives for work groups. These, in turn, feed into program objectives, clusters of programs, and hence to organizations. Finally, clusters of organizations and their objectives are intended to contribute to government-wide objectives (see, for example, British Columbia Government, 2003).

Normative Goals and Behavioral Goals in Organizations

Although strategic planning is widely accepted as sound management practice, it is essentially a set of tools that create normative goals for an organization. These goals are statements of what is desirable, what the organization ought to do, and hence, what the members of the organization ought to strive to do. Organization theorists, however, have observed that organizations appear, as well, to pursue **behavioral goals**, which may or may not coincide with statements enshrining the rational purposes of the organization.

Morgan (1997) elaborates a biological metaphor for organizations and, therein, notes that *if* organizations can be seen as "organisms," it makes sense to predicate of them goals that organisms share: namely, the will to survive and, in many cases, to grow.

The import of such "biological" goals accords with our experiences: many who have worked in the public sector would agree that organizations appear to "want" to grow and to survive. Suppose we take, as an example, an agency having a stated program goal of working with clients to reduce their dependency on the agency (i.e., a department dispensing public assistance payments may assert that its objective is to get its clients off of public assistance).

Suppose it succeeded. It would put itself out of business—perhaps a commendable eventuality, but one that is rarely observed. In fact, the goal of the program conflicts with an important behavioral goal: the survival of the agency.

This, perhaps, is an extreme example, but the broader point is that program objectives can conflict with other, perhaps implicit, organizational objectives. When that happens, the organization will tend to resist or deflect the program, and resources will be consumed as that struggle is resolved. Another way to look at this kind of situation is to say that programs exist in environments where values compete. Resolving competitions among values is the essence of organizational politics (Morgan, 1997).

STRENGTHS AND LIMITATIONS OF PROGRAM LOGICS ●

Conceptualizing programs as open systems that we can represent with logic models has the advantage of facilitating ways of communicating about programs that words alone cannot. Visual models are worth "a thousand words" and we have come to rely on logic models in our work as evaluators. Internationally, logic modeling is increasingly being viewed as a language that makes it possible to describe programs in universal terms (Montague, 2000).

But, embracing logic modeling means that we *expect* to see an open system when we work with a program. We work to fit programs into a logic modeling framework, and assume that we have done justice to the program itself. In Chapter 12, we discuss the role that expectations play in our work as evaluators. Our contention is that we are all affected by our expectations, and they can make a difference in what we "see" when we are doing our work. When we are describing programs, most of us expect to see an open system that can be depicted as a logic model. We proceed as if that assumption is true.

Program logics do three things. They categorize organizational work, they outline cause and effect linkages, and they distinguish what is in the program from what is in its environment. Program components facilitate our communicating about the program and often correspond to the way we have structured the organization that delivers the program. Program components can turn out to be branches or divisions in a department or, at least, be work units that are responsible for that component.

Organizational logic models and charts are useful as we focus on authority and responsibility. They encourage an administratively rational view of what we do when we implement programs. But, organizational categories can impede a capacity to respond to situations or clients who "fall between the cracks." We can find ourselves in situations where the model we have constructed serves as a filter between our perceptions and the actual situations that exist. Logic models, instead of being a means of understanding programs, may become ends in themselves and arrest our capacity to recognize the need for change.

There are some potential limitations with logic modeling as a way to understand programs. First, some programs do not lend themselves to modeling of this sort. Consider a program that operates in a highly turbulent environment. Change is the order of the day, and program managers are constantly working to figure out how the program can be adapted so that it can survive. Logic models are akin to photographs. They render a picture of a program at the point in time at which the evaluator has constructed the image. But, they can become obsolete and misleading if there is change in the program or in the relationship between the program and its environment.

A second and related point is that all program logic models are time-limited. Even programs that exist in relatively placid environments will change. This fact has implications for our efforts to construct performance measurement systems. Once the time and resources needed to construct a logic model are invested, it may be difficult to acknowledge that the picture will not represent reality a couple of years hence. We may find that the logic model we have used to identify and measure performance will at some point be an impediment to understanding what the program is currently about.

SUMMARY ●

Logic models are visual representations of programs that show how resources for a program are converted into activities, and hence into intended results. Logic models provide evaluators, program managers, and other stakeholders with a summary of a program that does two basic things. It divides the program delivery process into categories: inputs, components, implementation objectives, and results (outputs, linking constructs, and outcomes); and it displays the intended causal linkages among outputs and other program results. Based on an open systems metaphor, logic models can also distinguish between what is in the program and what is in the program's environment. Typically, program outcomes are expected to have impacts in the program's environment, consistent with the intended objectives.

Program logics are an important foundation for an evaluator's efforts to understand whether and in what ways the program was effective—whether the program actually produced the outcomes that were observed and whether those outcomes are consistent with the program objectives. They are also very useful for developing performance measurement systems to monitor program outputs and outcomes. They assist evaluators by identifying key constructs in a program that are candidates for being translated into performance measures.

Constructing logic models is an iterative process that relies on qualitative research methods. Evaluators typically consult documents that describe a program, consult stakeholders, and other sources of information about the program as it is being drafted, and then review the logic model with stakeholders to establish its adequacy.

Logic modeling has become an essential tool for evaluators and program managers alike. Because it is based on an open systems framework, it tries to "fit" programs and experiences with programs into this framework. Where programs are established and where program environments are stable, logic models are an efficient way to communicate program structure and objectives to stakeholders. In environments where there is a lot of uncertainty about program objectives or a lot of turbulence in the organization in which the program is embedded, logic models, with their "snapshot" perspective, may not capture the dynamic complexity that actually exists.

● DISCUSSION QUESTIONS

1. Read the program description of the COMPASS program in the appendix to this chapter and use the material in the chapter to build a model of the program. To make your learning more effective, *do not* look at the solution before you have completed the logic model. The following hints may be useful as you build the logic model.

Helpful Hints in Constructing Logic Models

- When your are constructing logic models, state your constructs as simply and as precisely as possible—keep in mind that later on these constructs will usually need to be measured and if they are not stated clearly, the measures may be invalid.
- Constructs cannot cause themselves in logic models—it is necessary for you to decide where a given construct fits in the logic, and having done so, do not repeat that construct at a later point in the model.
- Implementation objectives must focus on what needs to be done to get the program onto the "ground" and *not* on expected outcomes— if an employment program is training people so that they can find jobs, the implementation objective is "providing training" and the intended outcome is "employment."

2. Based on your experience of constructing the logic model for the COMPASS program, what advice would you give to a classmate who has not yet constructed his/her first logic model? What specific step-by-step guidance would you give (five key pieces of advice)? Try to avoid repeating what the chapter says logic models are, that is, components, implementation objectives, outputs, linking constructs, and outcomes. Instead focus on the *actual process* you would use to review documents, contact stakeholders, and other sources of information as you build a logic model. How would you know you have a "good" logic model when it is completed?

3. Logic models "take a picture" of a program that can be used in both program evaluations and performance measurement systems. What are the organizational and program conditions that make it possible to construct accurate logic models?

4. What are the organizational and program conditions that make it challenging to develop accurate and useful logic models?

5. Why is it that formulating clear objectives for programs is so challenging?

6. What are program technologies? How do program technologies affect program managers, particularly as they work with evaluators and with stakeholders who expect managers to be accountable for program results?

● APPENDIX 1: APPLYING WHAT YOU HAVE LEARNED IN CHAPTER 2: DEVELOPING A LOGIC MODEL FOR THE COMPASS PROGRAM

The COMPASS program was launched in Nova Scotia in 1995 as a joint initiative of the Nova Scotia Department of Community Services and Human Resources Development Canada. The program description below can be used as the basis for developing a program logic model of the program.

Using the written description of the COMPASS program below, *construct a program logic model* that includes the following:

- Program components
- Implementation objectives
- Outputs
- Linking constructs
- Intended outcomes

Be sure to indicate in your logic model how *each* component is linked to respective implementation objectives and how the implementation objectives are connected to *specific* outputs. In addition, show how *each* output is connected to *particular* linking constructs and how *each* linking construct is connected to *particular* outcomes.

Diagram your logic model. Put the model on one page and label it clearly. Keep in mind that there is a desirable balance between being explicit about important linkages, and putting so much in the model that it is cluttered. The solution to this modeling problem is included in this appendix.

Nova Scotia COMPASS Program: Program Description

Background to the COMPASS Program

The COMPASS Program is the Nova Scotia response to the joint federal-provincial Strategic Initiatives Program aimed at identifying innovative approaches to social security reform. The Strategic Initiatives Program, launched initially in March 1994 with a budget of CAD$800 million and subsequently reduced by half in the 1995 federal budget, supports innovative approaches to improve job opportunities, reduce barriers to employment, and curtail reliance on social security.

The COMPASS Program was developed by the provincial Department of Community Services in partnership with Human Resources Development Canada and in consultation with the Economic Renewal Agency and the Department of Education and Culture.

Program Context

Social assistance caseloads at both the provincial and municipal levels have been increasing dramatically over the past 5 years in Nova Scotia.

The COMPASS Program was developed in response to these rising caseloads and in particular to a number of trends associated with the growing dependency on social assistance. One trend concerns the increasing number of youth depending on municipal assistance. A second is the growing numbers of single parents appearing on the provincial family benefits caseload.

The COMPASS Program was specifically designed to provide provincial job-ready clients with access to employment. The province was aware that its continuum of service to provincial clients prior to COMPASS was limited on the employment side and set out to ensure that COMPASS would fill this gap.

Components of the COMPASS Program

The following are the four components of the COMPASS Program:

- Work Experience Option
- Transitional Training Option
- Enterprise Development Option
- Opportunity Fund

a. Work Experience Option

Recent years have seen a growing increase in the numbers of unemployed youth on the municipal social assistance caseloads. Many of these youth have been assessed as lacking both specific work skills and experience in a work environment.

The Work Experience Option was designed to address these needs by providing youth (between 18 and 30 years of age) in receipt of municipal assistance with an opportunity to gain work experience to enhance their employability.

The Work Experience Option provides youth with their first real job and an opportunity to feel positive about working every day. Through their participation, it is intended that youth will realize meaningful jobs are not easy to find and that they may require more training or education to obtain the job they want. The option is also useful for providing clients with the opportunity to explore different careers or to provide an opportunity to practice skills acquired but never implemented in a work situation. Finally, the Work Experience Option serves as an early intervention for a potentially high-risk client group.

Youth are paid an allowance of CAD$160 a week while on placement. Placements vary in length but may not exceed 26 weeks. Should clients not be hired on at the end of their placements, Employment Resource Centre (ERC) staff will continue to work with them as may be required, providing counseling, training in job search skills, or other job development activities.

b. Transitional Training Option

The Transitional Training Option was designed to assist skilled job-ready individuals obtain work experience to access employment. More specifically this component was designed to assist single parents in receipt of Family Benefits, persons with disabilities and displaced fishers/plant workers in receipt of social assistance.

Under this component, a wage subsidy is available to private sector employers to hire job-ready social assistance recipient (SAR) clients to enhance their employability. Employers contribute a minimum of 25% of the hourly wage. The maximum that COMPASS can contribute is CAN$5.62 an hour. Tenure is for a maximum of six months for full-time work.

c. Enterprise Development Option

Prior to the COMPASS Program, self-employment was not a viable program option for SAR clients. Although a self-employment program has been available through Human Resources Development Canada, the majority of seats are reserved for unemployment insurance recipients. As well, SAR clients have had no access to start-up equity prior to COMPASS.

COMPASS is breaking ground in presenting self-employment as a viable option for SAR clients across the province. Under the Enterprise Development Option, up to CAN$1.2 million has been allocated to assist SAR clients establish small business/micro enterprises. Of this amount, CAN$1 million was initially allocated to support the entrepreneurial training stream of the program with the remaining CAN$200,000 allocated to a Micro-Enterprise Loan Fund. On the basis of the first six month's activity, the Enterprise

Development Advisory Committee adjusted the allocation (CAN$800,000 to training and CAN$400,000 to the loan fund) to more closely reflect the take-up of the two streams.

Under Stream I, clients who have been assessed by the ERC and provincial counsellors as having the potential for self-employment are referred to an approved COMPASS funded training program to acquire core training in entrepreneurial skills and business development. The training occurs over a 20-week period, the first 4 weeks of which focus on core life skills. A completed business plan is one of the outputs of the training. The thrust of the training is to work with the clients to ensure that their business idea is viable and to provide them with the necessary skills to get their businesses up and running.

Stream II consists of the Micro-Enterprise Loan Fund. The fund is "a loan of last resort" and is intended to provide seed capital of up to CAN$2,000. In "exceptional cases" the advisory committee has the discretion to increase this amount up to CAN$5,000. The loan is to be repaid over a period of 3 years from the time of signed acceptance of the letter of offer.

Clients who meet the basic criteria for Stream II are referred by the ERCs to one of the nine Business Service Centres of the Nova Scotia Economic Renewal Agency established across the province. Business Service Centre officers meet with clients to review their Business Plans and, if found to be viable, assist the clients in preparing an application to the Micro-Enterprise Loan Fund.

d. Opportunity Fund

Prior to the introduction of the COMPASS Program, ERC staff had limited access to resources to purchase items that would enhance the employability of their clients. The Opportunity Fund was established as part of COMPASS to fill that gap. Through COMPASS, funds of up to CAN$200,000 have been made available to purchase specialty items/services such as textbooks, manuals, course fees, work boots, and safety equipment.

The Opportunity Fund is a loan of "last resort" and can be accessed only when all other options have been eliminated. It can be used both with ERC clients and clients who are involved in provincial vocational training and employment programs.

76

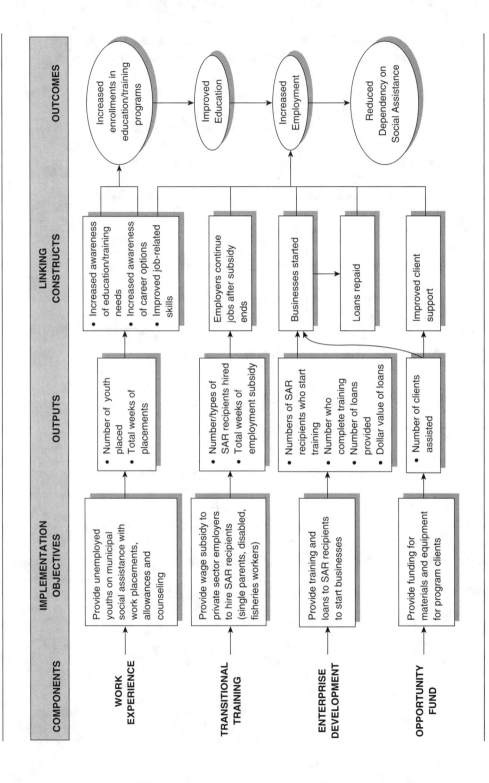

Figure 2A.1 Logic Model for Nova Scotia COMPASS Program

REFERENCES ●

Arthur, J. (1996). *Measuring client outcomes in alcohol and drug services.* Unpublished Master's thesis, University of Victoria, BC, Canada.

British Columbia Government. (2003). *Provincial government strategic plan 2004/2005 - 2006/2007.* Queen's Printer, Victoria, BC.

Coffman, J. (1999). *Learning from logic models: An example of a family/school partnership program.* Retrieved July 19, 2004, from http://www.gse.harvard.edu/~hfrp/pubs/onlinepubs/rrb/learning.html

Cook, T. D. (Ed.). (1994). *Meta-analysis for explanation: A casebook* (paper ed.). New York: Russell Sage Foundation.

Corbeil, R. (1985). *Mastering logic charts.* Unpublished workbook. Victoria, BC, Canada.

de Lancer Julnes, P., & Holzer, M. (2001). Promoting the utilization of performance measures in public organizations: An empirical study of factors affecting adoption and implementation. *Public Administration Review, 61*(6), 693–708.

Dinsdale, G., Cutt, J., & Murray, V. (1998). *Performance and accountability in non-profit organizations: Concepts and practice.* Unpublished manuscript, School of Public Administration, University of Victoria, BC, Canada.

Easton, D. (1967). *A systems analysis of political life.* New York: John Wiley & Sons.

House, E. R. (1984). How we think about evaluation. In R. F. Connor, D. G. Altman, & C. Jackson (Eds.), *Evaluation studies review annual* (vol. 9, pp. 1–12). Beverly Hills, CA: Sage.

Human Resources Development Canada. (2002). *Evaluation of the social development partnerships program.* Ottawa, ON: Government of Canada.

Mark, M. M., Henry, G. T., & Julnes, G. (2000). *Evaluation: An integrated framework for understanding, guiding, and improving policies and programs.* San Francisco: Jossey-Bass.

Martin, L. L., & Kettner, P. M. (1996). *Measuring the performance of human service programs.* Thousand Oaks, CA: Sage.

Michalopoulos, C., Tattrie, D., Miller, C., et al. (2002). *Making work pay: Final report on the self-sufficiency project for long-term welfare recipients.* Ottawa, ON: Social Research and Demonstration Corporation.

Montague, S. (2000). Focusing on inputs, outputs, and outcomes: Are international approaches to performance management really so different? *Canadian Journal of Program Evaluation, 15*(1), 139–148.

Morgan, G. (1997). *Images of organization* (2nd ed.). Thousand Oaks, CA: Sage.

Nagarajan, N., & Vanheukelen, M. (1997). *Evaluating EU expenditure programs: A guide.* Retrieved July 19, 2004, from http://europa.eu.int/comm/budget/evaluation/guide/guide00_en.htm

Nathan, R. P. (2000). *Social science in government: The role of policy researchers* (Updated ed.). Albany, NY: Rockefeller Institute Press.

Owen, J. M., & Rogers, P. J. (1999). *Program evaluation: Forms and approaches* (International ed.). London: Sage.

Patton, M. Q. (2001). Evaluation, knowledge management, best practices, and high quality lessons learned. *American Journal of Evaluation, 22*(3), 329–336.

Rossi, P. H., Freeman, H. E., & Lipsey, M. W. (1999). *Evaluation: A systematic approach* (6th ed.). Thousand Oaks, CA: Sage.

Rush, B., & Ogborne, A. (1991). Program logic models: Expanding their role and structure for program planning and evaluation. *Canadian Journal of Program Evaluation, 6*(2), 95–106.

Suchman, E. A. (1967). *Evaluative research: Principles and practice in public service & social action programs.* New York: Russell Sage Foundation.

Thompson, J. D. (1967). *Organizations in action: Social science bases of administrative theory.* New York: McGraw-Hill.

Treasury Board of Canada Secretariat. (2003). *Guidance: Departmental performance reports 2003-2004.* Retrieved July 22, 2004, from http://www.tbs-sct.gc.ca/rma/dpr/03-04/guidance/table-of-contents_e.asp

Tucker, L. (2002). *Use and the effects of using performance measures for budgeting, management, and reporting.* Case study: State of Texas. Retrieved July 21, 2004, from http://www.seagov.org/sea_gasb_project/state_tx.pdf

U.S. General Accounting Office. (2004, January). *Performance budgeting: Observations on the use of OMB's Program Assessment Rating Tool for the fiscal year 2004 budget.* Retrieved July 21, 2004, from http://www.gao.gov/new.items/ d04174.pdf

Weiss, C. H. (1972). *Evaluation research: Methods for assessing program effectiveness.* Englewood Cliffs, NJ: Prentice Hall.

Weiss, C. H. (1998). *Evaluation: Methods for studying programs and policies* (2nd ed.). Upper Saddle River, NJ: Prentice Hall.

CHAPTER 3

RESEARCH DESIGNS FOR PROGRAM EVALUATIONS

INTRODUCTION ●

Over the last 30 years, the field of evaluation has become increasingly diverse in terms of what is viewed as proper methods and practice. During the 1960s and into the 1970s most evaluators would have agreed that a good program evaluation should emulate social science research and, more specifically, methods should come as close to **randomized experiments** as possible. The ideal evaluation would have been one where people would have been randomly assigned either to a program or to a control group, the program was implemented, and after some predetermined period of exposure to the program, quantitative comparisons were made between the two groups. Program success was tied to statistically significant differences between program and control group averages on the **variable**(s) of interest.

Large-scale social experiments were funded, and evaluations were set up that were intended to determine whether the programs in question produced the outcomes that their designers predicted for them. Two examples of such experiments were the New Jersey Negative Income Tax Experiment and the Kansas City Preventive Patrol Experiment. In the New Jersey experiment, samples of low income families were randomly assigned to **treatment groups** where each group received a combination of minimum guaranteed family income plus a specific negative income tax rate. The latter worked like this: A family that was above the poverty line (say, $10,000 of income a year) would receive no benefit from the negative income tax part of the experiment. But if the family income fell below $10,000 per year, there would be payments that were related to how far below $10,000 the family earned. The lower the family income, the greater the payment—the greater the negative income tax (Pechman & Timpane, 1975).

The Kansas City Preventive Patrol Experiment (Kelling, 1974) was intended to test the hypothesis that the level of routine preventive patrol in a neighborhood would not affect the actual crime rate (measured by surveys of residents), the reported crime rate, or citizen perceptions of safety and security (measured by surveys of residents). Police beats were randomly assigned to one of three conditions: no routine preventive patrol (police would only enter the beat if there was a call for their services); normal levels of patrol; and two to three times the normal level of patrol. The experiment was run for a year, and during that time extensive measurements of key variables were made. The designers of the experiment believed that if the level of patrol could be shown to not affect key crime and citizen safety indicators, police departments across the United States could save money by modifying the levels of patrol that they deployed.

Typically, the programs were administered to samples of clients chosen randomly from a specific population, and clients for the **control groups** were chosen in the same way. The dominant belief in this approach was rooted in the assumption that randomized experiments are the best way to determine what effects programs have—they are the best test of the cause and effect relationship between the program and its intended outcomes.

In the 1970s and into the 1980s, as the field of evaluation diversified, some evaluators introduced and elaborated the use of **quasi-experiments** as ways of assessing cause and effect relationships in evaluations (Cook & Campbell, 1979). The costs and administrative challenges of doing randomized experiments prompted an interest in **research designs** that retained some of the features of experiments, but generally were more feasible to implement.

But, as we will see in Chapter 5, the biggest single change in the field during the 1970s and beyond was the introduction of **qualitative program evaluations**. The key difference between advocates of qualitative evaluations (Guba & Lincoln, 1981; Patton, 1997) and those who continued to assert the superiority of experiments and quasi-experiments was the qualitative evaluators' emphasis on words as a basis for the analyses in evaluations. Advocates of experiments and quasi-experiments tended to emphasize the use of quantitative techniques, often involving applications of statistics to numerical data.

Today, we continue to have this diversity of approaches to evaluation. As you can see just by reviewing the chapter titles and topics in this book, program evaluators can choose from a wide range of tools, representing differing views of the field. Knowing which ones are appropriate in a given situation is one of the skills that practitioners must master.

Although there are many different reasons for conducting evaluations, a principal one is to learn whether the program achieved its intended outcomes. As we saw in Chapter 1, the question of program effectiveness really is two separate questions:

Was the program responsible for (or the cause of) the observed outcomes?

Were the observed outcomes consistent with the expected outcomes?

In Chapter 2, we learned how to describe programs as logic models, and in doing so, how to make explicit the intended **causal linkages** in the program process. In this chapter, we focus on ways of testing the intended causal linkages in the program, beginning with the principal one: the connection between the program and the observed outcomes.

Figure 3.1   Did the Program Cause the Observed Outcomes?

WHAT IS RESEARCH DESIGN? ●

Research design is fundamentally about examining the linkage depicted in Figure 3.1.

Notice what we have done in Figure 3.1. We have taken the program, which we "unpacked" in Chapter 2 in logic models, and repacked it. The complexity of logic models has been simplified again so that the program is back in a box.

Why have we done this? Wouldn't it make more sense to *keep* the logic models we have worked on so far and test the causal linkages in such models? That way, we would be able to confirm whether the intended linkages between various outputs, linking constructs, and outcomes are supported by evidence gathered in the evaluation.

The reason we "repack" the logic models is that research designs are actually fairly blunt instruments in the tool kit of an evaluator. Research designs demand that we isolate *each* causal linkage that we wish to test. In order to examine a given cause and effect relationship, we must find ways of testing it while **holding constant** other factors that could influence it. A typical program logic will have a number of important causal linkages. In order to test these linkages using research designs we would need to isolate *each one* in turn to know whether that particular linkage is supported by evidence, holding constant the linkages in the rest of the logic model.

The problem is in finding ways of holding everything else constant while we examine each linkage in turn. In nearly all evaluations of programs, we simply do not have the time or the resources to do it. Thus, in thinking about research designs, we tend to focus on the *main* linkage, that being the one between the program as a whole (back in its box) and the observed outcomes.

Later in this chapter, we look at ways that have been developed to more fully test program logics. One approach that is quite demanding in terms of resources is to conduct an evaluation that literally tests all possible combinations of program components in an **experimental design** (Cook & Scioli, 1972). Another one that is more practical is to use several complementary research designs in an evaluation, and test different parts of the program logic with each one. These designs are often referred to as **patched-up research designs** (Poister, 1978), and usually, they do not test all the causal linkages in a logic model.

Research designs that fully test the causal links in logic models often demand more resources than are available to an evaluator. We will look at an example of such a research design later in this chapter to see what is involved in fully testing a logic model.

The Origins of Experimental Design

Experimental design originated in disciplines where it was essential to be able to isolate hypothesized cause and effect relationships. In agricultural research in the post–World War I period, for example, people were experimenting with different kinds of seeds to produce higher yields. There was keen interest in improving crop yields—this was a period when agriculture was expanding and being mechanized in the United States and Canada.

Researchers needed to set their tests up so that variation in seed types was the *only* factor that could explain the number of bushels harvested per acre. Typically, plots of a uniform size would be set up at an agricultural research station. Care would be taken to ensure that the soil type was uniform across all the plots and was generalizable to the farmlands where the grains would actually be grown.

Then, seed would be planted in each plot, carefully controlling the amount of seed, its depth, and the kind of process that was used to cover it. Again, the goal was to ensure that seeding was uniform across the plots. Fertilizer may have been added to all plots (equally) or to some plots to see if fertilizers interacted with the type of seed to produce higher (or lower) yields.

The seed plots might have been placed side by side or might have had areas of unplanted land between each. Again, that might have been a factor that was being examined for its effects on yield.

During the growing season, moisture levels in each plot would be monitored, but typically, no water would be provided other than rainfall. It was important to know if the seed would mature into ripe plants with the existing rainfall in that region. Because the seed plots were in the same geographic area, it was generally safe to assume that rainfall would be equal across all the plots.

Depending on whether the level of fertilizer and/or the presence of unplanted land next to the seed plots were also being deliberately manipulated along with the seed type, the research design might have been as

Plots with experimental seed	X	O_1
Plots with standard seed		O_2

Figure 3.2   Research Design to Test Seed Yields Where Seed Type Is the Only Factor Being Manipulated

simple as two types of plots: one type for a new "experimental" seed, and the other for an existing, widely used seed. Or, the research design might have involved plots that either received fertilizer or did not, and plots located next to unplanted land or not.

Figure 3.2 displays a research design for the situation where just the seed type is being manipulated. As a rule, the number of plots of each type would be equal. As well, there would need to be enough plots so that the researchers could calculate the differences in observed yields and statistically conclude whether the new seed improved yields. Statistical methods were developed to analyze the results of agricultural research experiments. Ronald A. Fisher, a pioneer in the development of statistical tools for small samples, worked at the Rothansted Experimental [Agricultural] Station in England from 1919 to 1933. His book, *Statistical Methods for Research Workers* (Fisher, 1925) is one of the most important statistics textbooks written in the 20th century.

In Figure 3.2, "X" denotes the factor that is being deliberately manipulated, in this case, the seed type. More generally, the "X" is the treatment or program that is being introduced as an innovation to be evaluated. The O_1 and O_2 are observations made on the **variable** that is expected to be affected by the "X." Treatments or programs have intended outcomes. An outcome that is translated into something that can be measured is a variable. In our case, O_1 and O_2 are measures of the yield of grain from each group of seed plots: so many bushels per acre (or an average for each group of plots).

Figure 3.3 displays the more complex research design involved when seed type, fertilizer, and cultivated (nonseeded) land nearby are all being manipulated. Clearly, many more seed plots would be involved—costing considerably more money to seed, monitor, and harvest. Correspondingly, the amount of information about yields under differing conditions would be increased.

Figure 3.3 is laid out to illustrate how the three factors (seed type, fertilizer, contiguous cultivated land) that are being manipulated would be "paired up" to fully test all possible combinations. In each of the cells of the

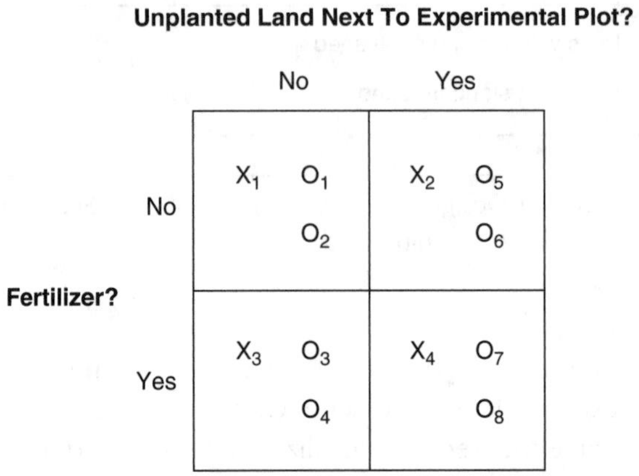

Unplanted Land Next To Experimental Plot?

		No	Yes
Fertilizer?	No	X_1 O_1 O_2	X_2 O_5 O_6
	Yes	X_3 O_3 O_4	X_4 O_7 O_8

Figure 3.3   Research Design to Test Seed Yields Where Seed Type, Fertilizer, and Contiguous Unplanted Land Are All Possible Causes of Grain Yield

figure, there are the original two types of plots: those with the new seed and those without. The plots where "X" has occurred in each of the four cells of the figure have the new type of seed, and the plots in each of the four cells that do not get "X" are planted with regular seed. Because each cell in Figure 3.3 represents a different treatment, each "X" has been subscripted uniquely. In effect, the simpler research design illustrated in Figure 3.2 has been reproduced four times: once for each of the combinations of the other two factors.

This agricultural experimental research design and others like it have been generalized to a wide range of program evaluation situations. In program evaluations, research designs work best where the evaluator is involved in both the design and implementation of a program, and there are sufficient resources to achieve a situation where the effects of the program can be determined while other factors are held constant. The paradigmatic situation is an experiment where, for example, clients are randomly assigned to the groups that do and do not get the program, and differences in client outcomes are measured. If the experiment has "worked," outcome differences can confidently be attributed to the program: we can say that the program *caused* the observed difference in outcomes, that is, the causal variable occurred before the observed effect, the causal variable co-varied with the effect variable, and there were no **plausible rival hypotheses**.

The difficulty in most program evaluation settings, of course, comes in ruling out the possible rival hypotheses. Lawrence Mohr (1995) makes this point when he states: "The crux of the analysis of the efficacy of a treatment or program with respect to a particular outcome . . . is *a comparison of what did*

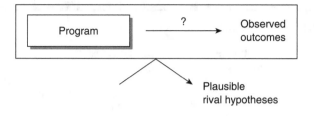

Figure 3.4   Visual Metaphor for a Defensible Research Design

appear after implementing the program with what would have appeared had the program not been implemented. Events in the what-would-have-happened category must obviously be troublesome" (p. 4, italics added).

Figure 3.4 suggests a visual metaphor for what research design strives to achieve. The causal linkage between the program and the observed outcomes is isolated so that other, plausible rival hypotheses are deflected. The line surrounding the program and its outcomes in the figure represents the barrier against rival hypotheses which is created by a defensible research design.

Achieving this isolation of the program/observed outcomes linkage is often associated with designing and implementing the program as a randomized experiment. Table 3.1 shows two different experimental designs: the first involves measuring the outcome variable(s) before the program is implemented, and then after. This **before-after design** is the classic experimental design and is often used where evaluators have sufficient resources and control to design and implement a before-after outcome measurement and data collection process. Having done so, it is possible to calculate the before-after changes in both the treatment and control groups, as well as ensuring that the pretest results indicate the two groups are similar before the program begins. The second design is called the **after-only experimental design** and does not measure program/control differences before the treatment begins; it generally works well where random assignment to treatment and control groups has occurred. Random assignment generally assures that the only difference between the two groups is the treatment. Both designs in Table 3.1 include "no program" groups to achieve the "all other things being equal" comparison, which permits us to see what differences, if any, the program makes.

The random assignment of cases/units of analysis to treatment and control groups is indicated in Table 3.1 by the letter "R" in front of both the treatment and control groups. This process is intended to create a situation where the clients in the program and no-program groups are equivalent in

Table 3.1	Two Experimental Designs		
Pre-Test Post-Test Design (Classic)			
R_1	O_1	X	O_2
R_2	O_3		O_4
Post-Test Only Design (After-Only)			
R_1		X	O_1
R_2			O_2

all respects, except that one group gets the program. Not pretesting the two groups *assumes* that they are equivalent. Where the numbers of clients randomly assigned are small, pretesting can establish that the two groups are really equivalent, at least in terms of the outcome measure(s).

When reviewing Table 3.1, keep in mind that the "X" designates the program or treatment of interest in the evaluation. The "O"s indicate measurements of the outcome variable of interest in the evaluation. The "R" indicates that we have assigned cases/units of analysis (usually people) randomly to the two groups. Where we measure the outcome variable before and after the program has been implemented, we are able to calculate the average change in the level of outcome. For example, if our program was focused on improving knowledge of parenting skills, we could measure the average gain in knowledge (the "after" minus the "before" scores) for the program group $(O_2–O_1)$ and the control group $(O_4–O_3)$. When we compare the average gain in outcome levels between the two groups after the program has been implemented, we can see what the incremental effect of the program was.

Where we do not have pretest measures of the outcome variable, as in the after-only experimental design, we would compare the averages of the program and control groups after the program was implemented $(O_1–O_2)$. One additional thing to keep in mind is that in experimental evaluations where we have several outcomes of interest, we have separate (but parallel) experimental designs for each variable. For example, if we were interested in evaluating the attitude changes towards parenting as well as the knowledge gains from a parenting program, we would have two research designs—one for each outcome. Likewise, where we have more than one treatment, or combination of treatments, each combination is designated by a separate X, subscripted appropriately.

The second design in Table 3.1 does have the potential of solving one problem that can arise in the first design. Sometimes, pretesting can be intrusive and have its own effect on the posttest measurement. Suppose you are evaluating a server intervention program that is intended to train employees who serve alcoholic beverages in bars and other establishments. One expected outcome might be improved knowledge of ways to spot customers who should not be served any more drinks, and how to say "no" in such situations.

If "knowledge level" was measured with the responses to a set of true-false statements before the training, and the same instrument was used after the training, we might expect higher scores simply because employees are familiar with the instrument. Using the first design might produce outcomes that are higher than they should be, misleading those who might want to generalize the results to other program situations.

WHY PAY ATTENTION TO EXPERIMENTAL DESIGNS? •

The field of program evaluation continues to debate the value of **experimental designs** (Shadish, Cook, & Campbell, 2002). On the one hand, they are generally seen to be costly, require more control over the program setting than is feasible, and are vulnerable to a variety of implementation problems. But for some evaluators, and some government jurisdictions, experimental designs continue to be "the gold standard" when it comes to testing **causal relationship**s (Lipsey, 2000; Office of Management and Budget, 2004).

Weisburd (2003), in a discussion of the ethics of randomized trials, asserts that the superior (internal) validity of randomized experiments makes them the ethical choice in criminal justice evaluations.

> At the core of my argument is the idea of ethical practice. In some sense, I turn traditional discussion of the ethics of experimentation on its head. Traditionally, it has been assumed that the burden has been on the experimenter to explain why it is ethical to use experimental methods. My suggestion is that we must begin rather with a case for why experiments should not be used. The burden here is on the researcher to explain why a less valid method should be the basis for coming to conclusions about treatment or practice. The ethical problem is that when choosing non-experimental methods we may be violating our basic professional obligation to provide the most valid answers we can to the questions that we are asked to answer. (p. 350)

Although Weisburd's view would be supported by some advocates of experimentation in evaluation, most practitioners recognize that there can be risks associated with randomized experiments. Shadish, Cook, and Campbell (2002) discuss the ethics of experimentation and point out that in the history of research with human participants, there are examples that have shaped our current emphasis on protecting the rights of individuals, including their right to informed consent, before random assignment occurs.

Among the best known American examples of controversial experiments was a series of studies conducted by Stanley Milgram (1983). His work focused on obedience to authority and was intended to see whether American university students who had agreed to be experimental subjects could be induced to behave in ways that would injure others, based on obeying the orders of the experimenter. Milgram demonstrated that his subjects regularly behaved in ways that they believed were harming a person whom they were ordered to punish in the experiments.

One feature of the obedience experiments was deception: the subjects did not know that the experimenter was colluding with the person who was supposedly being punished. Instead the subjects thought that they were actually administering electric shocks that were harmful or perhaps even fatal.

Deception has become a central concern with any research, but is highlighted in situations where people are randomly assigned and one group does not receive a possible benefit. In situations where the participants are disadvantaged (socially, economically, or mentally), even informed consent may not be adequate to ensure that persons fully understand the consequences of agreeing to random assignment. Shadish, Cook, and Campbell (2002) suggest strategies for dealing with situations where withholding treatment is problematic. For example, persons assigned to the control group can be promised the treatment at a later point.

We discuss the ethics of evaluation in Chapter 11 and suggest core values that need to be taken into account in conducting evaluations. The evaluation profession has guidelines for practice, but unlike most other professions, they are not enforceable.

Some program evaluators have argued that because opportunities to use experimental or even quasi-experimental designs are quite limited, the whole idea of making experiments the paradigm for program evaluations that examine causes and effects is misleading. We are setting up an ideal that is not achievable and expecting evaluators to deal with issues that they cannot be expected to resolve. As Berk and Rossi (1999) argue, "there is really no such thing as a truly perfect evaluation, and idealized textbook treatments of research design and analysis typically establish useful aspirations but unrealistic expectations" (p. 9).

The reality is that many situations in which evaluations are wanted simply do not permit the kind of control and resources that experiments demand, yet we do proceed with the evaluation, knowing that our findings, conclusions, and recommendations will be based in part on evidence that does not meet the standards implied by the experimental approach. Evidence is the essential core around which any program evaluation is built, but the constraints on resources and time available will usually mean that at least some issues that ideally should be settled with data from experiments will, in fact, be settled with professional judgments.

The key point is that there is value in knowing the requirements for experimental evaluations. Understanding the logic of experimental designs is important for understanding what we are claiming when we address issues of program effectiveness. Not being able to conduct an experiment does not change the need for us to appreciate what is involved in answering the question: Did the program cause the outcomes that we observe? We believe that most evaluators who are asked to address the issue of program effectiveness can tackle this question—perhaps not answering it definitively, but instead, reducing the uncertainty around the question.

USING EXPERIMENTAL DESIGNS TO EVALUATE PROGRAMS: ● THE ELMIRA NURSE HOME VISITATION PROGRAM

In the field of evaluation, there is a rich literature that chronicles the experiences of researchers and practitioners with studies where a core feature is the use of randomized experiments. Although the field has and continues to diversify in terms of criteria for judging appropriate evaluation designs, randomized experiments continue to be a key part of our profession. In the United States, there are currently several groups of researchers who are committed to using randomized experiments to understand how programs work to ameliorate social problems.

One such group is headed by David Olds, and since 1977 he and his colleagues have conducted three extensive randomized experiments in Elmira, New York, Memphis, Tennessee, and Denver, Colorado to determine whether home visits by nurses to first-time mothers prevent poor pregnancy outcomes, prevent subsequent health and development problems, and improve the mother's own life (Olds, Henderson, Kitzman, et al., 1998). The Elmira study was the first one conducted and we will summarize its main features to illustrate how randomized experiments have been used to address social policy issues.

The Elmira Nurse Home Visitation Program

Begun in 1977, the Elmira Nurse Home Visitation Program was implemented with pregnant women who had not given birth before and had one or more of the following characteristics: were young (19 years or younger), a single parent, or poor. These conditions, or combinations of them, were expected to result in the mothers being more likely to be smokers, drug or alcohol users, come from families where their own childhood experiences were difficult, have diet and exercise deficiencies, and in general not have the health, knowledge, skills, and attitudes that predicted successful birthing and caring for their first baby.

Random Assignment Procedures

A total of 400 women were randomly assigned to one of four groups. In group 1, 94 women were offered screening services for their babies at age 1 and again at age 2 to see whether the babies had developmental problems. In group 2, 90 were offered the screening services plus free transportation to their prenatal and postnatal medical appointments, up to age 2. In group 3, 100 women were offered screening, transportation and regular home visits by nurses during their pregnancy. And in group 4, 116 women got screening, transportation, and home visits both before they gave birth and for the first 2 years of their child's life. The first two groups were the "control group" in the main analyses, and groups 3 and 4 were the "experimental group," those mothers who received home visits by nurses.

The randomization procedure occurred in several stages. First, women who generally fit the criteria for "at risk" mothers were invited to participate in the study. Between 1978 and 1980, 500 women were interviewed and, of those, 400 agreed to participate in the study. Each participant was assigned to one of the four groups using a deck of cards—at the end of the intake interview, each woman was invited to draw a card from a deck that consisted of randomly shuffled cards with a number from 1 to 4 on each card. As the 400 participants were recruited, the deck was periodically rebalanced to overrepresent the groups that had fewer participants (Olds, Henderson, Chamberlin, & Tatelbaum, 1986).

Comparisons of the control and experimental (or program) groups at the point where they were recruited indicated that the two groups were equivalent in their background characteristics. Very few of the participants dropped out of the study before giving birth, meaning that the two groups continued to be equivalent in terms of background characteristics (Olds

Table 3.2	\multicolumn{2}{l}{Research Design for the Elmira Nurse Home Visitation Program}

$$R_1 \quad X_1 \quad O_1 \text{ to } O_n$$

$$R_2 \quad X_2 \quad O_2 \text{ to } O_n$$

$$R_3 \quad X_3 \quad O_3 \text{ to } O_n$$

$$R_4 \quad X_4 \quad O_4 \text{ to } O_n$$

Control Group

X_1 is developmental screening for the first born child, ages 0 to 2.

X_2 is screening plus transportation for pre- and postnatal appointments.

Program Group

X_3 is screening, transportation, and prenatal nurse home visits.

X_4 is screening, transportation, prenatal nurse home visits, and postnatal home visits.

et al., 1986). Table 3.2 shows the research design for the first 3 years of the Elmira study (prenatal to 2 years of age).

In the research design, there were many different variables to measure the effects of the treatments on the children and the mothers. These have been designated as the O_{1-4} to O_n series of variables for the four groups.

The Findings

The findings from the Elmira nurse home visitation experiment suggest that when the home-visited mothers are compared to those that did not get this program, home-visited mothers tended to change their attitudes and behaviors before giving birth: their diets improved; they smoked fewer cigarettes; they had fewer kidney infections; they took advantage of community services; and among young mothers aged 14 to 16, they tended to give birth to heavier babies.

In the first years of childhood, home-visited mothers tended to have fewer incidents of abused or neglected children, fewer injuries for their children, less severe injuries where there was physical abuse, fewer subsequent pregnancies, and more work force participation.

Although the initial plan was to follow the mothers and their children up to age 2, additional funding permitted the study to continue so that the parents and children have been followed to age 15. The longer time frame has meant that additional variables could be measured: home visited mothers had fewer subsequent births; longer times until the next birth after their first born child; less time on welfare; and among the children, fewer behavior problems due to substance abuse and fewer arrests.

A cost-benefit analysis comparing the two groups up to age 15 showed that the costs of the program were considerably less than the benefits to society from problems (and costs) prevented by the program, particularly for families where the mother had been low income and single at the time she was pregnant. The experimental design made it possible to say that the problems prevented (arrests, for example) were attributable to the home visitation program.

Policy Implications of the Home Visitation Research Program

The Elmira study is part of a sustained research program that has been focused on the question of how to reduce child and family problems associated with young women in difficult circumstances getting pregnant and giving birth. Over the past 28 years, randomized controlled trials of different combinations of program components have been conducted by Olds and others in different locations in the United States (Olds, O'Brien, Racine, et al., 1998). The commitment of the evaluation community to this issue suggests that the cumulative findings should be the basis for public policies. However, even with the sustained research and cumulative evidence that home visits by nurses do work, there are substantial barriers to the widespread dissemination of these programs.

First, critics point out that the findings from studies are not consistent. Given the volume of research, the variability in methods and rigor, some studies find no effects of nurse home visits, notwithstanding that the most rigorous ones consistently find positive effects from this intervention (Olds, Hill, Robinson, et al., 2000). Public policy makers can exploit this diversity in findings by suggesting that the results are not strong enough.

Second, advocacy groups tend to oversell the research results to make their case that public monies ought to be spent. Again, the issue that is ultimately raised by this approach is the credibility of the programs.

Third, these programs are costly. And in a political environment where the trend is to limit or even reduce social programs, any efforts to spend more are met with skepticism.

Finally, even the most rigorous randomized controlled trials of these programs run into a substantial problem in being able to translate program components that are tailored to an experiment into institutionalized programs. The variability in local circumstances means that even if funding is available, fidelity to the program components that have been tested is not assured. Even in the Elmira program, once the 2-year point after the births of the first-born children had occurred, changes in the leadership of the local funding agency resulted in substantial changes to the focus of the program. It took considerable pressure and negative publicity to steer the program back to its original focus (Olds, O'Brien, Racine, et al., 1998).

Olds and his colleagues sum up their own stance about the policy relevance of their research in the following way (Olds, O'Brien, Racine, et al., 1998):

> The divisiveness of current national policy debate and the understandable public skepticism about a social science which in the past has promised more than it can deliver, should send us neither into hiding while we wait for a better day nor into overzealous advocacy as we declare our political allegiances. Rather, we should seek to maintain both the integrity and utility of the scientific enterprise—both its capacity for producing new knowledge and the applicability of that knowledge to social problems that must be addressed in the present. This, at least, is the perspective we intend to adopt in our continuing effort to make what we learn through careful study about effective home visitation service relevant in the public domain. (p. 98)

ESTABLISHING VALIDITY IN RESEARCH DESIGNS ●

The example of spurious increases in knowledge due to testing and re-testing knowledge in a server-training program is one possible rival hypothesis that could serve as an alternate explanation for observed outcomes. Having a no-program group mitigates that problem, but it persists if efforts are made to use evaluation results as a benchmark for other, or future, program settings.

Over the last 35 years, several major contributions have been made to describe the ways that research designs can be weakened by validity problems. Cook and Campbell (1979) defined and described four different classes of validity problems that can compromise research designs in program evaluations.

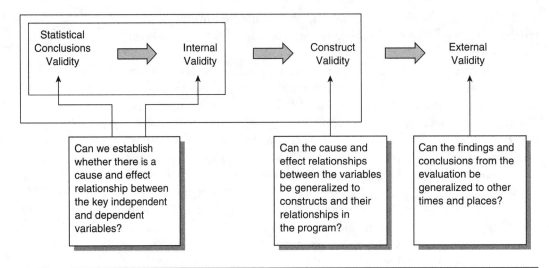

Figure 3.5   The Four Kinds of Validity in Research Designs

Figure 3.5 shows the four kinds of validity described by Cook and Campbell, and suggests ways that they can be linked. **Statistical conclusions validity** "feeds into" **internal validity**, and the two together support **construct validity**. All three support **external validity**. The questions in Figure 3.5 indicate the key issue that each kind of validity is intended to address. Notice that statistical conclusions validity and internal validity focus on the variables *as they are measured* in a program evaluation. Construct validity and external validity are both about generalizing; the former involves generalizing from the measured variables back to the constructs in the program model, and the latter is about generalizing the evaluation results to other situations.

In the following section, we will expand on these four kinds of validity, and the threats to each kind.

● DEFINING AND WORKING WITH THE FOUR KINDS OF VALIDITY

There are three conditions for establishing a causal relationship between two variables (Shadish et al., 2002).

1. **Temporal asymmetry**, that is, the variable that is said to be the cause precedes the variable that is the effect.

2. Co-variation, that is, as one variable varies, the other also varies either positively or negatively.

3. No plausible **rival hypotheses**, that is, no other factors that could plausibly explain the co-variation between the two variables.

These three conditions are **individually necessary and jointly sufficient** for establishing a causal relationship between two variables. The first tends to be treated at a conceptual level as well as at an empirical level. We hypothesize temporal asymmetry, and then look for ways of observing it in our program implementations. The second and third conditions are addressed by statistical conclusions validity and internal validity, respectively.

Statistical Conclusions Validity

Statistical conclusions validity is about establishing the existence and strength of the co-variation between the cause and the effect variables. To do so, one must pay attention to issues that fall within the purview of statistical analysis.

This book is not concerned with the details of statistical methods. Indeed, there are a large number of statistical tests that program evaluators can use, each having its own requirements for defensible applications. The appendix to this chapter offers a succinct summary of some of the statistical methods that are used by evaluators, as well as conditions under which they can be used.

For the purposes of this book, it is important to know that statistical conclusions validity involves the following (Cook & Campbell, 1979):

- Ensuring that sampling procedures are adequate, if they are used to gather data
- Ensuring that the appropriate statistical tests have been applied and that the outcomes have been interpreted correctly
- Ensuring that the measures that are used are sufficiently reliable

The third condition focuses on a general assumption that is usually made in applying most statistical tests—the theoretical models that underlie statistical tests assume that the measurement procedures that were used to gather the data are reliable (Berry, 1993).

Working With Internal Validity

Internal validity is about ruling out rival hypotheses. Beginning with Campbell and Stanley (1966), contributors to the program evaluation literature have categorized threats to the internal validity of research designs. We will learn about nine kinds of threats to internal validity. *Each one* is a *category* of possible problems and in a given situation there may be none, one, or even several factors that are a problem for a particular category.

Threats to Internal Validity

1. **History:** External events that coincide with the implementation of a program.

    Example: A state-wide Counter Attack program aimed at reducing accidents and injuries on state highways due to alcohol consumption is introduced in May. A state seat belt law is introduced in October of the same year. Because the seat belt law is intended to reduce accidents and injuries, it is virtually impossible to disentangle the outcomes of the two programs.

2. **Maturation:** As program participants grow older, their behaviors tend to change in ways that appear to be program outcomes.

    Example: A youth vandalism prevention program in a community is developed in a short period of time during a period of rapid population growth. The population matures roughly as a cohort. Children born in the community also mature as a cohort. If a program is developed to "combat a rising youth vandalism problem" when the average youth age is 12, by the time the average age is 16, the community may have outgrown the problem even without the program.

3. **Testing:** Taking the same posttest as had been administered as a pretest can produce higher posttest scores due to learning about the testing procedure.

    Example: Using a pre/postmeasure of knowledge in a server training program as described earlier, servers score higher on a test of "knowledge level" after the training, not because they have increased their knowledge during training, but simply because they are familiar with the test from when it was administered before the training.

4. **Instrumentation:** As the program is implemented, the way in which key variables are measured also changes.

> Example: A program to decrease burglaries is implemented at the same time that the records system in a police department is automated: reporting forms change, definitions of different types of crimes are clarified, and a greater effort is made to "capture" all crimes reported in the database. The net effect is to "increase" the number of reported crimes.

5. **Statistical regression:** Extreme scores on a pretest tend to have more measurement error. Thus, if program participants are selected because they scored low or high on the pretest, their scores will tend to regress toward the mean of the scores for all possible participants on the posttest.

> Example: People are selected for an employment skills training program on the basis of low scores on a self-esteem measure. On the posttest, their self-esteem scores increase.

6. **Selection:** Persons/units of analysis chosen for the program are different from those chosen for the control group.

> Example: A program to lower recidivism among youth offenders selects candidates for the program from the population in a juvenile detention center. The candidates are selected in part because they are thought to be reasonable risks in a half-way house living environment. If this group was compared to the rest of the population in the detention center (as a control group), differences between the two groups of youths, which could themselves predict recidivism, might explain program outcomes/comparisons.

7. **Mortality:** People/units of analysis "drop out" over the course of the program.

> Example: A program to rehabilitate chronic drug users may lose participants who would be least likely to succeed in the program. If the pretest "group" were simply compared to the posttest group, one could mistakenly conclude that the program had been successful.

8. **Ambiguous temporal sequence in the cause and effect variable:** It is not clear whether the key variable in the program causes the outcome, or vice versa.

> Example: A program that is intended to improve worker productivity hypothesizes that by improving worker morale, productivity

will improve. The data show that both morale and worker productivity improve. But the program designers may well have missed that improved morale is not the cause, but the effect. Or there is a reciprocal relationship between the two variables such that improvements in morale will induce improvements in productivity which, in turn, will induce improved morale, which in turn, improves productivity.

9. **Selection-based interactions:** Selection interacts with other internal validity threats so that the two (or more) threats produce joint effects (additive and otherwise) on variables of interest.

Example: A program to improve reading abilities in a school district is implemented so that program classes are located in higher-income areas and control classes in lower-income areas. Tests are given (pre, post) to both groups and the findings are confounded not only by selection bias, but also by the fact that higher-income children tend to mature academically more quickly.

Each of the nine categories of internal validity threats suggests possible solutions to the particular problems, although designing a study to sort out reciprocal or ambiguous causation can be challenging (Shadish et al., 2002). To avoid the intrusion of history factors, for example, anticipate environmental events that could coincide with the implementation of a program and, ideally, deploy a control group so that the history factors affect both groups, making it possible to sort out the incremental effects of the program.

The difficulty with that advice, or corresponding "solutions" to the other eight types of problems, is in having sufficient resources and control over the program design and implementation to structure the situation to effectively permit a "problem-free" research design.

Introducing Quasi-Experimental Designs: The Connecticut Crackdown on Speeding and the Neighborhood Watch Evaluation in York, Pennsylvania

Fundamentally, all research designs are about facilitating comparisons. Well-implemented experimental designs, because they involve random assignment of units of analysis to treatment and control groups, are constructed so that program and no-program situations can be compared "holding constant" other variables that might explain the observed differences in program outcomes. It is also possible to construct and apply research designs that allow us to compare program and no-program conditions in circumstances where random assignment does not occur. These **quasi-experimental** research designs typically are able to address one or more categories of internal validity threats, but not all of them.

Resolving threats to internal validity in situations where there are insufficient resources or control to design and implement an experiment usually requires the judicious application of designs that reduce or even eliminate threats that are most problematic in a given situation. Usually, the circumstances surrounding a program will mean that some potential problems are not plausible threats. For example, evaluating a 1-week training course for the servers of alcoholic beverages is not likely to be confounded by maturation of the participants.

In other situations, it is possible to construct research designs that take advantage of opportunities to use existing data sources, and combine these designs with ones involving the collecting of data specifically for the evaluation, to create **patched-up research designs** (Poister, 1978) that are combinations of quasi-experimental designs. Patched-up designs are usually stronger than any one of the quasi-experimental designs that comprise the patchwork, but typically still present internal validity challenges to evaluators.

The Connecticut Crackdown on Speeding

Donald T. Campbell and H. Laurence Ross (1968), in their discussion of the Connecticut Crackdown on Speeding, show how successive applications of quasi-experimental research designs can build a case for determining whether a program intervention is successful. They discuss three different ways of assessing whether the introduction of a state-wide crackdown on speeding motorists in 1955 resulted in a reduction of traffic-related fatalities.

The **before-after design** illustrated below shows that traffic fatalities in Connecticut had decreased from 325 in 1955 to 285 in 1956. In the diagram below, the "O" is reported traffic-related fatalities, measured before and after the crackdown on speeding was implemented. The crackdown was a policy change, not a program, and is the "X" in the diagram.

O X O

The reduction in fatalities could have been due to a number of other factors (threats to internal validity) including history (another policy or program might have been implemented then), maturation (automobile drivers on average might have been older and more safety-conscious), instrumentation (the way that reported accident statistics were gathered and recorded could have changed), and statistical regression (the reported accident level before the crackdown could have been unusually high, given the historical

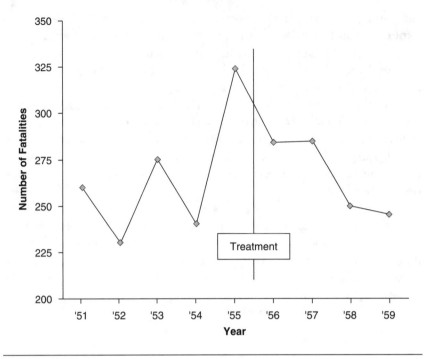

Figure 3.6   Reported Traffic Fatalities in Connecticut: 1951–1959

SOURCE: Campbell, D .T. & Ross, H. L. (1968, p. 42).

trend). So, to reduce the likelihood of those threats, Campbell and Ross introduced a **single time-series design**, which compared fatality levels for 5 years before and 4 years after the crackdown in Connecticut. This time series can be diagramed as follows:

OOOOOXOOOO

The data for 9 years, displayed in Figure 3.6, suggested that maturation and instrumentation did not appear to be rival hypotheses. There is no overall trend downward that might have suggested that Connecticut drivers were maturing, hence, less likely to take the risks associated with speeding. Although the before-after comparisons could have indicated that instrumentation was a problem, a review of data collection practices indicated that there was no change in the way accident data were gathered coincident with the crackdown. The data pattern displayed in Figure 3.6 could have been due to statistical regression—the jump in reported fatalities in 1954 was the biggest in the time series and would suggest the possibility of a movement in the next year back towards the mean of the pre-crackdown time series.

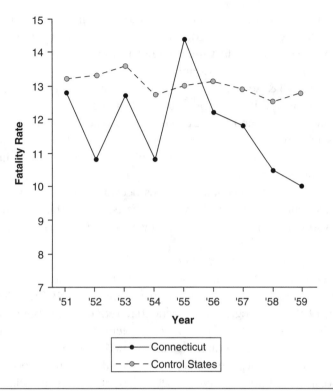

Figure 3.7   Reported Traffic Fatalities in Connecticut and Four Comparison
States: 1951–1959 (per 100,000 population)

SOURCE: Campbell, D .T. & Ross, H. L. (1968, p. 44).

To deal with possible history threats (some other event or events that coincided with the crackdown) and to deal with statistical regression, Campbell and Ross looked at traffic fatalities in the four surrounding states (New Jersey, New York, Massachusetts, and Rhode Island), both individually and combined. This **comparative time-series design** suggested that the crackdown in Connecticut did reduce fatalities.

The comparison with nearby states can be diagrammed as follows:

OOOOOXOOOO

OOOOO OOOO

Figure 3.7 displays two time series: one for Connecticut and the one for the four control states (New Jersey, New York, Massachusetts, and Rhode Island), averaged together. The patterns in the data suggest that the drop in reported fatalities in Connecticut was not replicated in the control states. So, although statistical regression continues to be a possible rival hypothesis, it needs to be weighted against the evidence from the control states. Further, the big jump in fatalities in Connecticut in 1954 also happened in Rhode Island (Campbell & Ross, 1968, p. 45), suggesting that large changes in year-over-year fatality levels were not unique to Connecticut.

Finally, Campbell and Ross examined the 9-year trends in several output variables in Connecticut—variables which might have been expected to be affected by the crackdown, and which would have contributed in turn to the intended outcome of reduced traffic fatalities. License suspensions for speeding increased sharply and speeding violations decreased. Both of these changes were consistent with expected program effects. In addition, an unintended effect was observed: arrests while driving with suspended licenses increased.

The York Neighborhood Watch Program

A second example of such a strategy was the evaluation of a neighbor-hood watch program in York, Pennsylvania, in the 1970s (Poister, McDavid, & Magoun, 1979). The program was intended to reduce crime, but was focused specifically on reducing burglaries. It was implemented in one area of the city and a no-program "control" area was established for comparison.

Reported burglaries were tracked at both the neighborhood and city-wide levels. In addition, a survey of all the block captains in neighborhood watch blocks was conducted to solicit their perceptions of the program, including estimates of resident attendance at neighborhood watch meetings. Finally, key environmental factors were also measured for the entire period, the principal one being the unemployment rate in the whole community.

Several research designs were embedded in the evaluation. At the level of the neighborhood watch blocks, the program was implemented and the block captains were interviewed.

$$X \quad O$$

where X = the neighborhood watch program and O = the measurement of block captain perceptions of program activity. This research design is called an **implicit design** (sometimes called a **case study design**), which is a weak design by itself because it does not include any comparisons that control for possible threats to internal validity.

Reported burglaries were compared between the neighborhoods that received the program and those that did not.

Program   OOOOOOOOXOXOXOXOXOXOXOXOXOXO

No Program   OOOOOOOOOOOOOOOOOOOOOOOOOOOO

where X = the neighborhood watch program and O = the reported burglaries in the program and no-program areas of the city. Notice that for the program area of the city, we show the "X"s and "O"s being intermingled. That shows that the program continued to operate for the full length of the time series, once it was implemented. This comparative time-series design is typically stronger than the implicit design because it includes a no-program group. Among the threats to this design is the possibility that the program group is not comparable to the no-program group (selection bias). That could mean that differences in reported burglaries are due to the differences in the two types of neighborhoods, and not the program.

Finally, reported burglaries were compared before and after the program was implemented, city-wide.

$$OOOOOOOOOOOXOXOXOXOXOXOXO$$

This single-time series design is vulnerable to several internal validity threats. In this case, what if some external factor or factors intervened at the same time that the program was implemented (history effects)? What if the way in which reported burglaries were counted changed as the program was implemented (instrumentation)? What if the city-wide burglary rate had jumped just before the program was implemented (statistical regression)?

In this evaluation, several external factors (unemployment rates in the community) were also measured for the same time period and compared to the city-wide burglary levels. These were thought to be possible rival hypotheses (history effects) that could have explained the changes in burglary rates.

Findings and Conclusions From the Neighborhood Watch Evaluation

The evaluation conclusions indicated that at the block level, there was some activity, but attendance at meetings was sporadic. A total of 62 blocks had been organized by the time the evaluation was conducted. That number was a small fraction of the 300-plus blocks in the program area. At the neighborhood level, reported burglaries appeared to decrease in *both* the program and no-program areas of the city. Finally, city-wide burglaries decreased shortly after the program was implemented. But given the sporadic activity in the neighborhood watch blocks, it seemed likely that some other environmental factor or factors had caused the drop in burglaries.

Figure 3.8 displays a graph of the burglaries in the entire city from 1974 through 1980. During that time, the police department implemented *two* programs: a neighborhood watch program and a team-policing program. The latter involved dividing the city into team policing zones and permanently assigning both patrol and detective officers to those areas.

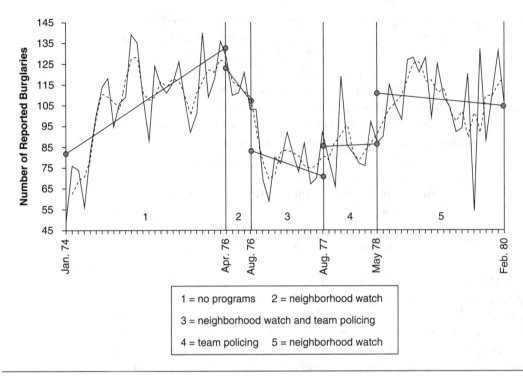

Figure 3.8   Burglary Levels in York, Pennsylvania: January 1974–February 1980

Figure 3.8 is divided into five time periods. The level of reported burglaries varies considerably, but by calculating a 3-month moving average (burglaries for January, February, and March would be averaged and that average reported for February; burglaries for February, March, and April would be averaged and that average reported for March, and so on), the graph is stabilized somewhat. The 3-month moving average is displayed as the dotted line.

As you can see by inspecting the graph, the police department initially implemented the neighborhood watch program, then shortly afterward moved to team policing as well. Both team-policing and the neighborhood watch program were in operation for period 3, then neighborhood watch was cancelled, but team policing continued (period 4). Finally, because the detective division succeeded in its efforts to persuade the department to cancel the team-policing program (detectives objected to being assigned to area-focused teams and wanted instead to operate city-wide), the police department restarted the neighborhood watch program (period 5).

Inspection of Figure 3.8 indicates that burglaries were increasing in the period prior to implementing the neighborhood watch program in 1976.

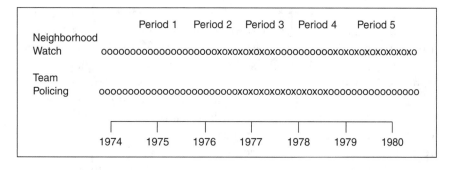

Figure 3.9   Implementation and Withdrawal of Neighborhood Watch and Team Policing

Burglaries dropped, but within 5 months of the neighborhood watch program being started up, team policing was implemented city-wide. When two programs are implemented so closely together in time, it is often not possible to sort out their respective outcomes—in effect, one program becomes a "history" rival hypothesis for the other. In this situation, the public response to a perceived burglary problem consisted of doing as much as possible to eliminate the problem. Although implementing the two programs may have been a good political response, it confounded any efforts to sort out the effects of the two programs, had the evaluation time frame ended in 1977.

By extending the time series as is shown in Figure 3.9, it was possible to capture two additional program changes: withdrawal of team policing in 1978 and the reinstatement of the neighborhood watch program at that point. Figure 3.9 depicts these program changes between 1974 and 1980. The neighborhood watch program is shown as a single-time series wherein the program is implemented (1976), withdrawn (1977–1978), and then implemented again (1978–1980). This on-off-on pattern facilitates being able to detect whether the program impacted on reported burglaries, notwithstanding some difficulties in drawing "borders" between the no-program and program periods. Because some neighborhood watch blocks could continue operating beyond the "end" of program funding in 1977, it is possible that some program outputs (block meetings, for example) for that program persisted beyond that point.

Team policing, being an organizational change, would likely experience some start-up problems (e.g., officers getting to know their assigned neighborhoods); but when it ended, there would be little carryover to the next segment of the time series.

It is clear from Figure 3.9 that when team policing and neighborhood watch operated together (period 3), the level of burglaries was lowest in the time series. When team policing operated alone (period 4), burglaries increased somewhat, but were still substantially lower than they were for either of the periods (2 and 5) when neighborhood watch operated alone.

Based on Figure 3.9 and the findings from evaluating neighborhood watch at the block and area of the city levels, it is reasonable (although not definitive) to conclude that the team-policing program was *primarily* responsible for reducing burglaries. Our conclusion is not definitive—very few program evaluation findings are—but it is consistent with the evidence and serves to reduce the uncertainty around the question of relative program effectiveness.

The evaluation of the York crime prevention programs employed several different research designs. Time-series designs can be useful for assessing program outcomes in situations where data exist for key program logic constructs before and after (or during) program implementation. The implicit design used to survey the block watch captains is perhaps the most vulnerable to internal validity problems of any of the possible research designs evaluators can use. For implicit designs, modeled as (X O), there is neither a pretest nor a control group, so we may not be able to reduce any of the uncertainty around the question of whether the program caused the observed outcomes.

But even in this situation (which is actually quite common for program evaluators), there are ways of strengthening your hand. Suppose you have been asked to evaluate a program that offers small businesses subsidies to hire people aged 17 to 24. The objectives of the program are to provide work experience, improved knowledge of business environments, and encouragement to either start their own business or pursue business-related postsecondary education.

As a program evaluator, it would be worthwhile having a **comparison group** who did not get the program, so that constructs like "increased knowledge of business practices" could be measured and the results compared. But that may not be possible, given resource constraints. Instead, you might still be expected to evaluate the program for its effectiveness and be expected to do so by focusing on the program alone.

One way to reduce uncertainty in the conclusions drawn is to acknowledge the limitations of an implicit (X O) design, but apply the design to *different stakeholder groups*. In the business experience program evaluation, it would make sense to survey (or interview) a sample of clients, a sample of employers, and the program providers. These three viewpoints on the program are complementary and allow the evaluator to triangulate the perspectives of stakeholders. In effect, the X O research design has been repeated for three different variables: client perceptions, employer perceptions, and program provider perceptions.

Triangulation is an idea that had its origins in the literature on measurement. As a measurement strategy, triangulation is intended to strengthen confidence in the validity of measures used in social research.

> Once a proposition has been confirmed by two or more independent measurement processes, the uncertainty of its interpretation is greatly reduced. The most persuasive evidence comes through a triangulation of measurement processes. If a proposition can survive the onslaught of a series of imperfect measures, with all their irrelevant error, confidence should be placed in it. (Webb, 1966, p. 3)

In our situation, triangulation that is focused on the question of whether the program was effective can at least establish whether there is a concurrence of viewpoints on this question, as well as other related issues. It does not offer a firm solution to the problem of our vulnerable research design, but it offers a workable strategy for increasing confidence in evaluation findings.

Construct Validity

Figure 3.5 indicates that construct validity is about being able to generalize from the variables and their observed relationships in a program evaluation back to the constructs and their relationships in the program logic. **Measurement** is about translating constructs into **observables**, and that at the "level" of observables, we work with variables, not the constructs themselves. In the evaluation of the York Neighborhood Watch Program, for example, the key dependent variable was the monthly total of burglaries that had been reported (and recorded) by the police department.

The evaluators wanted to learn whether the program was effective in *reducing burglaries* in York. The construct implicit in this question is the number of burglaries committed—we have measured the construct by *assuming* that reported burglaries is a valid measure of burglaries committed.

Basically, construct validity problems can result from the way *the program* has been **operationalized** in the evaluation, or the ways that *outputs, linking constructs, or outcomes* have been measured.

An example of a construct validity problem that involves the way the program has been operationalized can be taken from the Perry Preschool Program evaluation (Berrueta-Clement, 1984; Schweinhart, Barnes, & Weikart, 1993).

In the Perry Preschool experiment (1962–1967), children from families in Ypsilanti, Michigan, aged 3 and 4, were randomly assigned in cohorts to program and no-program groups. The purpose of the experiment was to see what the effects of preschool are on a set of variables related to academic achievement and social adjustment.

The authors reported that younger siblings of children who had been assigned to the program were included in the program when they reached preschool age (Schweinhart et al., 1993). The program included a classroom preschool component as well as weekly home visits. But for families with two or more preschool program children, it is possible that siblings reinforced each other and created a different at-home environment. Under this condition, does the program include the two intended components, or those plus sibling reinforcement? This ambiguity is a construct validity problem for the evaluation.

In an evaluation of a server training program, the evaluators assigned matched pairs of drinking establishments in Thunder Bay, Ontario, to program and no-program conditions (Gliksman, McKenzie, Single, et al., 1993). Managers in the establishments had been asked if they were willing to have their servers trained, but were cautioned not to tell their servers about the evaluation. Given the incentives for managers to "look good" or "cooperate," it is possible that some managers mentioned the evaluation to their servers. The construct validity problem created is: What is the program—server training, or server training plus the informal influence of bar managers?

Shadish, Cook, and Campbell (2002) point out that construct validity issues and internal validity issues are both about factors that confound our interpretation of the results we observe in an evaluation. Internal validity threats are sets of factors that could explain the outcomes we observe, *regardless of whether the program/treatment happened.* If a seatbelt law is implemented in a state at the same time that a program is implemented to crack down on speeding, a reduction in traffic fatalities could be due to the seatbelt law, regardless of whether the program was implemented.

Construct validity issues arise from the way the treatment/program was implemented. They cannot occur if there has not been a treatment or program in place. In the example from the Perry Preschool experiment mentioned above, the construct validity problems are created by the possibility that older siblings of preschool group children, having been to preschool themselves, act as home tutors and socializers for their younger siblings. This situation results in ambiguity in just what the program is: preschool (including home visits) or preschool plus sibling influence.

In an earlier formulation of threats to internal validity (Cook & Campbell, 1979), four threats were discussed that focused on the context for a treatment/program implementation. They could not have occurred if the treatment/program had not been implemented. Because they affect our ability to generalize from the observed findings back to the constructs in the program theory, they are included here as examples of construct validity threats.

1. **Diffusion of treatments:** People can sometimes communicate about their program experiences to members of the control group.

   Example: Two groups of employees in a company are selected to participate in a team-building experiment. One group participates in team-building workshops. The other group (who may have an opportunity to take the workshop later) serves as the control group. Employees communicate, and some of the skills are transferred informally.

2. **Compensatory equalization of treatments:** The group that is not supposed to get the program is offered the program to avoid criticisms of withholding something that could be beneficial.

   Example: One neighborhood/area gets a neighborhood watch program, other areas (that were control neighborhoods) insist on the right to the program in their areas so the program spreads.

3. **Compensatory rivalry:** The no-program group is given a "better" version of the existing program, which tends to cancel out the differences between the new program and the existing program.

   Example: Two versions of a course are offered in an undergrad program, one uses interactive multimedia tools while the other is taught in a conventional classroom format. The classroom instructor works twice as hard as she usually does, not to be outdone by this new instructional method.

4. **Resentful demoralization**: The control group perceives unfair treatment and reacts negatively.

   Example: Those persons not getting a program to test the effects of class size on learning (halving the size of classes) complain to the instructor and demand equal treatment. The administration refuses and students threaten to not take any of the remaining tests in the course.

Cook and Campbell (1979) and Shadish, Cook, and Campbell (2002) present lists of circumstances that might weaken construct validity in an evaluation. In summarizing the ways of minimizing this set of problems, they suggest that evaluators need to: make sure that constructs are clearly defined so that they can be measured appropriately; make sure that constructs are differentiated so that they do not overlap as measures are developed; develop "good" measures, that is, measures that produce unbiased information; and, ideally, use multiple measures of key constructs.

Shadish, Cook, and Campbell (2002) also point out that construct validity threats can be caused by the fact that people know they are a part of an evaluation process. Participant expectations can influence behaviors, confounding attempts to generalize the actual findings back to the constructs or program theory. One example of the way that participant behavior can confound an experiment is the **Hawthorne effect**.

In a worker-productivity experiment in the 1930s in the U.S., the experimenters discovered that being part of an experiment produced an effect regardless of the levels of the experimental variables being manipulated (Roethlisberger, Dickson, & Wright, 1939). No matter what conditions the experimenters varied (e.g., lighting level, speed of the assembly line, variability of the work), the results indicated that *any* manipulation increased productivity. This outcome is usually referred to as a Hawthorne effect, named after the location where the research occurred. Construct validity was compromised by the behavior of the workers.

External Validity

Figure 3.5 suggests that external validity builds on the "validities" that have been discussed thus far: statistical conclusions, internal, and construct validity. External validity is about generalizing the causal results of a program evaluation to other settings, other people, other program variations, and other times.

Fundamentally, we want to know whether the program, the evaluation results, the clients/participants and the setting(s) are *representative* of other circumstances in which we might wish to apply the program or the evaluation results. In the evaluation of the server-training program in Thunder Bay, Ontario, how confident are we that the results of that evaluation are generalizable, given that the drinking establishments were chosen for convenience by the evaluation team. Are the results unique to that setting, or would they be representative of what would happen if server training programs were implemented in other establishments in Thunder Bay, or other cities in the province of Ontario?

Shadish, Cook, and Campbell (2002) suggest five categories of external validity threats. In each one, the causal results obtained from a given evaluation are threatened by factors that somehow make the results unique. Four of the five threats are interaction effects that reflect their concern with generalizing to other units of analysis (typically people), other program variations, other outcome variations, and other settings.

1. Interaction between the causal results of a program and the people/participants.

   Example: The Perry Preschool experiment was conducted with a mix of boys and girls, but the program appeared to work much more effectively for girls (Schweinhart et al., 1993). Does that mean that efforts to implement the program elsewhere would encounter the same limitation? Why did the program not produce the same results for boys?

2. Interaction between the causal results of a program and the treatment variations.

   Example: The Perry Preschool experiment had two main components. Children in the program cohorts attended preschool 5 days a week during the school year for 2 years. In addition, there were home visits to the families each week to reinforce the school setting and establish acceptance of preschool in the family. This version of the program is expensive. Because it is called a "preschool program" other implementations of the program might skip the home visits. Would that undermine the program results?

3. Interaction between the causal results of a program and outcome variations.

   Example: A state operated program that is intended to train unemployed workers for entry-level jobs succeeds in finding job placements (at least 6 months long) for 60% of its graduates. A comparison with workers who were eligible for the program but could not enroll due to space limitations, suggests that the program boosted employment rates from 30% to 60%—an incremental effect of 30%. Another state is interested in the program but wants to emphasize long-term employment (2 years or more). Would the program results hold up if the definition of the key outcome were changed?

4. Interaction between the causal results of a program and the setting.

   Example: The Abecedarian Project (F. A. Campbell & Ramey, 1994) was a randomized experiment intended to improve the school-related and cognitive skills of children for poor families in a North Carolina community. The setting was a university town, where most families enjoyed good incomes. The project was focused on a segment of the population that was small relative to the rest of the community. The program succeeded in improving cognitive, academic, and language skills. But could these results, robust though they are, be generalized to other settings where poor, predominately minority families resided?

There is a fifth threat to external validity that also limits generalizability of the causal results of an evaluation. Shadish, Cook, and Campbell (2002) call this "context-dependent mediation." An example would be a situation where a successful crime prevention program in a community used existing neighborhood associations to solicit interest in organizing blocks as neighborhood watch units. Because the neighborhood associations were well established and well known, the start-up time for the crime prevention program was negligible. Members of the executives of the associations volunteered to be the first block captains, and the program was able to show substantial numbers of blocks organized within 6 months of its inception.

The program success might have been mediated by the neighborhood associations—their absence in other communities could affect the number of blocks organized, and the overall success of the program.

● TESTING THE CAUSAL LINKAGES IN PROGRAM LOGIC MODELS

Research designs are intended as tools to examine causal relationships. In program evaluation, there has been a general tendency to focus research designs on the linkage between the program as a whole and the observed outcomes. Although we learn by doing this, the current emphasis on elaborating program descriptions as logic models presents situations where our logic models are generally richer and more complex than our research designs. When we evaluate programs, we generally want to examine the linkages in the logic model, so that we can see whether (for example) levels of outputs are correlated with levels of linking constructs, and they, in turn, are correlated with levels of outcomes. Research designs are important in helping us to isolate each linkage so that we can assess whether the intended causal relationships are corroborated. But as we said earlier, designing an evaluation so that each program component can successively be isolated to rule out rival hypotheses is expensive and generally not practical.

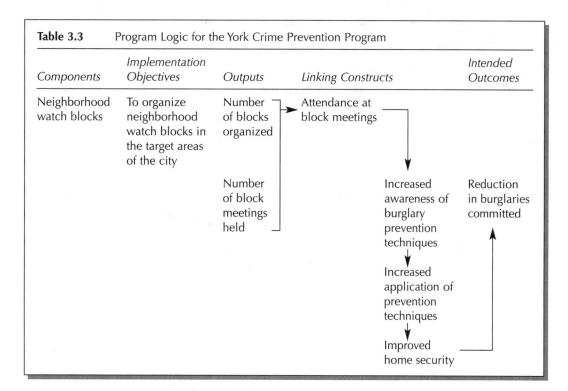

Table 3.3   Program Logic for the York Crime Prevention Program

Components	Implementation Objectives	Outputs	Linking Constructs	Intended Outcomes
Neighborhood watch blocks	To organize neighborhood watch blocks in the target areas of the city	Number of blocks organized	Attendance at block meetings	
		Number of block meetings held	Increased awareness of burglary prevention techniques	Reduction in burglaries committed
			Increased application of prevention techniques	
			Improved home security	

Having discussed some of the findings from the York crime prevention program, we can display the program logic in Table 3.3.

The intended outcome is a reduction in burglaries committed. To achieve that objective, the program logic specifies a series of steps, beginning with organizing city blocks into neighborhood watch blocks. If the program logic works as intended, then our program theory will have been corroborated in this implementation.

We can also summarize the research designs, the **units of analysis** and the key comparisons that were included in the York crime prevention program. Table 3.4 displays these features of the program.

By reviewing Figure 3.9, we can see that there were *four different* research designs in the evaluation and that *none* of them facilitated an examination of the whole logic model. Each design focused on one construct in the model, and the data collected permitted the evaluators to see *how that part of the logic model behaved.*

As an example, the block captain interviews focused on attendance at meetings, which is an important linking construct in the logic model. But they did not measure increased awareness of prevention techniques,

Table 3.4   Summary of the Overall Evaluation Design for the York Crime Prevention Program

Research Design	Unit of Analysis	Key Comparisons
Single time series	Time (monthly)	The number of blocks organized (an output variable) was recorded monthly by the police department and compared over time
Implicit design	Block captains	Block captains were interviewed at one point in time to obtain estimates of the numbers of meetings held (output), and resident attendance (linking construct)
Comparative time series	Time (monthly)	Reported burglaries (an outcome) were compared on a monthly basis in the two areas of York that were "program" and "control"
Single time series	Time (monthly)	Reported burglaries (an outcome) were compared each month city-wide before and after the program was implemented

increased applications of techniques, or improved home security. In fact, those three linking constructs were not measured in the evaluation. To do so would have required a survey of neighborhood residents, and there were insufficient resources to do that. Likewise, the time-series designs facilitated tracking changes in blocks that were organized, and reported burglaries over time, but were not set up to measure other constructs in the program logic.

In sum, each design addresses a part of the program logic, and helps us to see if those parts are behaving as the logic intended. But what is missing is a way to test the *connections* between constructs. Even if blocks are organized and neighbors attend block watch meetings, we do not know whether the steps leading to reduced burglaries have worked as intended.

Why not design program evaluations so that the whole logic model is tested? The main reason is cost. In the York crime prevention evaluation, a full test of the program logic in Table 3.3 (testing the intended linkages among outputs, linking constructs, and outcomes) would require that all the constructs be measured *using the same unit of analysis.*

Suppose that random samples of residents in the target and control neighborhoods were enlisted to participate in a 4-year study of crime prevention effectiveness in the city. Initially (2 years before the program started), each home (the unit of analysis) would be surveyed to find out if any had experienced burglaries in the past 12 months; householders would also be asked about their participation in any crime prevention activities, their awareness of burglary prevention techniques, and their existing home security measures.

This survey could be repeated (using the same sample) each year for 4 years (2 before the program, and 2 after implementation). After the program was implemented, the survey participants would also be asked whether they participated in the program, and if they did, how frequently they attended block watch meetings, whether the block watch meetings increased their awareness (they could be "tested" for their level of awareness), whether they were taking any new precautions to prevent burglaries, and finally, whether their homes had been burglarized in the previous 12 months.

Notice that this information "covers" the linkages in the logic model, and by comparing responses between the target and control neighborhoods, and responses within households over time, we could assess the causal linkages in all parts of the model. For example, comparisons between program and no-program residents after the program was implemented would indicate whether program residents were more likely to be aware of burglary prevention methods, more likely to apply such methods, and more likely to have more secure homes.

The Perry Preschool experimental study of the long-term effects of preschool on a sample of Black children in Ypsilanti, Michigan, has been able to follow the program and no-program children from age 3 to age 27 (Schweinhart et al., 1993). Over that time, multiple questionnaires were administered to participants and data gathered on a large number of separate measures of school achievement, marital and family relations, criminal behaviors, employment, earnings, and other personal, social, and economic variables.

The researchers have constructed and tested a logic model for the experiment. Key outcomes as of age 27 include: monthly earnings and lifetime arrests. These outcomes are explained in the model by years of schooling, school motivation in kindergarten and primary grades, IQ after 1 preschool year, whether the person had preschool or not, and finally, family characteristics at age 3 and 4 (Schweinhart et al., 1993).

Given sufficient money, time, and control over the program situation, it is possible to fully test program logics, as the Perry Preschool experiment demonstrates. Testing the program theory, using approaches that permit full tests of the logic model, is an important and growing part of the field (Bickman, 1987; Rogers, Hacsi, Petrosino, & Huebner, 2000). An additional example of this strategy is the evaluation of the WINGS project (Hennessy & Greenberg, 1999). The WINGS (Women In Group Support) project was focused on reducing sexual risk-taking among vulnerable women in the United States. The project was designed as a randomized experiment and implemented in multiple sites in 1992. The evaluators set up a logic model of the intended program theory that could be tested as a **recursive causal model**. In the program, women in the program groups were a part of 6 weekly sessions that provided information, modeled behaviors (including asking partners to use condoms), and provided opportunities for the participants to role play situations where the issue of safe sex would happen.

The evaluation demonstrated that the program was effective in reducing the infection rates from sexually transmitted diseases. The logic model was fully tested, to determine whether the intended linkages among constructs were corroborated by the data.

Donaldson's and Gooler's (2002) evaluation of the California Wellness Foundation's Work and Health Initiative introduces the role of evaluator judgment into the technical process of testing program theory. In an interview by Jody Fitzpatrick for the *American Journal of Evaluation,* Stewart Donaldson describes his approach to theory-driven program evaluation, based on two decades of experience (Fitzpatrick, 2002). Although there is clearly an important technical aspect to testing program theories, Donaldson implies that his own experience and judgment are also important parts of this process.

> Program theory clearly represents the presumed active ingredients of the program and how they lead to the desired outcomes. You're trying to consider whether the paths leading to this outcome have any chance at all for accounting for a significant portion of the variance in the desired outcome . . . I take each path and really think carefully about the timing and when it should occur and whether there are key moderators of that path. (p. 353)

When we look at the practice of evaluation, rarely do evaluators have the time, the resources, and the control to proceed as has been suggested above. Instead, we are usually expected to conduct program evaluations after the program has been implemented, using (mostly) existing data. These constraints usually mean that we can examine parts of program logics with evidence and other parts with our own observations and professional judgments.

● RESEARCH DESIGNS AND PERFORMANCE MEASUREMENT

The discussion so far has emphasized the connections between research designs and the central question of determining program effectiveness in program evaluations. Research design considerations need to be kept in focus if an evaluator wants to be in a position to construct and complete a credible evaluation.

Performance measurement parallels key parts of what we covered in Chapter 2. Identifying the key constructs that describe a program and outlining its intended cause and effect linkages are central to any effort to build a viable performance measurement system (Mayne, 2001).

Program evaluations are usually discrete projects: they have a beginning, middle, and an end point. Often, ad hoc committees of stakeholders are struck to oversee the process, and usually, the client(s) are involved in the evaluation from its first steps. Typically, program evaluations are commissioned because there are questions that need to be answered that cannot be addressed with existing administrative data sources and management information systems.

Performance measurement systems are often put together to serve the purposes of improving the efficiency and effectiveness of programs and accounting for program results to stakeholders (agency executives, elected officials, and legislatures) (Hatry, 1999). When performance measures are developed and implemented, it is usually with the intention of providing ongoing information on the outputs and outcomes of programs.

A key issue for any of us who are interested in developing and implementing credible performance measures is the expectation, on the one hand, that the measures we come up with will tell us (and other stakeholders) how well the observed outcomes approximate the intended program objectives, and on the other hand, constructing measures that *actually tell us what the program has accomplished.* This latter concern is, of course, our incrementality question: What differences did the program *actually* make? Answering it entails wrestling with the question of the extent, if any, to which the program *caused* the observed outcomes.

Performance measurement systems typically are not well equipped to tell stakeholders whether the observed outcomes were actually caused by the program. They can describe the observed outcomes, and they can tell us whether the observed outcomes are consistent with program objectives, but there is usually a shortage of information that would get at the question of whether the observed outcomes were the result of program activities (Newcomer, 1997).

If we think of performance measurement as a process of tracking programs over time, we can see that many of the measures built into such systems are, in fact, time series. Variables are measured at regular intervals and the changes in their levels are assessed. Often, trends and levels of performance variables are compared to targets or **benchmarks**. In some situations, where a change in program structure or activities has been implemented, it is possible to track the before-after differences, and at least see whether the observed changes in levels and trends are consistent with the intended effects.

In situations where we would want to use time-series data to look for effects that are consistent (or inconsistent) with intended outcomes, we need continuity in the way variables are measured. If we change the way the measures are taken, or change the definition of the measure itself (to improve its relevancy for current program and policy priorities), we may jeopardize its usefulness as a way to assess cause and effect linkages.

As we have seen in Chapter 2, logic models have two purposes: they describe a program (with components, elements, and implementation objectives), and they delineate intended cause and effect linkages (outputs, linking constructs, and outcomes). Outputs are typically thought of as measures of the work done, and we have conceptualized them as being part of the program, as distinct from being in its environment.

In program evaluations and in performance measurement systems, outputs are typically viewed as attributable to the program—one does not usually need an elaborate research design to test whether the outputs were caused by the program. That means that performance measures that focus on outputs typically can claim that they are measuring what the program actually produced.

Since outputs are generally more likely to be under the control of program managers, there is normally a much greater willingness to report outputs, as opposed to outcomes, as performance measures. As stakeholders insist that performance measures should focus on outcomes that are further along in the logic chain, there is typically much less willingness to "own" the results, that is, accept them as being due to program activities alone.

Thus, using performance measures as a way to determine what a program actually accomplished presumes that the causality issue has been settled. For outputs, that is less an issue than it is for outcomes. As we see in Chapter 10 the incentives for managers play a key role in how willing they are to participate in developing and using performance measures as ways of assessing what their programs accomplished.

● SUMMARY

This chapter focuses on the issue of how research designs can support program evaluators who want to assess whether a program has achieved its objectives. Examining whether the program was effective is a key question in most evaluations, regardless of whether they are formative or summative.

Although randomized experiments are often beyond the resources and control required to design and implement them well, the logic of experimental designs is important to understand if evaluators want to address questions of whether the program caused the observed outcomes. The three conditions for establishing causality—temporal asymmetry, co-variation between the causal variable and the effect variable, and no plausible rival hypotheses—are at the core of all experimental designs.

Through randomized experimental designs, whereby members are assigned randomly to either the program or the control group, it is possible to assert that the two groups are equal in all respects before the program begins. When the program is implemented, the only difference between the two groups is the program itself. This makes it possible to isolate the incremental effects of the program on the participants. Typically, in randomized experiments, we say that we have controlled for threats to the internal validity of the research design, although there can be internal validity problems with the implementation of experiments (Olds et al., 2000).

In assessing research designs, we should keep in mind that the four different kinds of validity are cumulative. Statistical conclusions validity is about determining whether the program and the outcome variable(s) co-vary. Co-variation is a necessary condition for causality. Internal validity builds on statistical conclusions validity and examines whether there are any plausible rival hypotheses that could explain the observed co-variation between the program and the outcome variable(s). Ruling out plausible rival hypotheses is also a necessary condition for causality. Construct validity generalizes the data-based findings back to the constructs and intended linkages in the logic model. Finally, external validity generalizes the results of the program evaluation to other times and places.

Departures from randomized experimental designs can also work well for determining whether a program caused the observed outcomes. One of the most common quasi-experimental designs is the time series, where a program is implemented part way through the time series. Single and multiple time series make is possible to selectively address internal validity threats. Deciding whether a particular threat to internal validity is plausible or not entails using professional judgment, as well as evidence from the evaluation itself.

When we develop a logic model of a program, we are specifying a theory of the program. Ideally, we want to test this theory in an evaluation. Most program evaluations do not permit this because the resources are not there to do so. Further, most evaluations use several different research designs, each having the capability of testing a part of the logic model, not all of it collectively.

Theory-driven program evaluations are designed so that it is possible to test the full logic model. By specifying a single unit of analysis that permits data collection for all the constructs in the model, statistical methods can be used to test each linkage in the model, controlling for the influences on that link, of other paths in the model. Although this approach to program evaluations typically requires extensive resources and control over the evaluation process, it is growing in importance as we realize that linking logic models to research designs that facilitate causal tests of linkages in the models is a powerful way to assess program effectiveness.

● DISCUSSION QUESTIONS

1. The following diagram shows several weak research designs that have been used in an evaluation. The "O" variable is the same for the entire diagram and is measured in such a way that it is possible to calculate an average score for each measurement. Thus, O_1, O_2, O_3, O_4, and O_5 all represent the same variable and the numbers in parentheses above each represent the average score for persons who are measured at that point. All the persons in group 1 are posttested; group 2 had been randomly divided into two subgroups and one subgroup has been pretested and the other one has not. Notice that all the members in group 2 got the program. Finally, for group 3, there is a pretest only.

   - Examine the averages that correspond to the five measurements and decide which threat to internal validity of the overall research design is clearly illustrated. Assume that attrition is not a problem, that is, all persons pretested are also posttested. Explain your answer, using information from Table 3A.1.

Table 3A.1   What Threat to Internal Validity Is Illustrated by This Patched-up Research Design?

Group 1		(6.0) X O_1		
Group 2	R	(4.0) O_2	X	(7.0) O_3
	R		X	(6.0) O_4
Group 3				(4.0) O_5

2. What is a key difference between internal validity and external validity in research designs?

3. What is the difference between testing and instrumentation as threats to the internal validity of research designs?

4. What is the difference between history and selection as threats to the internal validity of research designs?

5. A nonprofit organization in a western state has operated a 40-hour motorcycle safety program for the past 10 years. The program permits unlicensed, novice motorcycle riders to learn skills that are believed necessary to reduce accidents involving motorcyclists. Upon completing the 1-week course, trainees are given a standard state drivers test for motorcycle riders. If they pass, they are licensed to ride a motorcycle in the state. The program operates in one city and the training program graduates about 400 motorcyclists per year. The objective of the program is to reduce the number of motor vehicle accidents involving motorcyclists. Because the program has been targeted in one city, the effects would tend to be focused on that community. The key question is whether this course does reduce motorcycle accidents for those who are trained. Your task is to design an evaluation that will tell us whether the training program is effective in reducing motorcycle accidents. In designing your evaluation, pay attention to the internal and construct validities of the design. What comparisons would you want to build into your design? What would you want to measure to see whether the program was effective? How would you know if the program was successful?

Table 3A.2 Statistical Tools for Program Evaluation

Levels of Measurement	Describing One Variable	Generalizing One Variable	Describing Associations Between Two Variables	Generalizing Associations Between Two Variables	Describing Associations Among Three Variables	Generalizing Associations Among Three Variables
Nominal or nominal by nominal	• Mode • Variation ratio		• Phi • Cramer's V	• Chi square (null hypothesis of no association between the variables)	• Elaborating contingency tables and comparing values of phi or Cramer's V	• Pooled chi square
Ordinal or ordinal by ordinal	• Median • Semi-interquartile range		• Tau B • Tau C • Gamma • Spearman's rho	• Significance test (null hypothesis of no association between the variables)	• Elaborating contingency tables and comparing values of tau B or tau C or gamma	
Interval or interval by interval	• Mean • Standard deviation	• Confidence intervals • Single variable hypothesis tests	• Pearson's r	• Significance test for Pearson's r (null hypothesis of no slope on the regression line, i.e., b = 0)	Multiple regression analysis: • Multiple R • Unstandardized regression coefficients (b coefficients) • Standardized regression coefficients (beta coefficients)	Multiple Regression analysis: • Multiple F test • Partial t tests for the regression coefficients (unstandardized b coefficients)
Nominal by ordinal			• Phi • Cramer's V	• Chi square (Ho: no association)	• Elaborating contingency tables and comparing values of phi or Cramer's V	• Pooled chi square

(Continued)

Table 3A.2 (Continued)

Levels of Measurement	Describing One Variable	Generalizing One Variable	Describing Associations Between Two Variables	Generalizing Associations Between Two Variables	Describing Associations Among Three Variables	Generalizing Associations Among Three Variables
Ordinal by interval			• Tau B or tau C or gamma or Spearman's rho for cross tabs • One-way ANOVA: Eta (for situations where the **independent variable** is ordinal)	• Significance test (for cross-tabulations) • One-way ANOVA: F test (for situations where the independent variable is ordinal)	• Tau B or tau C or gamma for cross-tabs • Two-way ANOVA: partial Eta's and overall Eta (where the two independent variables are ordinal)	Two-way ANOVA: • Overall F test • Partial F tests for each independent variable • Tests for interactions
Nominal by interval			• Phi or Cramer's V or contingency coefficient for cross tabs • One-way ANOVA: Eta where the independent variable is nominal	• Chi square (for cross-tabs) • One-way ANOVA: F test (for situations wherein the independent variable is nominal)	• Multiple regression analysis (where the dependent variable is interval)—same statistics as interval by interval (above)	• Multiple regression—same statistics as interval by interval (above)
Dichotomous nominal by interval			• Pearson's r	• Two sample t test (where the independent variable is nominal) significance test for Pearson's r	• Two-way ANOVA where the interval variable is dependent (same statistics as above) • Multiple regression (where the interval variable is dependent)—same statistics as above	• Two-way ANOVA—same statistics as ordinal by interval • Multiple regression—same statistics as interval by interval

● REFERENCES

Berk, R. A., & Rossi, P. H. (1999). *Thinking about program evaluation* (2nd ed.). Thousand Oaks, CA: Sage.

Berrueta-Clement, J. R. (1984). *Changed lives: The effects of the Perry Preschool Program on youths through age 19.* Ypsilanti, MI: High/Scope Press.

Berry, W. D. (1993). *Understanding regression assumptions.* Newbury Park, CA: Sage.

Bickman, L. (Ed.) (1987). *Using program theory in evaluation* (New Directions for Program Evaluation, no. 33). San Francisco: Jossey-Bass.

Campbell, D. T., & Ross, H. L. (1968). The Connecticut crackdown on speeding: Time-series data in quasi-experimental analysis. *Law and Society Review, 3*(1), 33–54.

Campbell, D. T., & Stanley, J. C. (1966). *Experimental and quasi-experimental designs for research.* Chicago: Rand McNally.

Campbell, F. A., & Ramey, C. T. (1994). Effects of early intervention on intellectual and academic achievement: A follow-up study of children from low-income families. *Child Development, 65*(2), 684–698.

Cook, T. D., & Campbell, D. T. (1979). *Quasi-experimentation: Design and analysis issues for field settings.* Chicago: Rand McNally College Publishing.

Cook, T. J., & Scioli, F. P. J. (1972). A research strategy for analyzing the impacts of public policy. *Administrative Science Quarterly, 17*(3), 328–339.

Donaldson, S. I., & Gooler, L. E. (2002). Theory-driven evaluation of the Work and Health Initiative: A focus on winning new jobs. *American Journal of Evaluation, 23*(3), 341–346.

Fisher, R. A. (1925). *Statistical methods for research workers.* Edinburgh, UK: Oliver and Boyd.

Fitzpatrick, J. (2002). Dialogue with Stewart Donaldson. *American Journal of Evaluation, 23*(3), 347–365.

Gliksman, L., McKenzie, D., Single, E., Douglas, R., Brunet, S., & Moffatt, K. (1993). The role of alcohol providers in prevention: An evaluation of a server intervention program. *Addiction, 88*(9), 1195–1203.

Guba, E. G., & Lincoln, Y. S. (1981). *Effective evaluation.* San Francisco: Jossey-Bass.

Hatry, H. P. (1999). *Performance measurement: Getting results.* Washington, DC: Urban Institute Press.

Hennessy, M., & Greenberg, J. (1999). Bringing it all together: Modeling intervention processes using structural equation modeling. *American Journal of Evaluation, 20*(3), 471–480.

Kelling, G. L. (1974). *The Kansas City preventive patrol experiment: A technical report.* Washington, DC: Police Foundation.

Lipsey, M. W. (2000). Method and rationality are not social diseases. *American Journal of Evaluation, 21*(2), 221–223.

Mayne, J. (2001). Addressing attribution through contribution analysis: Using performance measures sensibly. *Canadian Journal of Program Evaluation, 16*(1), 1–24.

Milgram, S. (1983). *Obedience to authority: An experimental view.* New York: Harper & Row.

Mohr, L. B. (1995). *Impact analysis for program evaluation* (2nd ed.). Thousand Oaks, CA: Sage.

Newcomer, K. E. (Ed.) (1997). *Using performance measurement to improve public and nonprofit programs* (New Directions for Program Evaluation, no. 75, p. 102). San Francisco: Jossey-Bass.

Office of Management and Budget. (2004). *What constitutes strong evidence of a program's effectiveness?* Retrieved July 2004, from http://www.whitehouse.gov/omb/part/2004_program_eval.pdf

Olds, D. L., Henderson, C. R., Chamberlin, R., & Tatelbaum, R. (1986). Preventing child abuse and neglect: A randomized trial of home nurse visitation. *Pediatrics, 78*(1), 65–78.

Olds, D. L., Henderson, C. R., Kitzman, H., Eckenrode, J., Cole, R., & Tatelbaum, R. (1998). The promise of home visitation: Results of two randomized trials. *Journal of Community Psychology, 26*(1), 5–21.

Olds, D. L., Hill, P., Robinson, J., Song, N., & Little, C. (2000). Update on home visiting for pregnant woman and parents of young children. *Current Problems in Pediatrics, 30,* 109–141.

Olds, D. L., O'Brien, R. A., Racine, D., Glazner, J., & Kitzman, H. (1998). Increasing the policy and program relevance of results from randomized trials of home visitation. *Journal of Community Psychology, 26*(1), 85–100.

Patton, M. Q. (1997). *Utilization-focused evaluation: The new century text* (3rd ed.). Thousand Oaks, CA: Sage.

Pechman, J. A., & Timpane, P. M. (Eds.). (1975). *Work incentives and income guarantees: The New Jersey negative income tax experiment.* Washington, DC: Brookings Institution.

Poister, T. H. (1978). *Public program analysis: Applied research methods.* Baltimore: University Park Press.

Poister, T. H., McDavid, J. C., & Magoun, A. H. (1979). *Applied program evaluation in local government.* Lexington, MA: Lexington Books.

Roethlisberger, F. J., Dickson, W. J., & Wright, H. A. (1939). *Management and the worker: An account of a research program conducted by the Western Electric Company, Hawthorne Works, Chicago.* Cambridge, MA: Harvard University Press.

Rogers, P. J., Hacsi, T. A., Petrosino, A., & Huebner, T. A. (Eds.) (2000). *Program theory in evaluation.* (New Directions for Program Evaluation, no. 87). San Francisco: Jossey-Bass.

Schweinhart, L. J., Barnes, H. V., & Weikart, D. P. (1993). *Significant benefits: The High-Scope Perry preschool study through age 27.* Ypsilanti, MI: High/Scope Press.

Shadish, W. R., Cook, T. D., & Campbell, D. T. (2002). *Experimental and quasi-experimental designs for generalized causal inference.* Boston: Houghton Mifflin.

Webb, E. J. (1966). *Unobtrusive measures: Nonreactive research in the social sciences.* Chicago: Rand McNally.

Weisburd, D. (2003). Ethical practice and evaluation of interventions in crime and justice. *Evaluation Review, 27*(3), 336–354.

CHAPTER 4

MEASUREMENT IN PROGRAM EVALUATION

INTRODUCTION •

Program evaluation and performance measurement are both intended to contribute to **evidence-based decision making** in the performance management cycle. In Chapter 2, we discussed logic models as visual representations of programs or organizations and learned that describing program structures and specifying intended cause and effect linkages are the two main purposes for constructing logic models. Logic models identify key constructs that are a part of the program theory in those models. In program evaluations and in performance measurement systems, we need to decide which constructs will be measured; that is, which constructs will be translated into variables by procedures for collecting data.

Gathering evidence from a program evaluation or from performance measures entails developing procedures that can be used to collect information that is convincingly related to the issues and questions that are a part of a decision process. The procedures that are developed for a particular study or for an ongoing performance measurement system need to meet the substantive requirements of that situation and also need to meet the methodological requirements of developing and implementing defensible measures.

Measurement can be thought of in two complementary ways. Measurement is about finding/collecting data, often in circumstances where both time and other resources are constrained. Measurement is also about a set of methodological procedures that are intended to translate **constructs** into observables, producing valid and reliable data. In many program evaluations or performance measurement systems, the resource constraints will pose problems for those concerned with constructing defensible measurement methodologies.

This chapter will focus on measurement in program evaluation with emphasis on measuring program *outputs, linking constructs,* and *outcomes,* as well as measuring **environmental factors**, which can affect the program processes and offer alternate explanations (rival hypotheses) for the observed program outcomes. This approach provides us with a framework for developing our understanding of measurement in evaluations. The measurement methods that are discussed in this chapter can also be applied to needs assessments (Chapter 6).

As you read the chapter, keep in mind what Clarke and Dawson (1999) have to say about measurement in evaluations.

The evaluation enterprise is characterized by plurality and diversity, as witnessed by the broad range of data-gathering devices which evaluators have at their disposal. . . . It is rare to find an evaluation study based on only one method of data collection. Normally a range of techniques form the core of an overall research strategy, thus ensuring that the information acquired has . . . depth and detail. (p. 65–67)

Figure 4.1 links this chapter to the logic modeling approach introduced in Chapter 2. The program, including its outputs, is depicted as an open system, interacting with its environment. Program outputs cause linking constructs, where they are appropriate, which in turn cause outcomes. Environmental factors, which we introduced in Chapter 2, can affect the program and, at the same time, affect linking constructs and/or outcomes. In fact, environmental factors can affect the external validity of the program by mediating between outputs, linking constructs and outcomes (Shadish, Cook, & Campbell, 2002). Our goal is to be able to measure the outputs, linking constructs, and outcomes in a program logic model, and also measure environmental factors that constitute plausible **rival hypotheses** or mediating factors, explaining observed program outcomes.

Given the evaluation questions that motivate a particular program evaluation, and the **research designs** that are being used, some parts of the logic model will be more important to measure than others. If the evaluation focuses on program effectiveness, for example, we will want to measure the outcomes that are central to the intended objectives.

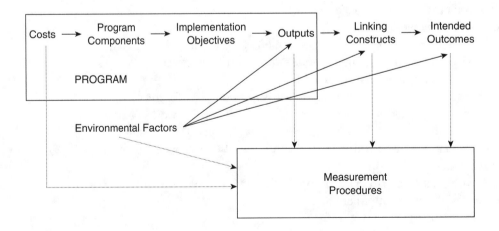

Figure 4.1   Measurement in Program Evaluation and Performance Measurement

Table 4.1	Research Design for the Vancouver Radar Camera Evaluation		
Knight Street	00000000	X0X0X0X0X0	00000000
Granville Street	00000000	000000000000	00000000
		Program Start	Program End

If our interest is whether the program is efficient, that is, what the relationships are between costs and outputs, we would measure outputs and also make sure that we had a robust way of estimating costs for the program or even for its key components.

MEASUREMENT PROCEDURES ●

If we focus on measuring program outputs and outcomes, the process of measurement really begins with building and validating the program logic model. Table 4.1 presents a research design for a **quasi-experimental** evaluation of photo radar cameras in Vancouver, Canada (Pedersen & McDavid, 1994). The program, as it was implemented in that setting, consisted of three components: radar camera enforcement, media publicity, and signage along the street where the cameras were being tested. The main objective was to reduce average vehicle speeds on the street where the program was implemented.

The radar camera program was implemented on Knight Street southbound for a period of 8 weeks (October to November 1990). A section of Granville Street southbound was used as a "**control**" street, and average vehicle speeds were measured (southbound and northbound) on both Knight and Granville Streets for 1 week prior to the intervention, throughout the intervention, and for 10 days after the program ended.

A key part of the program logic model is the first of two intended outcomes: reduced vehicle speeds. Measuring vehicle speeds was a key part of the program evaluation and was one of the dependent **variables** in the comparative time-series research design illustrated in Table 4.1.

ILLUSTRATING MEASUREMENT TERMINOLOGY ●

Table 4.2 is a logic model of the radar camera program. It categorizes the main activities of the program and classifies and summarizes the intended causal linkages among the outputs, the linking constructs, and outcomes.

Table 4.2   Program Logic of the Vancouver Radar Camera Intervention

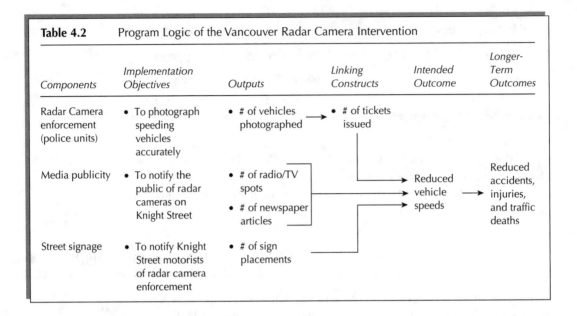

Each of the outputs, the linking constructs, and the intended outcomes is represented with words. The phrase "reduced vehicle speeds" tells us in words what we want the program to accomplish, *but it does not tell us how we will measure vehicle speeds.*

"Reduced vehicle speeds" is a construct in the logic model, as are the other outputs, linking constructs, and outcomes. Constructs are words or phrases that convey the meanings we have assigned to the constituents of the logic model.

Most of us have a reasonable idea of what it means to reduce vehicle speeds. But when you think about it, there are a number of different ways we could *measure* that construct. Measurement, fundamentally, is about translating constructs into observables. Or, in different words, measurement is about operationalizing constructs: translating them into a set of operations/physical procedures that we will use to tell (in our example) whether vehicle speeds have been reduced. It is worth remembering that a particular operational definition does not exhaust the possible ways we could have measured a construct. Often, we select one measurement procedure because of resource constraints, but being able to develop several measures of a construct is generally beneficial, since it makes triangulation of measurement results possible (Webb, 1966).

Some measurement procedures for a given construct are easier than others. Measurement procedures vary in terms of costs, the number of steps involved, and their defensibility. We will say more about the latter issue shortly.

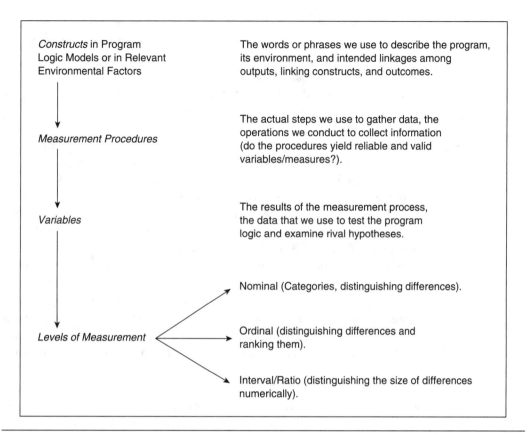

Figure 4.2   Measuring Constructs in Evaluations

Figure 4.2 offers a summary of key terminology used to describe measurement processes. Constructs are where we begin: we need to be clear, using words, about what we mean when we build a program logic model. Sometimes, we build logic models and end up with constructs that sound different but may *mean* the same thing upon closer reflection. Suppose we have a training program that is intended to improve job readiness in a client population of unemployed youth. One linking construct might be "improved job search attitude," and another "heightened self-esteem." It may not be practical to develop measurement procedures for each of these, since it is quite likely that a client who exhibits one would exhibit the other, and measures of one would be hard to differentiate from measures of the other.

Constructs are translated into variables via measurement procedures. Depending on the measurement procedures, the variables can be more or less *reliable*. **Reliability** generally has to do with whether a measurement result is repeatable (Goodwin, 2002), such as when we get the same (or a very similar) reading with our **measurement instrument** if we repeat the procedure in a given situation. If we are measuring the speed of a vehicle on Knight Street, a reliable measurement procedure would mean that we could measure and re-measure that vehicle's speed and get the same reading. We would say that getting the same speed on two different measurements would be a consistent result.

In addition to consistency, there are several other ways that we can assess reliability. In Chapter 5, we discuss ways that narrative (text from interviews with stakeholders, for example) can be coded in evaluations. If we conduct a **survey**, we may choose to include questions where the respondents can offer their own words. When we analyze these open-ended responses, one approach is to create categories that are intended to capture the key meanings of responses and allow us to group responses. Developing a coding scheme for open-ended questions involves considerable judgment. Checking to see whether the categories are successful in distinguishing among responses can be done by asking two or more independent persons to categorize the open-ended responses, using the coding categories (Armstrong, Gosling, Weinman, & Marteau, 1997). The extent to which their decisions are similar can be estimated by calculating an inter-coder reliability coefficient (Holsti, 1969).

The third and fourth types of reliability are more technical, and are applicable where researchers are developing an instrument (a set of survey items or a battery of questions) to measure some construct. For example, if a survey is being developed as part of an evaluation of a housing rehabilitation program, it may be desirable to develop **Likert-scale items** that ask people to rate different features of their neighborhood. A third type of reliability would focus on evaluators developing two sets of items, both of which are intended to measure resident perceptions of the neighborhood. These two parallel forms of the items are then tested in a pilot survey and are examined to see whether the results are consistent across the two versions of the perceptual measures. This is sometimes called a split-half reliability test (Goodwin, 2002).

The fourth way of assessing reliability is often used where a set of survey items are all intended to measure the same construct. For example, if a survey instrument focused on the perceived quality of police services in a community, respondents might be asked to rate different features of their police services, and their own sense of safety and security. To determine whether a set of survey questions was a reliable measure of the construct "quality of police services," we could calculate a measure called Cronbach's Alpha (Carmines & Zeller, 1979). This statistic is based on two things: the extent to which the survey items correlate with each other and the number of items. Cronbach's Alpha can vary between 0 (no reliability) and 1 (perfect reliability). Typically, we want reliability values of .80 or better, using this measure.

Reliability is not always easy to assess. In most program evaluations, the things we want to measure would not sit still while we re-measured a given construct—we typically get one opportunity to measure, and then we move on.

Sometimes, we may be able to use a measuring procedure (instrument) that we already know is reliable; that is, its ability to accurately reproduce a given measurement result is already known. In the radar camera intervention, the experimenters used devices called induction loops to measure the speed of each vehicle. An induction loop is buried in the pavement, and when a vehicle passes over it, the speed of the vehicle is measured and recorded. Because induction loops are widely used by engineers to measure both traffic volumes and speeds, they are generally viewed as a reliable way to measure vehicle speed.

Very few measurement procedures are completely reliable. The usual situation is that a measuring instrument will, if used repeatedly in a given situation, produce a range of results—some being "higher" than the true value and some "lower." The degree to which these results are scattered around the true value indicates how reliable the measure is. If the scatter is tightly packed around the correct value, the measure is more reliable than if there is a wide scatter. When we use statistics (for example, **correlations** to describe the co-variation between two variables), those methods generally assume that the variables have been measured reliably. In particular, if we have two variables, one of which is hypothesized to be the causal variable, we generally assume that the causal **variable** is measured without error (Pedhazur, 1997). Departures from that assumption can affect our estimates of the co-variation between the two variables, generally underestimating the co-variation. If the scatter tends to be higher (or lower) than the true value, then we say the measure is biased. **Bias** is a **validity** problem.

In our radar camera intervention, suppose that vehicle speeds were measured by a roadside radar device instead of in-ground induction loops. Suppose further that in setting the device up, the operator had not correctly calibrated it, so that it systematically over-estimated vehicle speeds. Even if the same speed value could be repeated reliably, we would say that that measure of vehicle speed was *invalid.*

More generally, validity has to do with whether we are measuring what we intend to measure: Is a given measure a "good" representation of a particular construct? Valid measures should be unbiased, and we can sometimes assess whether a measure is biased based on available information.

Validity has a judgmental component to it: Does a given measurement procedure make sense, given our knowledge of the construct and our experience with measures for other constructs? Suppose, for example, we are evaluating a community crime prevention program. The key objective might be to prevent crimes from happening through applying the technologies embedded in the program logic.

But directly measuring the numbers and types of crimes prevented is difficult. Direct measures might require us to develop ways of observing the community to determine how program outputs (neighborhood watch signs on the streets, for example) actually deter prospective criminals.

Usually, we don't do this—we do not have the resources. Instead, we rely on other measures that are available, and assume their validity. Instead of measuring crimes prevented, we might instead use police records of the numbers and types of crimes reported. The validity of such a measure assumes a systematic linkage between crimes prevented and crimes reported. In most evaluations we would not have independent evidence that such linkages exist—instead, we use our judgment and our knowledge of the situation to assess the validity of the measure. Such judgments are typically **face validity** judgments, made by the evaluator (or other experts) in a given situation. We are sometimes in a position to rely on published reports that use particular measures and discuss their presumed validity.

● MEASUREMENT VALIDITY

In Chapter 3, we introduced and discussed four different kinds of validity as they apply to research designs for program evaluations. As you may recall, construct validity pertains in part to the relationship between the variables or measures we manipulate and the constructs that they are intended to represent. One of the basic questions in construct validity is whether the measures that we use are valid representations of the constructs that are embedded in the program logic and research designs.

Clearly, construct validity and measurement validity are related. In fact, Trochim (2000) has suggested that measurement validity is a part of what is meant by construct validity. This view has been adopted in the most recent (1999) revision of the Standards for Educational and Psychological Testing (Goodwin, 2002). Broadly speaking, different forms of measurement validity can be thought of as different ways of getting at the question of the "fit" between variables and corresponding constructs. Trochim suggests that construct validity can be viewed as linking two levels of discourse in evaluations. One level is about the theory of the program, usually expressed in the logic model. At that level, we use language, models, and other verbal or visual ways to communicate what the program (and its environment) is about. The other level is about measures, variables, and observables.

For Trochim and for others (Cook & Campbell, 1979; Shadish et al., 2002), tying the two levels of discourse together means that we are doing two things: we are linking constructs to corresponding measures, and we are linking the observed patterns between and among variables to the predicted patterns among constructs in the program theory. Taken together, how well we succeed at these two sets of tasks determines the construct validity of our logic modeling, research design, and measurement process.

Measurement links individual constructs to the level of observables— the level at which all evaluations and performance measurements are conducted. The conventional types of measurement validity offer us ways of understanding different aspects of this process of tying the two levels together. Some types of measurement validity pertain to a single construct-variable pair and others pertain to expected connections between and among construct-variable pairs.

Types of Measurement Validity

Because measurement validity is really a part of construct validity, the types of measurement validity that we introduce here do not exhaust the types of construct validity. We can think of measurement validity as ways to improve construct validity, understanding that construct validity includes additional issues (Shadish et al., 2002).

We introduce and discuss three broad categories of measurement validity. Within each of these, we outline several specific types of validity that can be thought of as sub-categories. The first general validity category focuses on the relationship between a single measure and its corresponding construct. Within it, we discuss **face validity, content validity**, and **response process validity**. The second category focuses on the relationships between variables that are intended to measure one construct (**internal structure validity**). The third broad category focuses on relationships between one variable-construct pair and other such pairs. Within it, we will discuss **concurrent validity, predictive validity, convergent validity**, and **discriminant validity**.

Validity Types That Relate a Single Measure to a Corresponding Construct

Face Validity. This type of measurement validity is perhaps the most commonly applied one in program evaluations and performance measurement. Basically, the evaluator, or other stakeholders, makes a judgment about whether the measure has face validity with respect to the construct in question. As an example, suppose that the program logic for a meals-on-wheels program includes the intended outcome, "client satisfaction with the service." Using a survey-based question that asks clients of the program if they are satisfied with the service they receive is, on the face of it, a valid measure of the logic model construct "client satisfaction with the service."

Content Validity. This type of measurement validity also involves judgment, but here, we are relying on experts to offer their assessments of a measure (Goodwin, 2002).The issue is how well a particular measure of a given construct matches the full range of content of the construct. Suppose we think of the construct "head start programs." Given all that has been written and the wide range of programs that have been implemented that call themselves "head start programs," we might have a good idea of what a typical program is supposed to include. Further suppose that in a community where there is a substantial population of poorer families, a local nonprofit organization decides to implement its own version of a head start program, to give preschool children in those families an opportunity to experience preschool and its intended benefits. The "fit" between the general construct "head start programs" and its realization in this community would be a measure of the content validity of the local head start program.

Response Process Validity. This kind of validity was one of five categories created with the 1999 revisions to the Standards for Educational and Psychological Testing (American Educational Research Association, 1999). It focuses on the extent to which respondents to an instrument that is being validated demonstrate engagement and sincerity in the way that they participated. If an instrument was being developed to measure school-aged children's attitudes towards science and technology, we would want to know that the process of administering the instrument, and the ways that the children engaged with the instrument, suggest they took it seriously. Goodwin (2002) suggests that debriefing a testing process with a focus group is a useful way to determine whether the response process was valid.

Internal Structure Validity

Developing a measure often involves assessing a pool of items that are collectively intended to be a measure of a construct. In developing the items, the evaluator will use face validity and content validity methods to get an appropriate set of questions. But until they are tested on one or more samples of people who are representative of those for whom the measurement instrument was designed, it is not possible to know whether the items behave as if they are all measuring the same construct.

As an example, an evaluator working on a project to assess the effectiveness of a leadership training program on middle managers in a public sector organization develops an instrument that includes a pool of statements with which respondents are expected to agree or disagree. Among the statements are a set of eight that is intended to measure employee morale. The items have been selected from literature, with input from managers and some employees. A random sample of 150 middle managers takes a pilot version of the instrument and the evaluator analyzes the data to see if the set of eight items cohere, that is, are treated by the respondents as pertaining to their morale. Using a statistical technique called **confirmatory factor analysis** (Goodwin, 1997, 2002), it is possible to see whether the eight items cluster together and constitute one dimension in the data patterns. If one or more items do not cluster with the others, they can be discarded before the full survey is conducted.

Validity Evidence Based on Relationships with Other Variables

Concurrent Validity. Concurrent validity involves correlations between or among measures of a construct. Suppose that we have come up with a new measure for the level of blood cholesterol that involves painting a dye on a small area of the skin between the thumb and forefinger, waiting two minutes, and then shining an ultraviolet light on the dyed spot to see what color it is. The deeper the color of blue that emerges under ultraviolet light, the higher the cholesterol level in the bloodstream. The construct in this case (serum cholesterol level) is being measured by a novel, and comparatively straightforward, technique. One way to demonstrate its concurrent validity is to take a sample of persons who are willing to participate in this research and compare the estimated cholesterol levels given the test results from the "skin test" measure to independently measured serum cholesterol levels using a conventional blood test. Since the blood test method is generally accepted as a valid measure, it serves as a criterion against which our new method can be compared.

Predictive Validity. Two examples serve to illustrate predictive validity. In many graduate programs in Canadian and American universities, applicants are expected to take a standardized test called the Graduate Record Examination. The GRE is constructed so that higher scores are intended to indicate a higher aptitude on the skills that are tested. Research on what factors predict success in graduate programs has generally concluded that high

GRE scores predict higher grades in the programs (Kuncel, Hezlett, & Ones, 2001). The GRE has good predictive validity with respect to actual performance in graduate programs. A second example takes us back to cholesterol tests. One reason many medical practitioners advocate these tests is that they have good predictive validity, that is, high blood serum cholesterol levels is a good predictor of future arterial sclerosis.

Convergent Validity. This kind of measurement validity can be illustrated by a situation where an evaluator is assessing the effectiveness of an employment-training program and, as part of his methodology, has surveyed a sample of clients, asking them four questions that are intended to rate their overall satisfaction with the program. Past research has shown that satisfied clients are usually the ones that tend to be more committed to participating, that is, they attend the sessions regularly and learn the skills that are taught. In our situation, the evaluator also measures attendance and has access to records that show how well each person did in the training modules. As part of the analysis, the evaluator constructs an index of client satisfaction from the four questions, and discovers that persons who are more satisfied are also more likely to have attended all the sessions, and are more likely to have been rated by the instructors as having mastered the materials in the modules. The findings illustrate convergent validity, that is, correlations among measures of constructs that are expected to be related to each other.

Discriminant Validity. In the same employment-training program evaluation, scores on the client satisfaction index are compared to several variables in an intake questionnaire, completed by all clients. One of the intake variables is whether the person has any allergies (this is needed in case medical treatment is required). None of the literature on previous evaluations of these kinds of programs has suggested that having or not having allergies is correlated with client satisfaction. Not surprisingly, there is no such correlation in this case. Discriminant validity has been demonstrated in that measures for two constructs that are not supposed to be linked, in fact, are not correlated.

● LEVELS OF MEASUREMENT

Figure 4.2 indicated that variables that have been defined through measurement procedures can be classified according to their **levels of measurement**. The procedures that are used to collect data can be more or less sophisticated.

Nominal Level of Measurement

Classification is the most basic measurement procedure–we call it the nominal level of measurement. Suppose that one of the relevant environmental factors in a program evaluation was the previous work experience of program clients. In a job-training program, that might be an important alternative reason that explains client success, other than their participation in the program. We could measure previous work experience as a nominal variable: either the person did or did not have work experience (a yes/no variable). Nominal variables are widely used in evaluations because they entail the least demanding measurement procedures—basically, the evaluator needs to be able to classify situations so that for each person/case, the case will fall into one (but *only* one) category.

Nominal variables can have more than two categories. Suppose that an evaluator has interviewed a sample of program clients and has simply recorded their responses to several general questions about their experiences with the program. To see what kinds of patterns there are in these responses, the evaluator may want to develop a set of categories that are based on the actual responses themselves, but that can be used to classify the responses into groups of similar ones. The details of such a procedure are described in Chapter 5, but the evaluator is basically creating, from the client's own words, a nominal variable, which can be used in analyzing the information.

Nominal variables have two basic features: they permit the evaluator to classify every observation/response into *one,* and only one, category, and *all* the observations/responses must fit into the existing categories. In our example of the evaluator coding client responses, the challenge is to come up with categories/themes that do a good job of grouping all the client responses but do not leave the evaluator with a large percentage in a category that has to be labeled "miscellaneous" or "other."

Ordinal Level of Measurement

In the example of a job-training program, suppose we decided to measure previous work experience on a "less to more" basis. Program clients might be categorized as having "no previous work experience," "some previous work experience," and "a great deal of work experience." We could design the measurement procedures so that "some" and "a great deal" equated to ranges of months/years, but we might want to have special rules to take into account full or part-time work. The end result would be a variable that categorizes clients and ranks them in terms of previous work experience. We might have to make judgment calls for some borderline cases, but that would have been true for the previous "yes/no" version of this variable as well. In creating a variable that measures previous work experience on a less to more basis, we have constructed an **ordinal level of measurement**.

Interval and Ratio Levels of Measurement

Finally, we could use a measurement procedure that involved querying clients in some detail about their previous work experience: amounts, full time, part time (how many days per week). Then we could convert the information obtained from clients into a measure that *counts* the number of full-time equivalent months of previous work experience. The conversion process would necessitate rules for translating part-time into full-time equivalents.

The number of full-time equivalent months of a person's work experience is a ratio level of measurement. Although statistical methods used by evaluators do not generally distinguish between **interval** and **ratio levels of measurement**, it is useful for us to show the essential differences. In our example of the number of months of previous work experience, clients can have no previous work experience, or some number of months greater than zero. Because "zero" is a real or natural minimum for that measurement scale, it is possible for us to compare the amounts of work experience across clients. We could say, for instance, that if one client reported the equivalent of 6 months of work experience, and another reported 12 months, the ratio of work experience for the two would be 1 to 2. In different words, the more experienced client has twice as much work experience. Any time we can construct meaningful comparisons that give us ratios (twice as much, half as much, and so on), we are using a ratio level of measurement.

Notice what happens if we try to apply our ratios method to an interval variable. Recall our discussion of the New Jersey Negative Income Tax Experiment in Chapter 3 (Pechman & Timpane, 1975). The experimenters conceptualized family income relative to some poverty-related benchmark. The poverty benchmark became "0" income, for the experiment. If a family had more income in a given year than that benchmark, they would not receive any "negative income benefits." But, if a family's income fell below the benchmark value, they would be entitled to a benefit that increased, the lower their income fell below the poverty level. If we were comparing two families by constructing a ratio of their incomes using the poverty level benchmark as our 0 point, we would run into a problem. Suppose that one family earns $6,000 more that the benchmark and the other one earns $6,000 less than the benchmark. Since there is no natural 0 value in this experiment for income, we cannot construct a ratio of their incomes. We can add and subtract their incomes (we can do this for any interval measure) and that is sufficient to be able to use the most sophisticated statistical tools.

Typically, program evaluators use a mix of measures in an evaluation—some evaluations lend themselves to "counting" types of measures (interval and ratio), others do not. There is a philosophical issue embedded in how we measure in program evaluations. Some proponents of **qualitative evaluation methods** argue that words (e.g., narratives, detailed descriptions, discourse) are fundamentally more valid as ways of rendering experiences, viewpoints, and assessments of programs.

Proponents of **quantitative evaluation methods** tend to rely on numbers, hence, interval-level or ratio level measures of constructs. We will discuss the issue of objectivity in Chapter 11—whether it is possible and how evaluators might conduct their work to claim that they are being objective. Repeatability is a hallmark of scientific investigation and a key part of claims that evaluations can be objective. Advocates for objectivity point out that measurement procedures that yield numbers can be structured so that results are repeatable; that is, another evaluator could conduct the same measurement processes and ascertain whether the results are the same.

Interval-level variables also lend themselves to varied statistical manipulations, which can be very useful as evaluators try to determine the **incremental effects** of programs. If you can conduct a **multivariate statistical analysis** that includes both program measures and environmental variables as predictors of some outcome, it may be possible to assess the effect of the program on the outcome variable, controlling for the environmental variables in the analysis. In the appendix to Chapter 3 we introduced a chart that offers a summary of statistical techniques that are appropriate for different numbers of variables and different levels of measurement. Although this book does not focus on statistical methods per se, that appendix is a useful way to see how different statistical tools can be applied.

UNITS OF ANALYSIS •

In our discussion thus far, we have relied on a program logic approach to illustrate the process of identifying constructs that can be measured in an evaluation. Typically, we think of these constructs as characteristics of people or, more generally, the cases across which our measurement procedures reach. If we are measuring client characteristics as environmental variables that could affect client success with the program, we think of clients as the **units of analysis**.

In a typical program evaluation, there will often be more than one type of unit of analysis. For example, in an evaluation of a youth entrepreneurship program, clients may be surveyed for their assessment of the program; service providers may be interviewed to get their perspective on the program operation and service to clients; and business persons who hired clients might be interviewed by telephone.

Sometimes in program evaluations, the key constructs are expressed in relation to time. In our example of the radar camera program, average vehicle speed would be measured each day, then averaged up to a weekly figure for both Knight Street and Granville Street. The unit of analysis in this evaluation is time, expressed as weeks.

● SOURCES OF DATA IN PROGRAM EVALUATIONS AND PERFORMANCE MEASUREMENT SYSTEMS

Having described the program as a logic model, the constructs identified in the program process and outcomes become candidates for measurement. Typically, the research designs that are selected structure the comparisons in the evaluation and identify the constructs that will need to be measured. As we saw in Chapter 3, we rarely test all the linkages in the logic model.

Indeed, to fully test the program logic it would be necessary to have data on all key constructs so that predicted **co-variation** among them could be examined. Ruling out rival environmental hypotheses is also facilitated by being able to measure environmental factors and examine their actual influence on the program and its observed outcomes. Statistical techniques like **causal modeling** have been applied by some program evaluators to test program logics in this manner (Chen & Rossi, 1989; Costner, 1989; Petrosino, 2000; Shadish & Newman, 1995).

There are always limits to the amounts and kinds of data that can be gathered for a program evaluation. Program evaluators may find, for example, that in evaluating a community-based small business support program, some **baseline measures**, if taken in the community before the program was implemented, would have assisted in estimating the program's actual outcomes. But the data are not available, eliminating one program evaluation strategy (before-after comparisons) for assessing the program's incremental effects.

Existing Sources of Data

Existing sources of data, principally from agency records, governmental databases, client records, and the like, are used a great deal in program evaluations, needs assessments, and in constructing performance measures. It is important to keep in mind, whenever these sources are being relied on, that the operational procedures used to collect the information may not be known to the evaluator, and even if they are known, these data were collected with constructs in mind that may or may not coincide with key constructs in the evaluation at hand. Thus, the evaluator is always in the position, when using existing data sources, of essentially grafting someone else's intentions onto the evaluation design.

Further, existing data sources can be more or less complete, and the data themselves, more or less reliable. Suppose, for example, that the responsibility for recording client data in a family health center falls on two clerk-receptionists. Their days are likely punctuated by the necessity of working on many different tasks. Entering client data (recorded from intake interviews conducted by one or more of the nurses who see clients) would be one such task, and they may not have the time to check out possible inconsistencies or interpretation problems on the forms they are given by the nurses. The result might be a client database that appears to be complete and reliable, but on closer inspection has only limited utility in an evaluation of the program or the construction of performance measures.

Existing data sources present yet another challenge. Many public sector and nonprofit organizations keep reasonably complete records of program outputs. Thus, within the limits suggested previously, an agency manager or a program evaluator should be able to obtain measures of the *work done* in the program. Program managers have tended to see themselves being accountable for program outputs, so they have had an incentive to keep such records for their own use and to report program activities to senior managers, boards of directors, and other such bodies. But, increasingly, program evaluations and performance measurement systems are expected to focus on outcomes. Outcomes are further down the **causal chain** than are outputs, and program managers may experience some trepidation in gathering information on variables that they see as being outside their control.

Given the pressure to report on outcomes, one possible "solution" is to report outputs and *assume* that if outputs occur, outcomes will follow. Using measures of outputs instead of direct measures of outcomes is a process called **proxy measurement**: output measures become proxies for the outcome measures that are not available (Poister, 1978).

Because proxy measures entail an assumption that the outcomes they represent *will* occur, they are problematical. There may be independent evidence (from a previous evaluation of the program or from other relevant evaluations conducted elsewhere) that the outputs lead to the proxied outcomes, but one must approach such short cuts with some caution.

Managers who are expected to develop performance measurement systems for their programs are often in a position where, having identified the key constructs in a logic model that flows from program activities to outcomes, they have to struggle with the task of tying the model to measures that do not support the full model—the existing data available cannot be said to validly measure key constructs in the logic model.

Sources of Data Collected by the Program Evaluator

Most evaluations of programs involve collecting at least some data specifically for that purpose. There is a wide variety of procedures for measuring constructs "from scratch," and in this discussion, several of the main ones will be reviewed.

Perhaps the single most important source of data in many program evaluations is the evaluator. Program evaluations entail interacting with program managers, possibly clients, and other stakeholders, and reviewing previous evaluations. Much of this interaction is informal—meetings are held to review a draft logic model, for example, but each one creates opportunities to learn about the program and develop an experiential "database," which becomes a valuable resource as the evaluation progresses.

Program managers know from their own experiences that in the course of a day, they will conduct several informal "evaluations" of situations, resulting in some instances in important decisions. The data sources for many of these "evaluations" are a combination of documentary, observational, and interactive evidence, together with their own experiences (and their own judgment). Managers routinely develop working hypotheses or conjectures about situations or their interactions with people and "test" these hypotheses informally with subsequent observations, meetings, or questions. Although these "evaluation" methods are informal, and can have biases (e.g., not having representative views on an issue or weighing the gathered data inappropriately given subsequent events), they are the core of much current managerial practice. Henry Mintzberg (1990), who has spent much of his career observing managers to understand the patterns in their work, suggests that managerial work is essentially focused around being both a conduit for and synthesizer of information. Much of that information is obtained informally (e.g., by face-to-face meetings, telephone conversations, or casual encounters in and outside the workplace). Mintzberg (1990) suggests that effective managers ensure that they are well informed and rely on their own judgment.

Program evaluators can also develop some of the same skills managers use to become informed about a program and its context. Seasoned evaluators, having accumulated a wide variety of experiences, can often grasp evaluation issues, possible organizational constraints, and other factors bearing upon their work, in a short period of time. In Chapter 12, we will discuss the nature and practice of professional judgment in evaluation. We will talk about ways that evaluators can take advantage of opportunities to develop sound judgment in their work.

In the field of program evaluation there are important differences of opinion around the question of how "close" program evaluators should get to the people and the program being evaluated. Michael Scriven (1997) has taken the position that evaluators must be objective and, to be that, cannot "mix it up" with the program managers and other stakeholders. We will consider his viewpoint in Chapter 11. Michael Patton (1997), on the other hand, advocates a utilization-focused evaluation approach, which emphasizes the importance of interacting with stakeholders. We will consider some key elements of Patton's approach in Chapter 5.

Surveys as a Data Source in Evaluations

In addition to the evaluator's own observations and informal measurements, program evaluations usually entail systematic data-collection efforts, and a principal means of gathering information is through surveys of program stakeholders: clients, service providers, or other stakeholders. Surveys are also a principal means of collecting information in needs assessments.

Fundamentally, surveys are intended to be measuring instruments that elicit information from respondents. Typically, a survey will have measures for a number of different constructs. In some evaluations, survey-based measures of all the key constructs in a logic model are obtained. If this is feasible, it is then possible to consider using multivariate modeling techniques (causal modeling) to examine the strength and significance of the linkages among the variables that correspond to the logic model constructs (Shadish et al., 2002).

Surveys generally involve some kind of interaction between the evaluator and the respondent, although it is possible to conduct surveys of units of analysis that are inanimate: a neighborhood housing rehabilitation program evaluation might include a visual survey of a random sample of houses to assess how well they are being maintained.

Surveys that focus on people typically are intended to measure constructs that are a key part of a logic model, but these are *mental* constructs. A program in a government ministry that is intended to implement an electronic case management system for all clients might be evaluated in part by surveying the affected employees before and after the changeover to

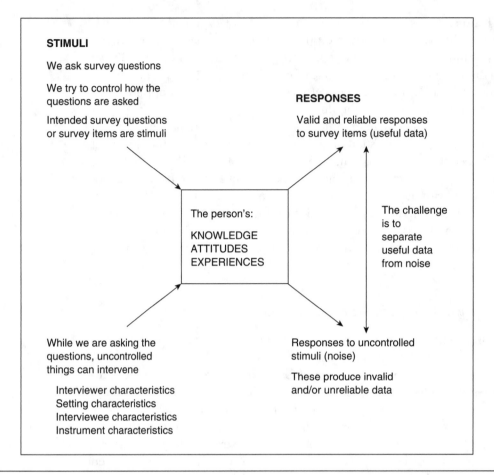

Figure 4.3   Measuring Mental Constructs

electronic client files to see how the change affected their perceptions of their work and the timeliness of their responses to clients.

One construct in the logic model of such a program might be "employee morale." If the electronic file changeover is smooth, morale should improve, given improved access to relevant information as cases are being processed and updated.

Figure 4.3 displays a model of the survey process, drawing attention to the issues of survey validity and reliability. On the upper left-hand side of the model, the survey questions we ask are the intended stimuli, which, if all goes well, elicit (on the upper-right side) the responses that become our data.

Our problem in most surveys is that other factors become stimuli as the survey questions are posed. These, in turn, produce responses that are, from the evaluator's perspective, mixed with the responses to the intended questions. Distinguishing the responses to survey questions from the responses to unintended stimuli is the essence of the reliability and validity challenge of using surveys. Possible sources of stimuli are included in Figure 4.3: characteristics of the interviewers (relevant whether telephone, in-person, or group interviews are conducted); setting characteristics (where the interview is conducted—asking employees about their relationships with their fellow workers might elicit different responses if the interviews are conducted by telephone after hours as opposed to interviews conducted at work); interviewee characteristics (elderly respondents could be hard of hearing, for example); and instrument characteristics (beginning a survey by asking respondents to provide personal demographic information may be seen as too personal, resulting in more cautious responses to other questions).

Among the four types of unintended stimuli, the easiest ones to control are those pertaining to the design of the instrument itself. Suppose, for example, that we are conducting a mailed survey of forestry consultants who have agreed to participate as stakeholders in the planning phase of a program to use chemicals and/or other means to control insects and unwanted vegetation in newly reforested areas. In the survey are a series of statements with which the respondent is expected to agree or disagree.

One such statement is as follows:

Improved pre-harvest planning, quicker reforestation, and better planting maintenance would reduce the need for chemical or mechanical treatments.

Please circle the appropriate response.

Strongly Agree		*Neither*		*Strongly Disagree*
1	2	3	4	5

The main problem with this statement is that any response is ambiguous. We cannot tell *what* the respondent is agreeing or disagreeing to since there are several distinct ideas included in the statement. In short, the question is not a valid measure because we cannot tell which construct is being measured. This problem can be remedied by making sure that only one idea is included in a given statement. Since these statements are commonly used in surveys, this rule of simplicity is quite useful.

Schwarz and Oyserman (2001) have discussed in some depth the problems that can arise from asking respondents for information in interviews and surveys. They focus specifically on asking questions about past behaviors—a common way to establish baselines in program evaluations. In needs assessments, surveys are often used to elicit from clients their patterns of usage for programs or services.

Recall of uses of services such as visits to health or social service agencies, waiting times before a practitioner is available, and the kinds of services delivered, can be an important part of estimating the adequacy of service coverage, and the use patterns of different subpopulations of clients. Schwarz and Oyserman (2001) have outlined validity and reliability problems that can arise where surveys or interviews ask respondents to recall their behavior. They point out that in many evaluations, the time and resources allocated for instrument design are insufficient to address substantial problems that researchers have identified for surveying. For example, when respondents are asked how many times they visited their doctor in the past year, they may have different understandings of what a visit is (some may include telephone conversations, others not), who their doctor is (family doctor, specialists, other health practitioners [dentists, chiropractors]); how many times they visited (memory decays with time; visits that occurred before the year began could be included), and how many times are desirable (some older respondents may deliberately underestimate visits to avoid perceptions that they are using services too frequently).

To improve the likelihood that valid and reliable information about past behaviors will be elicited by surveying, Schwarz and Oyserman (2001) outline for survey designers the five key tasks that survey participants go through in responding to a question, including: "Understanding the question; recalling relevant behavior; inference and estimation; mapping the answer onto the response format; and, 'editing' the answer for reasons of social desirability" (p. 129).

The following list of eight key points on questionnaire design is adapted from the concluding remarks of their report:

1. Once the instrument is drafted, answer every question yourself. If you find questions difficult or confusing, respondents will as well.

2. Respondents will use the instrument to make sense of the questions. Some features may elicit responses that are invalid. Features of the instrument to pay attention to include: the response alternatives offered; the time period for recalling behaviors; the content of related/preceding questions; the title of the questionnaire; and the sponsor of the study.

3. Consult models of well-formed questions.

4. Pilot your questions to see how respondents interpret the wordings of questions and the anchors in rating scales. Use the pilot to check the graphic layout of the instrument.

5. Familiarize yourself with the basics of how people recall and report events—the psychology of responding to survey questions.

6. Give respondents enough time and be prepared to remind them that accuracy is important.

7. Consider using events calendars. These are tables that have time intervals across the top (months, for example) and categories of possible events along the left side (visits to health care facilities might be an example).

8. Train interviewers so that they know the intended meanings of questions in the instrument. (Schwarz & Oyserman, 2001, p. 154–155)

In designing, conducting, and processing data from surveys, evaluators can control some sources of validity and reliability problems more easily than others. There are tradeoffs among different ways of administering surveys. In-person interviews afford the most flexibility and can be used to check whether respondents understand the questions, might want to offer alternative responses, and, for unstructured questions, give respondents an opportunity to express themselves fully. Telephone surveys are somewhat less flexible, but still afford opportunities to confirm question understandings, etc. Mailed surveys are the least flexible, requiring a questionnaire design that is explicit, easy to follow, and unlikely to mislead respondents with the wording of questions.

Clearly, there are cost issues in choosing a survey method. In-person interviews generally are the most costly, followed by telephone surveys, then mailed questionnaires. As well, in-person and telephone surveys generally enjoy higher response rates than mailed surveys.

Sometimes, a mixed strategy is appropriate—in-person interviews with program managers, mailed surveys to clients with telephone follow-up to a sample of nonrespondents to check whether the returned mailed questionnaires are representative of the population being surveyed.

Table 4.3 summarizes some of the principal sources of validity and reliability problems in conducting and processing survey results. Some of these problems are easier to control than others. Although controlling instrument design can eliminate some problems, the evaluator needs to pay attention to the entire surveying process to reduce sources of noise that interfere with interpretations of data. Among the most effective ways to manage surveys, training interviewers (including role playing interview situations) and pretesting instruments yield the greatest validity and reliability benefits.

Careful survey design takes time. It is essential that the designer(s) knows what constructs are to be measured with the survey, and that information guides developing the contents of the instrument. A common experience for

Table 4.3 Examples of Validity and Reliability Issues Applicable to Surveys

Validity: Bias	Source of the Problem	Reliability: Random Error
Race, gender, disability, appearance, age interjections, interviewer reactions to responses	Interviewer	Inconsistency in the way questions are worded/spoken
Old age, handicaps, suspicion	Respondent	Wandering attention
Biased questions, response set, question order	Instrument	Single measures to measure client perceptions of the program
Privacy, confidentiality, anonymity	Interviewing situation/interviewing method/setting	Noise, interruptions
Biased coding, biased categories (for qualitative or quantitative data)	Data processing	Coding errors, inter-coder reliability problems

program evaluators is to be developing a survey instrument with an evaluation steering committee and realize that each person has his or her own "pet questions" or issues. Sometimes, the process of drafting the instrument will itself stimulate additional question items. Evaluators need to continually ask: Why are we including that? What is that intended to measure? How will that question help us to answer the evaluation questions that motivate the project?

Although there is no one format that fits all survey designs, a general sequence of question types that applies in many evaluations is as follows:

- Begin the survey with factual, "warm-up" questions. Ask respondents to relate how they became connected to the program, how long they have been connected, in what ways, etc.
- Ask about program-related *experiences*. Again, begin with factual questions. If the survey is intended for program clients, the instrument can be structured to "walk the client through" the program process.
- As program-related experiences are related, it may be appropriate to solicit evaluations of *each* phase of the process. For example, if clients of a debtor assistance program are being surveyed, the first set of program experience questions might focus on the initial interview. Once the facts of that interview have been recounted, it may be

appropriate to ask for a rating of that experience. Then, as the client is "walked through" the program, each phase can be rated.

- Global ratings of a program should always come *after* ratings of specific experiences/steps in the process. If global ratings are solicited first, there are two risks: the initial global rating will "color" (create a halo effect) subsequent ratings; and second, a global rating without soliciting ratings of specific experiences first is less likely to be based on a full recall of program experiences—in short, it will be less valid.

- Demographic information should be solicited near or at the end of the survey. Typically, demographic information (gender, age, education, income) will be viewed by some as personal and, hence, not shareable with the interviewer. *Any* effort to solicit such information should be differentiated from the rest of the survey, and the respondent should be informed that these data are optional, and if any question is viewed as too personal, not to respond to it. Typically, response rates are somewhat better if these questions are solicited after the respondent has completed the rest of the survey.

- Instruments that have been drafted should be **pre-tested** before they are used in a program evaluation. Typically, the instrument design will be an amalgam of several viewpoints or contributors. Making sure that the questions are clear, and that the instrument as a whole "works" is an essential step in increasing the validity and reliability of the measures. Consider this example, which comes from an actual survey conducted (but not pre-tested) in a U.S. suburban area. The topic of the survey was the possible amalgamation of several smaller police departments into one, larger department.

Question 8: Do you think that your police services would improve if your police department and *all other police departments* [emphasis in the original] in the West Shore area combined into one department?

_____ Yes   _____ No   _____ Undecided

Question 9: Have you discussed this question of police consolidation with friends or neighbors?

_____ Yes   _____ No   _____ Undecided

Question 10: Are you for or against combining your police department with police departments in *surrounding* municipalities?

_____ Yes   _____ No   _____ Undecided

The problem with Question 10 was not discovered until the survey had been mailed to 1,000 randomly selected homes in 15 communities. Fortunately for the project, many respondents detected the problem and simply circled "for" or "against" in the question. But some did not, diminishing the value of the entire survey.

● USING SURVEYS TO ESTIMATE THE INCREMENTAL EFFECTS OF PROGRAMS

In Chapter 3, we examined the issue of incrementality in some depth. The fundamental question in many program evaluations is often what difference, if any, the program made.

Sometimes, in the context of an evaluation, it is appropriate to ask the clients of a program to estimate the incremental effect of the program for themselves. Although such estimates are usually subject to recall and other kinds of biases, it is usually worthwhile taking advantage of a survey to pose questions of this sort.

Bamberger, Rugh, Church, and Fort (2004) develop an interesting approach to evaluation they call "Shoestring Evaluation." The basic idea of their approach is to recognize that in many evaluation situations, the most appropriate methodologies are just not feasible. Evaluators must learn to make do, by using approaches that are within small budgets and do not require data collection efforts that are not feasible. Although they include Shadish et al.'s (2002) four kinds of validity as a part of assessing the methodological adequacy of a particular evaluation approach, they are clear that evaluation practice often diverges from the expectations established by evaluation methodologists.

One technique Bamberger and his colleagues discuss is using surveys and interviews to establish baselines retrospectively. Although respondent recall is clearly an issue, there are patterns in recall bias (tending to telescope events forward into the time frame of the recall query, for example) that make it possible to adjust recall of events so that they are less biased. Schwarz and Oyserman (2001) suggest ways that surveys can be designed so that recall biases are minimized.

In a client survey of a provincial government debtor assistance program, which had, as its objective, counseling clients to avoid personal bankruptcy, respondents were asked a series of questions about their experiences with the program (Rogers, 1983). These questions led up to the following question:

What would you have done if this counseling service had not been available? (Choose the most likely one)

a. Contacted your creditors and tried to work out your problems with them

b. Tried to work out the problems yourself

c. Try to get assistance and advice from another provincial ministry

d. Get advice from a friend or acquaintance

e. Applied for bankruptcy

f. Other (please specify)

_____

_____

The number and percentage of clients who selected option "e" above would be one measure of the incremental effect of the program—if the program had not been available, they would have chosen the option the program was designed to avoid. Although this measure of incrementality will be subject to recall bias, it was part of an evaluation that used three different lines of evidence (client survey, service provider survey, and organizational managers' interviews) to elicit information on the extent to which the program succeeded in steering clients away from bankruptcy.

SURVEY DESIGNS AND RESEARCH DESIGNS ●

Designing a survey is a demanding task. This chapter has suggested some important issues, but it is *not* intended as a detailed guide on this topic. Textbooks that focus on surveying provide more information and are worth consulting as needed (Alreck & Settle, 1995; Babbie, 2001; Rea & Parker, 1997). Some sources are quite skeptical of the use of surveys for measuring constructs that focus on reported behaviors and attitudes (Schwarz, 1999). In addition, it is worth remembering that other extant surveys can be a useful source of ideas for constructing a survey-based measuring instrument, particularly if the surveys have been previously validated.

There is an important difference between survey designs and research designs. Surveys are a way to measure constructs in a program evaluation.

They are, fundamentally, measuring instruments. As such, they are not intended to be research designs. The latter are much broader and will typically include several complementary ways of measuring constructs. Fundamentally, research designs focus on the linkages between the program and its observed outcomes and are intended to get at the question of whether the program caused the observed outcomes. Surveys can be used to measure constructs in a wide variety of research designs. For example, in a quasi-experimental evaluation of a program called the Kid Science Program (Ockwell, 1992), intended to improve children's attitudes toward science and technology, 10 classes of children ages 10 to 12 who participated in the 1-day program at a community college were matched with 10 classes of students who were on the waiting list to participate. All the students in the 20 classes were surveyed before and after the visit to see what differences the visit made on their attitudes.

The survey was a key part of the overall program evaluation, but the research design was clearly independent of the survey as a measuring instrument. The research design for the student surveys is a before-after nonequivalent control group design that measures children's attitudes before and after the program, for both the program and the control groups. Using the terminology introduced in Chapter 3, the research design would look like the diagram below, where the dotted line between the program and control groups indicates nonrandom assignment of classrooms to the program and control groups.

$$O_1 \; X \; O_2$$
$$\text{-----------------------------}$$
$$O_3 \quad O_4$$

In the same evaluation, the teachers in the 10 participating classes were surveyed (interviewed in person) after the visit to get at their perceptions of the effectiveness of the program. The research design for this part of the evaluation was an implicit design. Surveys were used for both designs, illustrating clearly the difference between surveys as measuring instruments and research designs as the basis for the comparisons in evaluations.

● VALIDITY OF MEASURES AND THE VALIDITY OF CAUSES AND EFFECTS

Consider this example: Police departments routinely record the numbers of burglaries reported within their jurisdiction. Reported burglaries, and all the operational steps involved in actually getting those numbers together, are usually considered to be a measure of the number of burglaries actually committed in the community.

But are reported burglaries a valid measure of burglaries committed? Evidence from criminal victimization surveys suggests that in most communities, residents tend to report fewer burglaries than actually occur. In other words, reported burglaries tend to underestimate all burglaries in the community.

A more valid measure might be based on periodic community surveys of householders. Carefully constructed instruments could elicit burglary experiences in the previous 12 months, details of each experience, and whether the police were called. That might be a more valid measure, but it is costly. As well, questions must be carefully worded so as to avoid the issue of "telescoping" the recall of events that occurred outside of the designated time frame into the last 12 months (Schaeffer & Presser, 2003).

The main point is that there are several possible alternative ways of measuring burglaries committed in the community. Each one has different validity and reliability problems.

Now suppose the police department designs and implements a program that is intended to reduce the number of burglaries committed in the community. The program includes components for organizing neighborhood watch blocks, having your property identified, and a series of TV advertisements about ways of "burglar-proofing" your home. The police department wants to know if its program made a difference: Did the program cause a reduction in burglaries committed in the community? Answering that question is very different from answering the question about alternative ways of validly measuring burglaries committed.

Answering the question of whether the program caused a reduction in burglaries and doing a defensible job of answering that question has to do with the validity of causes and effects (the subject of Chapter 3).

SUMMARY •

Measurement is the process of translating constructs into procedures for collecting data. The translation process can produce nominal, ordinal, interval, or ratio measures of constructs. Assessing the validity of measures is an essential part of determining whether they are adequate for a program evaluation or a performance measurement system. Since we tend to rely on data that already exist from organizational records (particularly in constructing performance measures) the validity of these measures can be challenging to assess. It is not uncommon for existing data sources to be connected with constructs where the match is at best approximate.

We can distinguish different kinds of validity for measures that we use in evaluations. All of them pertain to one or more aspects of the construct validity of the measures. Thus, measurement validity is part of the construct validity of measures. Among the types of measurement validity, one pertains to content of the measure itself. The question that this content validity addresses is: To what extent is the measure a complete or convincing representation of the corresponding construct? We tend to rely on stakeholder or expert judgment as ways of estimating content validity.

A second type of measurement validity focuses on the measurement process itself: Did the respondents take the questions seriously? Did the measures "work" from a process standpoint? A third type of validity estimates how well the measure performs empirically when it is used for a group of respondents. This type of validity is relevant where evaluators have constructed an instrument that includes multiple items measuring one construct. If the instrument is tested with a sample of respondents, do they tend to answer the questions so that their responses are highly correlated with each other?

A fourth type of validity compares the variable of interest to other variables to see whether the new measure correlates with other measures that it should correlate with, does not correlate with those where no relationship would be expected, and can predict other variables that should be connected to it in the future.

When evaluators collect their own data, surveys are often used as a measuring instrument. Surveys can be very useful in evaluations or for performance measurement systems. Constructing and administering surveys to minimize the "noise" that can occur from failures to anticipate the ways people will react to instrument design, for example, is key to making surveys worthwhile.

Measurement is perhaps the most undervalued aspect of evaluations. As practitioners, we tend to rely on our own judgment in constructing and in using measures. Particularly in performance measurement systems, there is a tendency to rely very heavily on data that already exist, without taking the time to find out whether the data have been gathered in a reliable way. Experiences with auditing performance measurement systems suggest that even in systems that have taken the time to integrate performance measurement into the planning and budgeting cycle, there are substantial problems with the reliability of the data (see, for example, Texas State Auditor's Office, 2002). If performance measures are not reliable, then it is less likely that they will be used.

DISCUSSION QUESTIONS •

1. What is the basic difference between the reliability and validity of measures?

2. What is the difference between face validity and content validity?

3. What is the difference between concurrent validity and predictive validity?

4. Would you agree that ordinal measures have all the characteristics of nominal measures? Why?

5. Are surveys research designs? Why?

6. What is the difference between the validity of measures and the validity of causes and effects?

7. Design four Likert-scale items (with five points ranging from strongly disagree to strongly agree) that collectively measure patron satisfaction with restaurant dining experiences. Discuss the face validity and the content validity of these measures with a classmate.

REFERENCES •

Alreck, P. L., & Settle, R. B. (1995). *The survey research handbook: Guidelines and strategies for conducting a survey* (2nd ed.). New York: McGraw-Hill.

American Educational Research Association. (1999). *Standards for educational and psychological testing*. American Psychological Association; National Council on Measurement in Education. Washington, DC: American Educational Research Association.

Armstrong, D., Gosling, A., Weinman, J., & Marteau, T. (1997). The place of inter-rater reliability in qualitative research: An empirical study. *Sociology—the Journal of the British Sociological Association, 31*(3), 597–606.

Babbie, E. R. (2001). *The practice of social research* (9th ed.). Belmont, CA: Wadsworth Thomson Learning.

Bamberger, M., Rugh, J., Church, M., & Fort, L. (2004). Shoestring evaluation: Designing impact evaluations under budget, time and data constraints. *American Journal of Evaluation, 25*(1), 5–37.

Carmines, E. G., & Zeller, R. A. (1979). *Reliability and validity assessment.* Beverly Hills, CA: Sage.

Chen, H.-T., & Rossi, P. H. (1989). Issues in the theory-driven perspective. *Evaluation and Program Planning, 12*(4), 299–306.

Clarke, A., & Dawson, R. (1999). *Evaluation research: An introduction to principles, methods, and practice.* London, Thousand Oaks, CA: Sage.

Cook, T. D., & Campbell, D. T. (1979). *Quasi-experimentation: Design & analysis issues for field settings.* Chicago: Rand McNally College Publishing.

Costner, H. L. (1989). The validity of conclusions in evaluation research: A further development of Chen and Rossi's theory-driven approach. *Evaluation and Program Planning, 12*(4), 345–353.

Goodwin, L. D. (1997). Changing conceptions of measurement validity. *Journal of Nursing Education, 36*(3), 102–107.

Goodwin, L. D. (2002). Changing conceptions of measurement validity: An update on the new standards. *Journal of Nursing Education, 41*(3), 100–106.

Holsti, O. R. (1969). *Content analysis for the social sciences and humanities.* Reading, MA: Addison-Wesley.

Kuncel, N. R., Hezlett, S. A., & Ones, D. S. (2001). A comprehensive meta-analysis of the predictive validity of the Graduate Record Examinations: Implications for graduate student selection and performance. *Psychological Bulletin, 127*(1), 162–181.

Mintzberg, H. (1990). The manager's job—Folklore and fact. *Harvard Business Review, 68*(2), 163–177.

Ockwell, P. (1992). *An evaluation of the Kid's Science Program run by the Science and Technology Division of Camosun College.* Unpublished Master's report, University of Victoria, British Columbia, Canada.

Patton, M. Q. (1997). *Utilization-focused evaluation: The new century text* (3rd ed.). Thousand Oaks, CA: Sage.

Pechman, J. A., & Timpane, P. M. (Eds.). (1975). Work incentives and income guarantees: The New Jersey negative income tax experiment. Washington, DC: Brookings Institution.

Pedersen, K. S., & McDavid, J. C. (1994). The impact of radar cameras on traffic speed: A quasi-experimental evaluation. *Canadian Journal of Program Evaluation, 9*(1), 51–68.

Pedhazur, E. J. (1997). *Multiple regression in behavioral research: Explanation and prediction* (3rd ed.). Fort Worth, TX: Harcourt Brace College Publishers.

Petrosino, A. (2000). Answering the why question in evaluation: The causal-model approach. *Canadian Journal of Program Evaluation, 12*(1), 1–25.

Poister, T. H. (1978). *Public program analysis: Applied research methods.* Baltimore: University Park Press.

Rea, L. M., & Parker, R. A. (1997). *Designing and conducting survey research: A comprehensive guide* (2nd ed.). San Francisco: Jossey-Bass.

Rogers, P. (1983). *An evaluation of the debtor assistance program.* Unpublished Master's report, University of Victoria, British Columbia, Canada.

Schaeffer, N. C., & Presser, S. (2003). The science of asking questions. *Annual Review of Sociology, 29*(1), 65–88.

Schwarz, N. (1999). Self reports: How the questions shape the answers. *American Psychologist, 54*(2), 93–105.

Schwarz, N., & Oyserman, D. (2001). Asking questions about behavior: Cognition, communication, and questionnaire construction. *American Journal of Evaluation, 22*(2), 127–160.

Scriven, M. (1997). Truth and objectivity in evaluation. In E. Chelimsky & W. R. Shadish (Eds.), *Evaluation for the 21st century: A handbook.* Thousand Oaks, CA: Sage.

Shadish, W. R., Cook, T. D., & Campbell, D. T. (2002). *Experimental and quasi-experimental designs for generalized causal inference.* Boston: Houghton Mifflin.

Shadish, W. R., & Newman, S. L. (1995). Developing the guiding principles. *New Directions for Program Evaluation, 66,* 3–18.

Texas State Auditor's Office. (2002). *An audit report on fiscal year 2001 performance measures at 14 entities.* Austin: Texas State Auditor's Office.

Trochim, W. M. K. (2000). The research methods knowledge base. Retrieved August 22, 2004 from http://www.socialresearchmethods.net/kb/index.htm

Webb, E. J. (1966). *Unobtrusive measures: Nonreactive research in the social sciences.* Chicago: Rand McNally.

CHAPTER *5*

APPLYING QUALITATIVE
EVALUATION METHODS

James C. McDavid
Irene Huse
Laura R. L. Hawthorn

INTRODUCTION ●

This chapter introduces a different approach to program evaluation—one that has emerged in parallel to the more structured, quantitative approach that has been elaborated in Chapters 2, 3, and 4. This chapter will show how **qualitative evaluation methods** can be incorporated into the range of options available to evaluators and their clients, and will offer some comparisons between the two approaches. In general, qualitative approaches, such as interviews and **focus group** discussions, are more open-ended than quantitative methods, and are most valuable in collecting and analyzing data that do not readily reduce into numbers. Qualitative methods are particularly useful for exploratory research and participatory, or empowerment, research. Empowerment research (see Fetterman, 1996) involves significant collaboration between the evaluator and stakeholders during most or all of the steps in the evaluation process, from the planning and design to the final interpretation and recommendations. Chapter 11 includes further information on empowerment evaluation in its discussion of the relationship between program evaluation and performance management.

It is worth reiterating that the key issues in deciding on which method or methods to use for any evaluation are the context of the situation and the evaluation questions that need to be addressed. Qualitative methods can be used in various stages of an evaluation:

- Determining the focus of the evaluation
- Evaluating the implementation or the process of a program
- Determining improvements and changes to a program

To introduce qualitative evaluation methods, it is important to first elaborate on the diversity of approaches even within the theory and practice of qualitative evaluation. Qualitative evaluation approaches differ from each other on at least two important fronts: their philosophical beliefs about how and what we can know about the kinds of situations evaluators typically face (these are called **epistemological beliefs**); and their methodologies, that is, the ways that evaluations are organized and conducted. In this chapter, we will learn about some of the key philosophical differences among qualitative evaluation approaches, but will spend more time focusing on the ways that qualitative methods can be used.

● COMPARING AND CONTRASTING DIFFERENT APPROACHES TO QUALITATIVE EVALUATION

When qualitative evaluation approaches emerged as alternatives to the then-dominant **social scientific approach** to evaluation in the 1970s, proponents of these new ways of evaluating programs were a part of a much broader movement to re-make the foundations and the practice of social research. Qualitative research has a long history, particularly in disciplines like anthropology and sociology, and there have been important changes over time in the ways that qualitative researchers see their enterprise. There is more diversity within qualitative evaluation approaches than within quantitative approaches:

> A significant difference between qualitative and quantitative methods is that, while the latter have established a working philosophical consensus, the former have not. This means that quantitative researchers can treat methodology as a technical matter. The best solution is one which most effectively and efficiently solves a given problem. The same is not true for qualitative research where proposed solutions to methodological problems are inextricably linked to philosophical assumptions and what counts as an appropriate solution from one position is fatally flawed from another. (Murphy, Dingwall, Greatbatch, et al., 1998, p. 58)

Denzin and Lincoln (2000) summarize the history of qualitative research in their introduction to the *Handbook of Qualitative Research.* They offer an interpretation of the history of qualitative research in North America as comprising seven moments, beginning with traditional anthropological research (1800s to about 1950), characterized by lone anthropologists spending time in other cultures and then rendering their findings in "objective" accounts of the values, beliefs and behaviors of the natives. In their definition of qualitative research Denzin and Lincoln (2000) include the following:

> Qualitative research is a situated activity that locates the observer in the world . . . qualitative researchers study things in their natural settings, attempting to make sense of, or to interpret, phenomena in terms of the meanings people bring to them. (p. 3)

and later continue:

> Qualitative researchers stress the socially constructed nature of reality, the intimate relationship between the researcher and what is studied,

and the situational constraints that shape inquiry. Such researchers emphasize the value-laden nature of enquiry. They seek answers to questions that stress *how* social experience is created and given meaning. In contrast, quantitative studies emphasize the measurement and analysis of causal relationships between variables, not processes. (p. 8)

Understanding the Issue of Paradigms

In the field of program evaluation, academics and practitioners in the 1970s were increasingly under pressure to justify the then-dominant social science-based model as a way of thinking about and conducting evaluations. Questions about the relevance and usefulness of highly structured evaluations (often experiments) were being raised by clients and by academics alike. An alternative paradigm was emerging, based on different assumptions, different ways of gathering information, different ways of interpreting that information, and finally, different ways of reporting evaluation findings and conclusions.

On the whole, the qualitative research approach embraces a different view of research than the **positivist** "rational" approach traditionally taken by most quantitative researchers. Thomas S. Kuhn (1962), in his revolutionary book *The Structure of Scientific Revolutions,* asserted that when scientists "discover" a new way of looking at phenomena, they literally *see the world in a different way.* He developed and popularized the notion of a **paradigm**, a self-contained perceptual and theoretical structure akin to a belief system. Although Kuhn was describing the change in world view that happened in physics when Einstein's relativity theory began its ascendancy at the turn of the 20th century, he used language and examples that invited generalizing to other fields. In fact, because his book was written in a nontechnical way, it became a major contribution to the widespread and continuing process of questioning the foundations of our knowledge in the social sciences and humanities.

Paradigms, for Kuhn, were at least partly **incommensurable.** That is, adherence to one paradigm, and its attendant way of seeing the world, would not be translatable into a different paradigm. Proponents of dissimilar paradigms would experience an inability to communicate with their counterparts. They would talk past each other, because they would use different words and literally see different things even when they were pointing to the same object.

Norwood Russell Hanson (1958) illustrated this problem with a series of optical illusions and puzzles. A well-known one is reproduced here to make several points about seeing. Figure 5.1 is a line drawing of a person. Look

Figure 5.1 The Old Woman or the Young Woman?

carefully at the drawing and see if you can discern the old woman's face. She is facing the left side of the page, her nose is prominent, her lips are a line, and she is wearing a hood that exposes the hair on her forehead.

Now, look at the same drawing again and see if you can discern the young woman's face. She is facing away from you, and is showing you the left side of her face. You cannot see her nose, but you can see her left ear. Her jaw and her hair are visible as is her necklace. A hood that loosely covers her head exposes the left side of her face.

Generally, people find it easier to "see" the old woman. But once you "see" the young woman, she will be there every time you look for her.

The point of Figure 5.1 is to show you that the same information can be interpreted in different ways. Further, the patterns of information are incommensurable with each other. When you are seeing the old woman, you are not (at that moment) seeing the young woman. Kuhn was arguing that scientists, too, can "see" the same information and interpret it differently.

In the qualitative methods arena, one key paradigm, often called the **constructivist** paradigm, has been developed and articulated by a number of central contributors. Egon Guba and Yvonna Lincoln (2001) are two well-known scholars who have made major contributions to constructivist research approaches. In explaining relativism, one of the fundamental assumptions of the constructivist paradigm, they state that "human (semiotic) sense-making that organizes experience so as to render it into apparently comprehensible, understandable, and explainable form, is *an act of construal and is independent of any foundational reality.* Under relativism there can be no 'objective' truth" (p. 1, original emphasis).

Guba and Lincoln are among those who have stated a belief that the "scientific" approach to program evaluation and the constructivist approach are fundamentally different, and at least partially incommensurable, paradigms:

> It is not appropriate to "mix and match" paradigms in conducting an evaluation, for example, utilizing both scientific (positivist) and constructivist propositions within the same study. This is not a call for "purity" nor is it intended to be exclusionary. It is simply a caveat that mixing paradigms may well result in nonsense approaches and conclusions. (Guba & Lincoln, 2001, p. 2)

However, other influential evaluators, including Michael Patton, while acknowledging the continued existence of paradigmatic differences between social scientific evaluators and some qualitative research approaches, have argued that the philosophical differences have been at least substantially submerged by more practical concerns:

> The trends and factors just reviewed suggest that the paradigms debate has withered substantially. The focus has shifted to methodological appropriateness rather than orthodoxy, methodological creativity rather than rigid adherence to a paradigm, and methodological flexibility rather than conformity to a narrow set of rules. (Patton, 1997, p. 295)

For Patton, pragmatism makes it possible to overcome the differences between paradigms:

> I believe that the flexible, responsive evaluator can shift back and forth between paradigms within a single evaluation setting. In doing so, such a flexible and open evaluator can view the same data from the perspective of each paradigm and can help adherents of either paradigm interpret data in more than one way. (Patton, 1997, p. 296)

Patton's view is by no means universal, however. The recent discussions about the primacy of randomized control trials (RCTs) within the evaluation community (Scriven, in progress, unpublished) are connected with the ongoing debate about the relative merits of qualitative and quantitative evaluation approaches.

The Pragmatic Approach

In evaluation, although there continue to be debates that stem from differences in epistemological beliefs and methodological approaches, there is a general movement toward Patton's more pragmatic pluralism. Even in a textbook that has been considered to be a benchmark for **positivist** and **post-positivist** approaches to evaluation (Cook & Campbell, 1979), the more recent edition (Shadish, Cook, & Campbell, 2002) mentions the value of qualitative approaches as complements to the experimental and quasi-experimental approaches that comprise the book.

Mark, Henry, and Julnes (2000) have offered a theoretical approach to evaluation that they claim will settle the qualitative/quantitative dispute in the profession. Their approach relies on combining what they call "**sense-making**" with commonsense realism:

> Neither qualitative nor quantitative methods are superior in all evaluation situations. Each inquiry mode corresponds to a functional aspect of sensemaking and each can be addressed through qualitative and quantitative methods. More generally, commonsense realism, integrated with sensemaking, offers a potent grounding for a lasting peace following the paradigm wars. (p. 335)

The paradigm debate is not dead yet. But its role in the practice of program evaluation has diminished considerably. Most program evaluators have taken the position that qualitative and quantitative methods do not carry the freight of different philosophical traditions, or if they do, methodological pluralism is the solution of the day. Ernest House (1994), for example, has argued that the quantitative-qualitative dispute is dated and has stated that he does not believe that the two methods represent distinct paradigms that incorporate incommensurate worldviews. The two methods can be applied regardless of whether one believes that we share the same reality, or that each person has a reality that is ultimately known only to the perceiver.

● QUALITATIVE EVALUATION METHODS: SOME BASICS

What is qualitative evaluation? How is it distinguished from other forms of program evaluation? How do qualitative evaluators do their work?

These questions are practical ones, and the main focus of this section will be to offer some answers to them. It is worth saying, however, that qualitative evaluation methods have developed in many different ways and that there are a number of different textbooks that offer evaluators ways to design, conduct, and interpret evaluations that rely on qualitative data (for example Denzin & Lincoln, 2000; Patton, 1997). Patton, in the Evaluation Checklists Project (The Evaluation Center, 2001), maintains "Qualitative methods are often used in evaluations because they tell the program's story by capturing and communicating the participants' stories." They normally encompass interviews, focus groups, narrative data, field notes from observations, and other written documentation. Often, the sample size is quite small. In contrast, quantitative evaluations use numbers gathered from measures over comparatively large samples, and use statistical procedures for describing and generalizing the patterns between and among variables.

An evaluation may be entirely conducted using a qualitative approach, but it will depend on the context and needs of the evaluation. Sometimes initial qualitative exploratory work is followed by a quantitative approach, particularly when an evaluator is developing survey questions. Developing logic models is a qualitative process that relies on interpreting documents, interviewing stakeholders, and putting together a visual representation of a program. Sometimes qualitative findings are collected and/or presented along with quantitative data, such as that gathered from a survey with both closed-ended and open-ended questions. Finally, qualitative research sometimes occurs *after* quantitative research has been completed, such as when an organization is determining how to follow up on survey results that indicate a program needs changes.

There are strong practical reasons to view qualitative evaluation methods as complementary to **quantitative methods**. Indeed, as Reichardt and Rallis (1994) and many others have argued, using two methods can be better than one. In her discussion of the "paradigm wars," for example, Datta (1994) concludes:

> [T]he differences [between the qualitative and quantitative paradigms] are less sharp in practice than in theoretical statements. The best examples of both paradigms seem actually to be mixed models. . . . Perhaps this is not surprising. . . . Most evaluations are conducted under many constraints. These include relatively short time frames, relatively little money, often intractable measurement challenges. . . . In most circumstances, evaluators have to do the best they can and need more, not fewer, approaches on which they can draw. (p. 67)

Some qualitative research crosses the bridge between qualitative and quantitative methods. If we survey clients of a program and ask them to tell us, in their own words, about their experiences with the program, we end up with narrative. Likewise, if we interview persons who delivered the program, questioning them in depth about their perceptions of the program environment, including the clients they serve, we again have narrative. To use this in the evaluation, we need to sort it, organize it, and interpret it. To categorize it we may conduct a **thematic analysis**, looking for groups of similar word or statement clusters.

If we look at the qualitative findings from the client survey, our thematic analysis would give us categories of different themes, and could give us the numbers of times we detected client responses that fit each theme. If we had done our work well, the themes we discerned would cover the range of issues raised by clients in their responses to the survey question and tell us how often each issue was raised.

But we have also created a *nominal variable,* that is, a variable that has mutually exclusive and jointly exhaustive categories. Having done so, we could report the frequencies and relative percentages of themes. We could even cross-classify the frequencies of themes with other nominal variables (the gender of our clients, for example). If we cross-classified themes by other variables, we might even go so far as to test for the statistical significance of the associations: for example, was the distribution of themes significantly different for men than it was for women?

Nominal variables can, in some situations, be just as amenable to statistical manipulations as "higher" levels of measurement—the statistical tools are different, but we are still adding, subtracting, and counting as we do the analysis.

Key Differences Between Qualitative and Quantitative Evaluation Approaches

Table 5.1 is a listing of some of the differences that have been cited between qualitative and quantitative evaluation approaches. The two lists are intended to convey principles and practices that evaluators might use to distinguish the two approaches. It is worth noting that the differences in Table 5.1 are not absolute. Because our views of the roles of qualitative and quantitative evaluation continue to change, it is possible to find advocates for and examples of the view that qualitative data can be the main source of information in randomized experiments (Miles & Huberman, 1994).

Table 5.1 suggests an image of qualitative program evaluation that, although it does not convey the differences among qualitative approaches, does highlight some common central features.

Table 5.1   Key Differences Between Qualitative and Quantitative Evaluation Approaches

Qualitative Evaluation *Is Often Characterized by*	Quantitative Evaluation *Is Often Characterized by*
• Inductive approach to data gathering, interpretation, and reporting	• Research hypotheses and questions that are tested in the evaluation
• Holistic approach: finding **gestalts** for the evaluation results	• Finding patterns that either corroborate or disconfirm particular hypotheses and answer the evaluation questions
• **Verstehen:** understanding the subjective lived experiences of program stakeholders (discovering their truths)	• Understanding how social reality, as observed by the evaluator, corroborates or disconfirms hypotheses and evaluation questions
• Using natural language throughout the evaluation process	
• In-depth, detailed data collection	• Emphasis on measurement procedures that lend themselves to numerical representations of variables
• Use of case studies	
• The evaluator as the primary measuring instrument	• **Representative samples** of stakeholder groups
• A **naturalistic approach:** does not explicitly manipulate the setting	• Use sample sizes with sufficient statistical power to detect expected outcomes
	• Measuring instruments that are constructed with a view to making them reliable and valid
	• Evaluator control and ability to manipulate the setting, which improves the internal validity, the statistical conclusions validity, and the construct validity of the research designs

In qualitative evaluation, emphasis is placed on the uniqueness of human experiences, eschewing efforts to impose categories or structures on experiences, at least until they are fully rendered in their own terms.

Qualitative program evaluation tends to build from these experiences upwards, seeking patterns but keeping an open stance toward the new or unexpected. The **inductive approach** starts with "the data," namely, narratives, direct and indirect (unobtrusive) observations, interactions between stakeholders and the evaluator, documentary evidence, and other sources of information, and then *constructs* an understanding of the program. Putting together the themes in the data, weighting them, verifying them with stakeholders, and finally, preparing a document that reports the findings and conclusions is part of a **holistic approach** to program evaluation. A holistic approach also, like an empowerment approach, entails taking into account and reporting different points of view on the program, its operations, and its effects on stakeholders. Thus, an evaluation is not just conducted from the program manager's standpoint, but takes into account clients' viewpoints, as well as other stakeholders' views. Later in this chapter we will provide further suggestions for structuring a qualitative evaluation research project. Many of the major steps in implementing a qualitative evaluation design are, however, fundamentally parallel to the steps in a quantitative evaluation design:

1. Data collection

2. Analysis of the data

3. Writing of the report

4. Dissemination of the report

5. Making changes, based on the evaluation

Similarly, the major questions to be addressed in designing the evaluation have parallels with those in a quantitative evaluation approach. These are summarized in Table 5.2.

Qualitative evaluations tend to be **naturalistic**, that is, they do not attempt to control or manipulate the program setting. Instead, the evaluator works with the program as it is and works with stakeholders as they interact with or perform their duties in relation to the program or with each other. Naturalistic also means that natural language is used by the evaluator—the same words that are used by program stakeholders. There is no separate "languages of research design," for example, and usually no separate language of statistics.

In qualitative evaluations, the evaluators themselves are the principal measuring instrument. There is no privileged perspective in an evaluation. It is not possible for an evaluator to claim objectivity. Evaluator observations, interactions, and renderings of narratives and other sources of information are a critical part of constructing patterns, and creating an evaluation report. A principal means of gathering data is face-to-face interviews/conversations. Mastering the capacity to conduct interviews and observations while recording the details of such experiences is a key skill for qualitative program evaluators.

In contrast, quantitative evaluation tends to emphasize hypotheses (embedded in the program logic, for example) or evaluation questions, which generally reflect a limited number of possible stakeholder perspectives. Typically, a key evaluation question is whether the program produced/caused the observed outcomes, that is, *was the program effective?*.

Table 5.2   Summary of Key Questions in Conducting Qualitative Evaluation Assessments and Evaluation Studies

---

1. Who are the client(s) for the evaluation?

2. What are the questions and issues driving the evaluation?

3. What resources are available to do the evaluation?

4. What has been done previously?

5. What is the program all about?

6. What kind of environment does the program operate in and how does that affect the comparisons available to an evaluator?

7. Which research design alternatives are desirable and appropriate?

8. What information sources are available/appropriate, given the evaluation issues, the program structure and the environment in which the program operates?

9. Given all the issues raised in points 1 to 8, which evaluation strategy is least problematical?

10. Should the program evaluation be undertaken?

Quantitative evaluation is concerned with validity and, in particular, threats to internal validity that would undermine efforts to assess the incremental outcomes of the program. Concerns with internal validity and statistical conclusions validity, in particular, usually mean that quantitative evaluators prefer having some control over the program design, program implementation, and the evaluation processes. Naturalistic settings, however, present limited opportunities in terms of controlled research design options, and render efforts to attribute causal relationships problematical, at least from an internal validity perspective.

Mohr (1999) agrees that most evaluators have tended to question the value of qualitative methods for determining causality in evaluations. The basic problem is that our conventional notion of causality, discussed in Chapter 3, requires some kind of comparison to see what would have happened without the program. In other words, seeing whether the program caused the actual observed outcomes involves establishing what the pattern of outcomes would have been *without* the program. The logic of this process is that if X (the program) caused Y (the observed outcome), then both X and Y occurred, and if X had not occurred, then neither would Y have occurred. If there are no rival hypotheses to cause Y to occur in the absence of X, the counterfactual condition can be demonstrated, and we can conclude that X caused Y.

Mohr suggests that we invoke an alternative model of causality—the modus operandi approach introduced by Scriven (1976). This alternative model is demanding—it requires that there be "physical" connection between the causal variable (X) and Y. The example Mohr uses is from medicine: it is possible to work backwards from a set of symptoms to determine whether a patient had a recent heart attack. A blood test will show whether there are enzymes present that are uniquely associated with a heart attack. Where it is possible to connect the cause physically to the effect, it is not necessary to have a counterfactual comparison—a single case permits us to determine causality.

In principle then, even single cases, which typify some qualitative evaluations, can contribute to understanding cause and effect linkages. Although establishing physical causality in program evaluations can be daunting, Mohr notes that the weakest research designs get used in evaluations often, and we seem to be able to learn by using them:

> Furthermore, ex-post facto or observational studies are well known to be pathetic in terms of internal validity, yet they are in constant use, probably because there is often so much to be learned from them nevertheless. (Mohr, 1999, p. 80)

Parenthetically, it is worth noting that among quantitative evaluators there has been considerable debate over the importance of internal validity. Cook (1991), for example, has argued that internal validity has to be a central concern of evaluations because assessing whether the program really did cause the observed outcomes is a key part of knowing whether the program worked and, thus, whether to ask how the findings and conclusions might be generalized.

Lee Cronbach, on the other hand, has argued that although internal validity should not be ignored, the key issue is external validity: the issue being the generalizability of the evaluation results (Cronbach in Cook, 1991). Cronbach looked at this issue from a practitioner's standpoint and concluded that pinning down causal linkages was so difficult that it tended to absorb far too much of an evaluator's efforts, when the real issue was what could be generalized from a given evaluation and used elsewhere. In their recent book, Shadish, Cook, and Campbell (2002) have come some ways toward Cronbach's position. No longer is internal validity the arbiter of the value of an evaluation. External validity is more prominent now in their schema of four kinds of validities, and they agree with Cronbach that external validity can be viewed independently of internal validity in a program evaluation.

Later in this chapter we will examine the sets of validity issues applicable to qualitative evaluation research, and compare them to the four kinds of validity discussed in Shadish, Cook, and Campbell (2002).

STRUCTURING QUALITATIVE ●
PROGRAM EVALUATIONS

The issue of how much structure to impose on a qualitative evaluation is contentious. At one end of the spectrum, some evaluators advocate an unstructured approach, which does not depend on stakeholders articulating their evaluation questions and expectations for the evaluation process in advance. Evaluators in such settings explore stakeholder viewpoints, and as information is gathered, inductively construct issues that are supported by evidence. These issues, in turn, can be used to guide further data collection. The goal is a full, authentic representation of issues and views that stakeholders have contributed.

The other end of the spectrum is perhaps more common. Program evaluators can construct **conceptual frameworks,** which then guide the evaluation, including what data to collect, who to interview, and what to ask them.

One way to look at the issue of structure is in terms of more specific topics:

1. Identifying evaluation questions and issues

2. Identifying research designs and comparisons

3. Sampling methods

4. Data collection instruments

5. Collecting and coding qualitative data

In expanding these steps, we provide some examples, including a qualitative research study related to home nursing visitation programs. In that study, Byrd, (1999) describes an 8-month field study of 53 home visits to at-risk infants by one nurse. Another study, by McNaughton (2000), provides a synthesis of seventeen qualitative studies of nurse home visitation. These studies relate to the experimental research conducted by Olds and his colleagues in the United States (see Chapter 3). Olds' work has emphasized understanding whether home visits by nurses improve the well-being of children and mothers, but does not describe the actual processes that nurses use in developing relationships with their clients.

Identifying Evaluation Questions and Issues in Advance

In all program evaluation situations, time and money are limited. Usually, evaluations are motivated by issues or concerns raised by program managers or other stakeholders. Those issues constitute a beginning agenda for the evaluation process. The evaluator will usually have an important role in defining the evaluation issues and may well be able to table additional issues. But it is quite rare for an evaluation client or clients to support a fully exploratory evaluation.

In the Byrd (1999) study, the researcher explained that "[t]he processes public health nurses use to effectively work with . . . families are not adequately described" and that "[d]escribing and interpreting the process of home visiting can contribute to the development and refinement of theory that provides a meaningful framework for practice and for studies examining the efficacy of these challenging home visits" (p. 27). The key issue was that although many quantitative studies had done comparisons between program groups and control groups in terms of nurse home visitations, there had been no in-depth studies that looked at what typically occurs during nurse home visits. The study, then, set out to address that gap.

The McNaughton (2000) study was conducted later and looked at the by-then larger selection of qualitative studies on the effects of nurse home visitations. The issue was that before the qualitative studies had been done, "it was difficult to determine aspects of nursing interventions that were or were not effective" (p. 405). The goal of this qualitative research was to gather and analyze qualitative nurse home visitation studies "to provide an organized and rich description of public health nursing practice based on identification of common elements and differences between research reports" (p. 405).

Identifying Research Designs and Appropriate Comparisons

Qualitative data collection methods can be used in a wide range of research designs. Although they can require a lot of resources, qualitative methods can be used even in fully randomized experiments, where the data are compared and analyzed with the goal of drawing conclusions around the program's incremental outcomes.

More typically, the comparisons are not structured around experimental or even quasi-experimental research designs. Instead, **implicit designs** are often used. The emphasis, then, is on what kinds of comparisons to include in the data collection and analysis.

Miles and Huberman (1994) indicate that there are two broad types of comparisons, given that you have collected qualitative data. One is to focus on single cases and conduct analyses on a case-by-case basis. Think of a case as encompassing a number of possibilities. In an evaluation of the Perry Preschool experiment (see Chapter 3), the individual children in the study (program and control groups) were the cases. In an evaluation of a Neighbourhood Integrated Service Teams (NIST) Program in Vancouver, Canada, each NIST was a case; there were a total of 15 in the city and all were included in the evaluation (Talarico, 1999). In the Byrd (1999) study of home nursing, the case was the one nurse who was observed for 8 months, and the researcher aggregated the results from observing 53 home visits, in order to produce "a beginning typology of maternal-child home visits" (p. 31).

Cases are, in the parlance of Chapter 4, **units of analysis.** When we select cases in a qualitative evaluation, we are selecting units of analysis.

Cases can be described in depth. Events can be reconstructed as a chronology. This is often a very effective way of describing a client's interactions with a program, for example. Cases can include quantitative data. In the NIST evaluation, it was possible to track and count the levels of activities for each NIST from the program's inception in 1996 to the evaluation in 1999.

The second kind of comparison using cases is *across* cases. Selected program participants in the Perry Preschool experiment, for example, were compared using qualitative analysis. Each person's story was told, but his or her experiences were aggregated: men versus women, for example. In the McNaughton (2000) study, the research reports were analyzed individually first ("within case" analysis), and then later the author conducted cross-case analysis that "consisted of noting commonalities and differences between the studies" (p. 407).

Cases can be compared across program sites. If a program has been implemented in a number of geographic locations, it might be important to conduct **case studies** of clients (for example) in each area, and then compare client experiences across areas. There is no reason why such qualitative comparisons could not also be complemented by quantitative comparisons. Program sites might be compared over time on the levels of program activities (outputs) and client satisfaction, and provide perceptions of program outcomes.

Identifying Appropriate Samples

Qualitative **sampling strategies** generally deliberately select cases. Contrast this approach with a quantitative evaluation design that emphasizes random samples of cases. Typically, the total number of cases sampled is quite limited, so the selection of cases becomes critical. Note that the Byrd (1999) study followed just *one* nurse over an 8-month period. The author notes "[i]nitially, observations with several nurses were planned, but well into the fieldwork, it became evident that prolonged full engagement with just one nurse was critical" (p. 28). Initially "the researcher observed all home visits scheduled during the fieldwork day" but as the patterns of the process emerged, "the emphasis shifted to describing and elaborating the patterns, so the researcher asked to accompany the nurse on visits anticipated to follow a specific pattern" (p. 28).

Table 5.3 is a typology of sampling strategies developed by qualitative researchers. This version of the table is adapted from Miles and Huberman (1994, p. 28). The 16 types of sampling strategies summarized in Table 5.3 are similar to the 14 purposeful sampling strategies summarized in Patton's (2003) Qualitative Evaluation Checklist.

Among the strategies identified in Table 5.3, several tend to be used more frequently than others. **Snowball sampling** relies on a chain of informants, who are themselves contacted, perhaps interviewed, and asked who else they can recommend, given the issues being canvassed. Although this sampling procedure is not random and may not be representative, it usually yields informed participants. One rough rule of thumb to ascertain when a snowball sample is "large enough" is to note when themes and issues begin to repeat themselves across informants.

The Byrd (1999) study would probably be considered an example of a case of *intensity* sampling, one that "manifests the phenomenon intensely, but not extremely." The author makes the point in her article that the nurse may not have been typical, so while the case was an in-depth one, it could not be assumed that the typology would apply broadly to all, or even most, nurses conducting home visits.

Sampling *politically important* cases is often a component of qualitative sampling strategies. In a qualitative study of stakeholder viewpoints in an intergovernmental economic development agreement, the 1991–1996 Canada/Yukon Economic Development Agreement (McDavid, 1996), the evaluator initially relied on a list of suggested interviewees, which included public leaders, prominent business owners, and the heads of several interest group organizations (the executive director of the Yukon Mining Association, for example). Interviews with those persons yielded additional names of persons who could be contacted, some of whom were interviewed, and others who were willing to suggest further names (McDavid, 1996).

The McNaughton (2000) study would be considered an example of **criterion sampling**, and their selection from the qualitative nursing studies was partly described as follows:

Studies included in the analysis were written in English and were published articles or doctoral dissertations reporting original research. Only research investigating home visits between PHNs [public health nurses] and mothers of young children was included. In addition, only reports

Table 5.3   Sampling Strategies for Qualitative Evaluations

Type of Sampling	*Purpose*
Maximum variation	Documents variation and identifies important common patterns
Homogeneous	Focuses, reduces, simplifies, facilitates group interviewing
Critical case	Permits logical generalization and maximum application of information to other cases
Theory based	Finding examples of a theoretical construct and thereby elaborating and examining it
Confirming and disconfirming cases	Elaborating initial analysis, seeking exceptions, looking for variation
Snowball or chain	Identifies cases of interest from people who know people who know what cases are information-rich
Extreme or deviant case	Learning from highly unusual manifestations of the phenomenon of interest
Typical case	Highlights what is normal or average
Intensity	Information-rich cases that manifest the phenomenon intensely, but not extremely
Politically important cases	Attracts desired attention or avoids attracting undesired attention
Random purposeful	Adds credibility to sample when potential purposeful sample is too large
Stratified purposeful	Illustrates subgroups; facilitates comparisons
Criterion	All cases that meet some criterion; useful for quality assurance
Opportunistic	Following new leads; taking advantage of the unexpected
Combination or mixed	Triangulation, flexibility, meets multiple interests and needs
Convenience	Saves time, money, and effort, but at the expense of information and credibility

using qualitative design and published after 1980 were reviewed. Seventeen studies were retrieved and 14 included in the final analysis. (p. 406)

Opportunistic sampling takes advantage of the inductive strategy that is often at the heart of qualitative interviewing. An evaluation may start out with a sampling plan in mind (picking cases that are representative of key groups or interests) but as interviews are completed, a new issue may emerge that needs to be explored more fully. Interviews with persons connected to that issue may need to be conducted.

Mixed sampling strategies are common. As was indicated for the Canada/Yukon Economic Development Agreement project, an initial sample that was dominated by politically important persons was combined with a snowball sample. In pursuing mixed strategies, it is valuable to be able to document how sampling decisions were made. One of the criticisms of qualitative samples is that they have no visible rationale—they are said to be drawn capriciously and the findings cannot be trusted. Even if sampling techniques do not include random or stratified selection methods, documentation can blunt criticisms that target an apparent lack of a sampling rationale.

Structuring Data Collection Instruments

Typically, qualitative data collection components of program evaluations are structured to some extent. It is very unusual to conduct interviews, for example, without at least a general agenda of topics. Additional topics can emerge, and the interviewer may wish to explore connections among issues that were not anticipated in the interview plan. But the reality in program evaluations is that resource constraints will mean that interviews are focused and at least semi-structured.

Qualitative survey instruments generally use **open-ended questions**, unlike highly structured quantitative surveys, which typically have a preponderance of **closed-ended questions.** Table 5.4 is a summary of the open-ended questions that were included in all interviews conducted for the Canada/Yukon Economic Development Agreement project (McDavid, 1996).

Each interview took at least 1 hour and most lasted 2 or more hours. The open-ended questions in Table 5.4 were structured to follow the evaluation questions that the overall program evaluation was expected to answer. The stakeholder interviews were intended to provide an independent source of information on the evaluation questions, and the findings from the interviews were integrated into the overall evaluation report (McDavid, 1996).

Each interview was tape-recorded, and the researcher conducting the interviews used the tapes to review and fill in his interview notes for each interview. Because the questions were organized around key issues in the overall evaluation, the analysis of the interview data focused on themes within

Table 5.4 Open-Ended Questions for the Economic Development Agreement Stakeholders Project

---

Canada/Yukon Economic Development Agreement Evaluation Stakeholder Interview Questions (Note date, time, location, name/position of interviewee)

Begin by introducing interviewer, reviewing purpose of interview, and requesting permission to tape-record interview. Note: Only the interviewer will have access to the tape of the interview.

1. What is your interest in and/or involvement in the 1991–1996 Economic Development Agreement between the Government of Canada and the Yukon Territorial Government?
   - Which of the six Cooperation Agreements are you involved with?
   - Were you involved in the previous EDA (1985–1989)?
   - How?

2. How much (if any) do you know about the EDA and the six Cooperation Agreements?
   - Do you have a good idea of how the whole EDA works?
   - Which of the CAs are you familiar with?
   - (If appropriate) How did you find out about the EDA?

3. How would you assess the strengths and weaknesses of the organization and the administration of the EDA and the CAs?
   - Particular strong points?
   - Particular weak points?
   - Suggestions for improving the organization/administration?
   - Barriers to improving the organization/administration?

4. Can you think of any EDA-funded projects that you would consider to be successful? Which projects, if any, were/are they? Why were they successful? Are there specific reasons for their success?

5. Which EDA projects, if any, would you consider to be failures, that is, they did not achieve any of their objectives? Are there specific reasons for their failure?

6. Now, understanding that we are asking for your overall assessment, how effective has each CA been in meeting its objectives (refer to list of Cooperation Agreement objectives, if necessary)?
   - Which CA(s) has/have been the most successful? Why?
   - Which CA(s) has/have been the least successful? Why?

   (Note: Some interviewees may not be familiar with all CAs.)

7. How effective would you say the EDA as a whole has been in meeting its overall objectives (refer to them in the list, if necessary)?
   - What part(s) of the EDA has/have been most successful in achieving these objectives?
   - What part(s) has/have been least successful?
   - Why?

(Continued)

Table 5.4 (Continued)

8. Have there been any unintended impacts from EDA-funded activities? Can you suggest some examples?

9. Has the EDA been equitably accessible to all Yukoners? Which CAs have been relatively more accessible? Which CAs have been less accessible? (Note: If necessary, remind interviewees that the four Federally-mandated target groups [Aboriginal peoples, women, handicapped persons, and visible minorities] are included in this question.)

10. If you were in a position to offer your advice on re-designing the EDA, which CAs would you change to make them more effective (more likely to achieve their objectives)?

    • What changes would you make?
    • What particular parts (elements) would you add or delete?

    (Note: Some interviewees may not be able to answer this question, given their knowledge of the EDA.)

11. Thinking of economic development as a broad issue for the Yukon, are there other kinds of programs, besides an EDA, that might be more cost-effective in achieving the objectives of the current EDA (refer to list of objectives again, if necessary)?
    • What would this/these program(s) look like?

12. Any other comments?

Thank you for your time and input into the evaluation of the EDA.

SOURCE: McDavid, 1996.

each of the issue areas. In some interviews, information relevant to one question was offered in response to another and was reclassified accordingly.

Structuring data collection instruments has several limitations. By setting out an agenda, the qualitative evaluator may miss opportunities to follow an interviewee's direction. If qualitative evaluation is, in part, about reconstructing others' lived experiences, structured instruments, which imply a particular point of view on what is important, can largely limit opportunities to empathetically understand stakeholders' viewpoints.

It may be appropriate in an evaluation to begin qualitative data collection without a fixed agenda, to learn what the issues, concerns, and problems are so that an agenda can be established. In the Byrd (1999) observational study of home nurse visitations, the data collection description provides an interesting example of a relatively fluid project, as the collection of data was guided as the analysis was occurring:

Simultaneous data collection, field note recording, and analysis were focused on the nurse's intentions, actions, and meanings as she anticipated, enacted, or reflected on her visits. In an effort to understand her thinking as she did home visiting, this nurse was informally interviewed before and after home visits. Interviews included probing questions to promote self-reflection. The investigator explored the nurse's perceptions of the reasons for the home visit; concerns before, during, and after the home visit; and her perceptions of what she was trying to accomplish, as well as the anticipated consequences of her actions. (p. 28)

A practical limitation on the use of unstructured approaches is their cost. Often, evaluation budgets do not permit us to solely conduct unstructured interviews, and then consume the resources needed to organize and present the findings.

Collecting and Coding Qualitative Data

A principal means of collecting qualitative data is interviews. Although other ways are also used in program evaluations (e.g., documentary reviews/ analyses, open-ended questions in surveys, direct observations), face-to-face interviews are a key part of qualitative data collection options.

Table 5.5 summarizes some important points to keep in mind when conducting face-to-face interviews. The points in Table 5.5 are not exhaustive, but are based on the writers' experiences of participating in qualitative interviews and qualitative evaluation projects. Patton (2003) includes sections in his Qualitative Evaluation Checklist that focus on field work and open-ended interviewing. Patton's experience makes his checklists a valuable source of information for persons involved in qualitative evaluations.

Table 5.6 offers some helpful hints about analyzing qualitative data, again, principally from face-to-face interviews. As Patton (2003) reiterates in the Qualitative Evaluation Checklist, it is important that the data are effectively analyzed "so that the qualitative findings are clear, credible, and address the relevant and priority evaluation questions and issues" (p. 10).

Coding of the data gathered in the Byrd (1999) study was described as follows:

initial analysis focused on fully describing the process of home visiting. Later, coding and theoretic memos—both analytic techniques from grounded theory—were used. Coding is assigning conceptual labels to

Table 5.5   Some Basics of Face-to-Face Interviewing

General Points to Keep in Mind

- Project confidence and be relaxed—you are the measuring instrument, so your demeanor will affect the entire interview.

- Inform participants—make sure they understand why they are being interviewed, what will happen to the information they provide, and that they can end the interview or not respond to specific questions as they see fit (informed consent).

- Flexibility is essential—it is quite possible that issues will come up "out of order" or that some will be unexpected.

- Listening (and observing) are key skills—watch for word meanings or uses that suggest they differ from your understanding. Watch for non-verbal cues that suggest follow-up questions or more specific probes.

- Ask for clarifications—do not assume you know or that you can sort something out later.

Conducting the Interview

- Ask questions or raise issues in a conversational way

- Show you are interested, but non-judgmental

- Look at the person when asking questions or seeking clarifications

- Consider the cultural appropriateness of eye contact

- Pace the interview so that it flows smoothly

- Note taking is *hard work:* the challenge is to take notes, listen, and keep the conversation moving
  - Consider having one researcher conduct the interview while another takes notes
  - Note key phrases
  - If you can, use a tape-recorder
    - Issues of confidentiality are key
    - If you are trying to record sentences verbatim to use as quotes, you may need to stop the conversation for a moment to write

- Your recall of a conversation decays quickly so *take time* to review your notes, fill in gaps, and generally make sense out of what you wrote, *before* you conduct the next interview.

- Pay attention to the context of the interview—are there situational factors (location of the interview, interruptions or interactions with other people) that need to be noted to provide background information as qualitative results are interpreted?

Table 5.6   Helpful Hints as You Analyze Qualitative Data

Getting Started

- Why did you conduct the interviews? How do the interviews fit into the program evaluation?

- What evaluation issues were you hoping could be addressed by the interview data?

- Can the relevant evaluation issues be organized or grouped to help you sort narrative and notes into themes and sub-themes? Can your interview data be categorized by evaluation issue?

- Always do your work in pencil so you can revise what you have done.

Analyzing the Data

- If you have tape-recorded the interviews, you should listen to the tapes as you review your interview notes to fill in or clarify what was said. Some analysts advocate transcribing tapes and using the transcripts as your raw data. That takes a lot of time, and in many evaluations is not practical. It does, however, ensure the accuracy and completeness of the data that you will be analyzing. Keep in mind that even verbatim transcriptions do not convey all the information that was communicated as part of interviews.

- If you have not taped the interviews, read all your interview notes, jotting down ideas for possible themes as penciled marginal notes.

- Pay attention to the *actual words* people have used—do not put words in interviewees' mouths.

- There is a balance between looking for themes and categories and imposing your own expectations. When in doubt, look for *evidence* from the interviews.

- Thematic analysis can be focused on identifying words or phrases that summarize ideas conveyed in interviews. For example, interviews with government program evaluators to determine how they acquired their training identified themes like: university courses; short seminars; job experience; and other training.

- Thematic analysis can be focused on identifying statements (subject/verb) in a narrative. It may be necessary to set up a database that translates a narrative into a set of equivalent statements that can be analyzed.

Re-read the interviews. Which of the preliminary themes still make sense? Which ones are wrong? What new themes emerge?

- What are the predominant themes? Think of themes as ideas: they can be broad (in which case lots of different sub-themes would be nested within each theme) or they can be narrow, meaning that there will be lots of them.

(Continued)

Table 5.6 (Continued)

- Are your themes different from each other (they should be)?
- Have you captured all the variation in the interviews with the themes you have constructed?
- How will you organize your themes? Alternatives might be by evaluation issue/question; or by affect, that is, positive, mixed, negative views of the issue at hand.
- List the themes and sub-themes you believe are in the interviews. Give at least two examples from the interviews to provide a working definition of each theme or sub-theme.
- Read the interviews again, and this time try to fit the text/responses into your thematic categories.
- If there are anomalies, adjust your categories to take them into account.
- There is almost always an "other" category. It should be no more than 10% of your responses/ coded information.
- Could another person use your categories and code the text/responses the way you have? Try it for a sample of the data you have analyzed.
- Calculate the percentage of agreements out of the number of categorizations attempted. This is a measure of inter-coder reliability.
- Are there direct quotes that are appropriate illustrations of key themes?

incidents or events. Memos link observations and help investigators make inferences from the data. For each visit, data were coded as descriptors of the phases of the process, consequences of the process, or factors influencing the process. Similarities and differences in the process, potential consequences, and influencing factors across visits were then compared. Distinct patterns of home visiting emerged from this analysis. (p. 28)

To illustrate, the thematic coding and analysis of "the phases of the process" in this study resulted in the following model of the nurse home visitation process (p. 28):

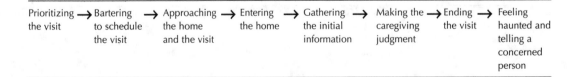

Prioritizing the visit → Bartering to schedule the visit → Approaching the home and the visit → Entering the home → Gathering the initial information → Making the caregiving judgment → Ending the visit → Feeling haunted and telling a concerned person

THE CREDIBILITY AND GENERALIZABILITY ●
OF QUALITATIVE FINDINGS

Debates about sampling and the ability to generalize from one's data are important—they are at the core of a frequent criticism of qualitative methods. Qualitative evaluators emphasize the value of in-depth, case-based approaches to learning about a program. Given that each stakeholder will offer his or her own world view, it is essential to "peel away" the layers of the onion to gain a full and authentic rendering of a person's experiences, views, and assessments of the program. As Kushner (2000) argues, since "the worth of a program is . . . subject to situational interpretation and contested meaning," program evaluators should seek to "document the lives and work of people and to use that as context within which to 'read' the significance and meaning of programs" (pp. xiv, 11).

This takes time and considerable effort. The main source of data is narrative, and analyzing these data is also time-consuming. In sum, there is a tradeoff between depth (and increasing the validity of the data) and breadth (increasing the representativeness of the data). Qualitative methods focus on fewer cases, but the quality and completeness of the information is viewed by proponents as outweighing any disadvantages due to lack of representativeness.

A challenge for evaluators who use qualitative methods is to establish the credibility and generalizability of their findings, that is, their believability and hence usefulness for stakeholders. Relying on analyses of cases can produce rich, detailed information, but if we cannot address possible concerns about the representativeness of the findings, or the methods used to produce them, our work has not been productive.

Maxwell (2002), in a synthesis of various approaches to validity in qualitative research, outlines five types of understanding and validity that typify qualitative research. His efforts were stimulated by the following observation:

Proponents of quantitative and experimental approaches have frequently criticized the absence of "standard" means of assuring validity, such as quantitative measurement, explicit controls for various validity threats, and the formal testing of prior hypotheses. (p. 37)

Maxwell outlines a typology that exemplifies the ways that qualitative researchers conceptualize validity. The typology (Table 5.7) includes: descriptive validity, interpretive validity, theoretical validity, generalizability, and evaluative validity. Maxwell's epistemological stance is critical realism, that is, he believes that there is a reality external to our perceptual knowledge of it, but

Table 5.7   Comparing Qualitative Validity with Experimental and Quasi-Experimental Validity

Types of Validity in Qualitative Research[a]	Definitions of Qualitative Validities	Related to the Following Types of Validity in Experimental/ Quasi-Experimental Research[b]
Descriptive Validity	The factual accuracy of the account (consensus of researchers—intersubjective agreement on the existence of physical and behavioral events); can include descriptive statistics (e.g., frequencies)	Statistical conclusions validity (specifically the reliability of measures)
Interpretive Validity	The meanings of actions or behaviors from participants' perspectives.	No correspondences with validity categories in the Shadish, Cook, and Campbell typology
Theoretical Validity	Focus is on the researcher's constructs—both individual constructs and causal relationships among constructs	Construct validity (how well do the factual accounts link with researcher constructs that interpret them, and how well do factual patterns correspond to relationships among constructs?)
Generalizability • Internal • External	Generalizing to other persons, organizations or institutions within the community Generalizing to other communities, groups or organizations	• Statistical conclusions validity (inferential statistics) • External validity (do the causal relationships hold for variations in persons, treatments, settings and outcomes?)
Evaluative Validity	Judging the appropriateness of actions or events from a values perspective	No correspondence with validity categories in Shadish, Cook, and Campbell

a. From Maxwell, 2002.

b. From Shadish, Cook, and Campbell, 2002.

we cannot know that reality directly. His validity categories are not intended to be a filter to assess or judge the quality of a study. Unlike positivists or post-positivists who use validity categories to discriminate among methods for doing research (randomized control trials are generally superior from an internal validity perspective, for example), Maxwell sees types of validity as fallible and not proscribing particular methodologies. The relevance of validities depends of the circumstances of a given research study.

Table 5.8   Ways of Testing and Confirming Qualitative Findings

1. **Check the cases for representativeness** by comparing case characteristics to characteristics of people (units of analysis) in the population from which the cases were selected.

2. **Check for researcher effects** by asking whether and how the evaluator could have biased the data collection or how the setting could have biased the researcher.

3. **Triangulate data sources** by comparing qualitative findings with other sources of data in the evaluation.

4. **Weigh the evidence** by asking whether some sources of data are more credible than others.

5. **Check outliers** by asking whether "deviant" cases are really that way or, alternatively, the "sample" is biased and the outliers are more typical.

6. **Use extreme cases** to calibrate your findings, that is, assess how well and where your cases sit in relation to each other.

7. **Follow up surprises,** that is, seek explanations for findings that do not fit the overall patterns.

8. **Look for negative evidence,** that is, findings that do not support your own conclusions.

9. **Formulate If/Then statements** based on your findings to see if interpretations of findings are internally consistent.

10. **Look for intervening variables** that could explain key findings—if you have information on these variables, can you rule their influences out, based on your findings?

11. **Replicate findings** from one setting to another one that should be comparable.

12. **Check out rival explanations** using your own data, your judgment, and the expertise of those who know the area you have evaluated.

13. **Get feedback from informants** by summarizing what they have contributed and asking them for their **concurrence** with your summary.

SOURCE: Miles and Huberman (1994, pp. 263–277).

Some, though not all, of the types have commonalities with the types of validity described by Shadish, Cook, and Campbell (2002). But Maxwell stresses that even where there are commonalities, it is important to keep in mind that they do not indicate a shared epistemology with positivist and post-positivist researchers.

Miles and Huberman (1994) have identified 13 separate ways that qualitative data and findings can be queried to increase their robustness. Table 5.8 lists, adapts, and summarizes these checks, together with a brief explanation of what each means.

Although these 13 points all offer complementary ways to increase our confidence in qualitative findings, some are more practical than others. In program evaluations, two of these are more useful:

- Triangulating data sources
- Getting feedback from informants

Feedback from informants goes a long way toward establishing the validity of qualitative data. It does not tell you how representative your cases are, but it does tell you whether you have rendered the information so that it accords with the views of those providing it—that is key to authentically representing their world views.

Triangulation of data sources is important to establish whether findings from qualitative analyses accord with those from other data sources. Typically, complementary findings suggest that the qualitative data are telling the same story as are other data. If findings diverge, then it is appropriate to explore other possible problems: representativeness of the cases, researcher bias, and weighing the evidence are reasonable places to begin.

● CONNECTING QUALITATIVE EVALUATION METHODS TO PERFORMANCE MEASUREMENT

Performance measurement has tended to rely on quantitative measures for program constructs. Program or organizational objectives are stated, annual performance targets are established in many performance measurement systems, and the data that are gathered are numerical. Numbers lend themselves to visual displays (graphs, charts) and are relatively easy to interpret (trends, levels). But, for some government agencies and nonprofit organizations, the requirement that their performance be represented in numbers forces the use of measures that are not seen by agency managers to reflect the key outcomes. Nonprofit organizations that mark their progress with clients by seeing individual lives being changed often do not feel that numerical performance measures weigh or even capture these outcomes.

Sigsgaard (2002) has summarized an approach to performance measurement that is called the Most Significant Change (MSC) approach. Originally designed for projects in developing nations, where aid agencies were seeking an alternative to numerical performance measures, the MSC approach applies qualitative methods to assessing performance. It has something in common with the Shoestring Evaluation approach (Bamberger, Rugh, Church, & Fort, 2004)—both are designed for situations where evaluation resources are very limited, but there is a need to demonstrate results and do so in ways that are defensible.

Sigsgaard (2002) describes how a Danish international aid agency (Mellemfolkeligt Samvirke) adopted the MSC approach as an alternative to the traditional construction of quantitative logic models of projects in developing countries. The main problem with the logic modeling approach was the inability of stakeholders to define objectives that were amenable to quantitative measurement.

The MSC approach involves an interviewer or interviewers (who have been briefed on the process and intent of the approach) asking persons who have been involved in the project (recipients/beneficiaries of the project) to identify positive or negative changes they have observed over a fixed time, for one or more domains of interest. Examples of a domain might be health care in a village involved in an aid project, or farming in a rural area where a project had been implemented. By eliciting both positive and negative changes, there is no bias towards project success. Then these same persons are asked to indicate which change is the most significant and why.

By interviewing different stakeholders, a series of change-related stories are recorded. Although they might not all relate to the project or to the project's objectives, they provide authentic views on how participants in the MSC interviews see their world and the project in it.

The performance stories are reviewed by different governance levels (boards) in the donor organization (within and outside the country), and from among them, the most significant stories (ultra-most significant) are selected along with reasons for their choices. Essentially, the set of performance stories are shared and discussed and finally winnowed to a smaller set. Performance stories are verified by additional investigation and are then used to guide any changes that are implied by the results that are communicated via the stories.

Sigsgaard (2002) sums up the experience of his aid organization with the MSC approach to qualitative performance measurement:

> The process of verification, and the curiosity aroused by the powerful data collected, will stimulate the country offices as well as the partners to supplement their knowledge through use of other, maybe more refined and controlled measures. The MSC system is only partially participatory. Domains of interest are centrally decided on, and the sorting of stories according to significance is hierarchic. However, I believe that the use of and respect for people's own indicators will lead to participatory methodologies and "measurement" based on *negotiated indicators* where all stakeholders have a say in the actual planning of the

development process. Some people in the MS [Mellemfolkeligt Samvirke] system have voiced a concern that the MSC method is too simple and "loose" to be accepted by our source donor, Danida, and our staff in the field. The method is not scientific enough, they say. My computer's Thesaurus program tells me that science means knowledge. I surely can recommend the Most Significant Change method as scientific. (p. 11)

● THE POWER OF CASE STUDIES

One of the great appeals of qualitative evaluation is the ability to render experiences in convincing detail. Narrative from even a single case, rendered to convey a person's own words and feelings, is a very powerful way to draw attention to an issue or a point of view.

Most of us respond favorably to stories, to narratives that chronicle the experiences of individuals. In the context of program evaluations, it is often much easier to communicate key findings by using case examples. For many clients, tables do not convey a lot of intuitive meaning. Graphs are better; but narratives, in some cases, are best. Patton (2003), in his checklist for qualitative evaluations, suggests that qualitative methods are best suited for telling stories:

> Qualitative methods are often used in evaluations because they tell the *program's story* by capturing and communicating the *participants' stories*. Evaluation case studies have all the elements of a good story. They tell what happened when, to whom, and with what consequences. (p. 2)

In the mass media, typically news stories focus on individuals, and a single well-stated opinion or carefully presented experience can have important public policy implications. The tragic death of a single child in British Columbia, Canada in 1994 at the hands of his mother became the basis for the Gove Commission (Gove, 1995) and, ultimately, the reorganization of all existing child protection functions into the provincial Ministry for Children and Families in 1996.

In program evaluations, case studies often carry a lot of weight, simply because we can relate to the experiences of individuals more readily than we can understand the aggregated/summarized experiences of many. Even though single cases are not necessarily representative, they are often treated as if they contained *more data* than just one case. For program evaluators, there is both an opportunity and a caution in this. The opportunity is to be able to use cases and qualitative evidence to render evaluation findings more credible and, ultimately, more useful. But the caution is to conduct qualitative evaluations (or the qualitative components of multi-source evaluations) so that they are *methodologically defensible* as well as being persuasive.

SUMMARY ●

Qualitative evaluation methods are an essential part of the range of tools that evaluators call upon in their practice. Since the 1970s, when qualitative evaluation methods were first introduced as an alternative to the then orthodox experimental/quasi-experimental paradigm, the philosophical underpinnings and methodological requirements for sound qualitative evaluation have transformed the evaluation profession. Debates continue about the relative merits of positivistic and constructivist approaches to evaluation, but many evaluators have come to the view that pragmatically, it is desirable to mix qualitative and quantitative methods—they have complementary strengths and the weaknesses of one approach can be mitigated by calling upon the other approach.

Qualitative approaches to evaluation are themselves diverse. Some proponents share the same epistemological beliefs as do practitioners who rely on quantitative methods—that there is a reality we share and can know (to varying degrees) through our efforts to measure aspects of it. Other qualitative evaluators have embraced one or another phenomenological approach—the underlying assumptions about the way we know do not include the belief that there is one (social) reality we share. Rather, each of us has our own "world" and the challenge for evaluators is to develop methods to learn about each person's world, render what has been learned in ways that are authentic, and find ways of working with those perspectives in the evaluation process.

Qualitative evaluation often relies on case studies—in-depth analyses of individuals (as units of analysis) who are stakeholders in a program. Case studies, rendered as stories, are an excellent way to communicate the personal experiences of those connected with a program. We, as human beings, have tended to be storytellers—indeed, stories and songs were the ways we transmitted culture before we had written language. Case studies convey meaning and emotion, rendering program experiences in terms we can all understand.

Although performance measurement has tended to rely on quantitative indicators to convey results, there are alternatives that rely on qualitative methods to elicit performance stories from stakeholders. In settings where data collection capacities are very limited, qualitative methods offer a feasible and effective way to describe and communicate performance results.

● DISCUSSION QUESTIONS

1. What is a paradigm? What does it mean to say that paradigms are incommensurable?

2. What is Patton's pragmatic approach to evaluation?

3. What are the key characteristics of qualitative evaluation methods?

4. What does it mean for an evaluation to be naturalistic?

5. What is snowball sampling?

6. Suppose that you have an opportunity to conduct an evaluation for a state agency that delivers a program for single mothers. The program is intended to assist pregnant women with their first child. The program includes home visits by nurses to the pregnant women and regular visits for the first 2 years of the child's life. The objective of the program is to improve the quality of parenting by the mothers, and hence, improve the health and well-being of the children. The agency director is familiar with the quantitative, experimental evaluations of this kind of program in other states and wants you to design a qualitative evaluation that focuses on what actually happens between mothers and children in the program. What would your qualitative evaluation design look like? What qualitative data collection methods would you use to see what was happening between mothers and children? How would you determine whether the quality of parenting had improved as a result of the program?

● REFERENCES

Bamberger, M., Rugh, J., Church, M., & Fort, L. (2004). Shoestring evaluation: Designing impact evaluations under budget, time and data constraints. *American Journal of Evaluation, 25*(1), 5–37.

Byrd, M. E. (1999). Questioning the quality of maternal caregiving during home visiting. *Image: Journal of Nursing Scholarship, 31*(1), 27–32.

Cook, T. D. (1991). Clarifying the warrant for generalized causal inferences in quasi-experimentation. In M. W. McLaughlin & D. C. Phillips (Eds.), *Evaluation and education: At quarter century. Yearbook of the National Society for the Study of Education* (90th ed., Pt. 2, pp. xiv, 296). Chicago: National Society for the Study of Education; University of Chicago Press.

Cook, T. D., & Campbell, D. T. (1979). *Quasi-experimentation: Design & analysis issues for field settings.* Chicago: Rand McNally College Publishing.

Datta, L. E. (1994). Paradigm wars: A basis for peaceful coexistence and beyond. *New Directions for Program Evaluation, 61,* 53–71.

Denzin, N. K., & Lincoln, Y. S. (Eds.). (2000). *Handbook of qualitative research* (2nd ed.). Thousand Oaks, CA: Sage.

Fetterman, D. M., (1996). Empowerment evaluation: An introduction to theory and practice. In D. M. Fetterman, S. Kaftarian, & A. Wandersman (Eds.), *Empowerment evaluation: Knowledge and tools for self-assessment and accountability.* Thousand Oaks, CA: Sage.

Gove, T. J. (1995). *Report of the Gove Inquiry into Child Protection in British Columbia: Executive Summary.* Retrieved August 5, 2004, from http://www.qp .gov.bc.ca/gove/gove.htm

Guba, E. G., & Lincoln, Y. S. (1989). *Fourth generation evaluation.* Newbury Park, CA: Sage.

Guba, E. G., & Lincoln, Y. S. (2001). *Guidelines and checklist for constructivist (a.k.a. Fourth Generation) evaluation.* Retrieved August 4, 2004, from http://www .wmich.edu/evalctr/checklists/constructivisteval.pdf

Hanson, N. R. (1958). *Patterns of discovery: An inquiry into the conceptual foundations of science.* Cambridge, UK: Cambridge University Press.

House, E. R. (1994). Integrating the quantitative and qualitative. *New Directions for Program Evaluation, 61,* 13–22.

Kuhn, T. S. (1962). *The structure of scientific revolutions.* Chicago: University of Chicago Press.

Kushner, S. (2000). *Personalizing evaluation.* London, Thousand Oaks, CA: Sage.

Mark, M. M., Henry, G. T., & Julnes, G. (2000). *Evaluation: An integrated framework for understanding, guiding, and improving policies and programs.* San Francisco: Jossey-Bass.

Maxwell, J. A. (2002). Understanding and validity in qualitative research. In A. M. Huberman & M. B. Miles (Eds.), *The qualitative researcher's companion* (pp. 37–64). Thousand Oaks, CA: Sage.

McDavid, J. C. (1996). Summary report of the 1991–1996 Canada/Yukon EDA evaluation. Ottawa, ON: Department of Indian and Northern Affairs.

McNaughton, D. B. (2000). A synthesis of qualitative home visiting research. *Public Health Nursing, 17*(6), 405–414.

Miles, M. B., & Huberman, A. M. (1994). *Qualitative data analysis: An expanded sourcebook* (2nd ed.). Thousand Oaks, CA: Sage.

Mohr, L. B. (1999). The qualitative method of impact analysis. *American Journal of Evaluation, 20*(1), 69–84.

Murphy, E., Dingwall, R., Greatbatch, D., Parker, S., & Watson, P. (1998). Qualitative research methods in health technology assessment: A review of literature. *Health Technology Assessment, 2*(16), 1–274.

Patton, M. Q. (1997). *Utilization-focused evaluation: The new century text* (3rd ed.). Thousand Oaks, CA: Sage.

Patton, M. Q. (2003). *Qualitative evaluation checklist.* Retrieved August 4, 2004, from http://www.wmich.edu/evalctr/checklists/qec.pdf

Reichardt, C. S., & Rallis, S. F. (1994). Qualitative and quantitative inquiries are not incompatible: A call for a new partnership. *New Directions for Program Evaluation, 61,* 85–91.

Scriven, M. (1976). Maximizing the power of causal investigations: The modus operandi method. In G. V. Glass (Ed.), *Evaluation studies review annual* (Vol. 1, pp. 108–118). Beverly Hills, CA: Sage.

Scriven, M. (In progress, unpublished). *Causation.* New Zealand: University of Auckland.

Shadish, W. R., Cook, T. D., & Campbell, D. T. (2002). *Experimental and quasi-experimental designs for generalized causal inference.* Boston: Houghton Mifflin.

Sigsgaard, P. (2002). MCS approach: Monitoring without indicators. *Evaluation Journal of Australasia, 2*(1), 8–15.

Talarico, T. (1999). *An evaluation of the Neighbourhood Integrated Service Team program.* Unpublished Master's thesis, University of Victoria, British Columbia, Canada.

The Evaluation Center. (2001). *The evaluation checklist project.* Retrieved July 13, 2004, from http://www.wmich.edu/evalctr/checklists/qec/index.htm

CHAPTER 6

ASSESSING THE NEED FOR PROGRAMS

INTRODUCTION ●

Needs assessments, which measure the nature and extent of the need for a program, are an important part of the range of evaluation-related activities that are conducted in the public and nonprofit sectors. Organizations can use the information produced by needs assessments to identify social problems that might be ameliorated by designing and implementing appropriate policies and programs. As well, evaluators can include needs assessments in their evaluations of ongoing programs to see whether they are still relevant.

Assessing the current need for programs is clearly a part of the core evaluation issues that are expected to be addressed in many government program evaluations. The assessment of need is a key component of the *Program Assessment Rating Tool* (PART), used since 2002 by the United States government to review and evaluate federal programs as part of the budget process (Office of Management and Budget, 2002). The PART is a series of approximately 30 questions designed to gather information about program efficiency and effectiveness. Question 2 asks "Does the program address a specific interest, problem *or need?*" (Office of Management and Budget, 2002, p.6, italics added). The PART explains the purpose of question 2 as being "to determine whether the program addresses a specific interest, problem or need that can be clearly defined and *presently exists.*" A *"Yes"* answer would require the existence of a relevant and clearly defined interest, problem or need that the program is designed to address. A *Yes* answer would also require that the program purpose *is still relevant to current conditions*" (Office of Management and Budget, 2002, p.6, italics added).

In the *Evaluation Policy of the Treasury Board of Canada* (Treasury Board of Canada Secretariat, 2001), the Canadian Federal Government identifies program relevance, along with results and cost-effectiveness, as the three evaluation issues that should be addressed in evaluation or performance measurement frameworks. Program relevance is defined in a related Treasury Board of Canada publication this way:

> *Relevance to Canadians*—Does the program area or activity continue to produce results that reflect government priorities *and the current needs of Canadians?* What targets have been set, and have they been achieved? (Treasury Board of Canada Secretariat, 2003, p. 2, italics added)

The performance management cycle introduced in Chapter 1 includes several activities that relate to assessing needs. Figure 6.1 is a simplified version of the performance management cycle which highlights activities that

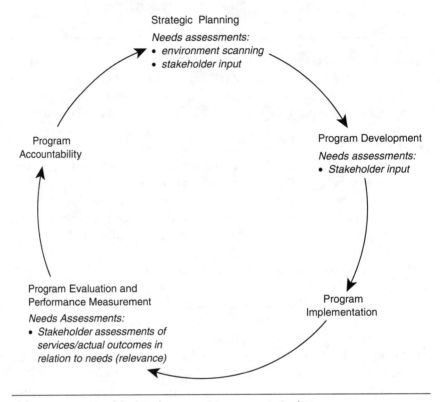

Figure 6.1   Simplified Performance Management Cycle

are central to public and nonprofit managers. Several of the major activities are associated with possible applications of needs assessment tools.

Beginning with the strategic planning phase of the cycle, most strategic plans are developed or modified in the light of information put together to represent the organization's strengths, weaknesses, opportunities, and threats. Among the sources of information would be current and possibly future stakeholders, including clients or customers. Setting strategic goals in public sector and nonprofit organizations is usually constrained by existing mandates. As well, relationships with stakeholders must be maintained, as they will typically be ongoing over time.

Environmental scans are often used to gauge the relationships between the organization and its external stakeholders. Typically, environmental scans offer comparisons that indicate trends that could affect the way the organization wishes to direct or redirect its efforts. As an example, an environmental scan for a university strategic planning process might concentrate on the following:

- Demographic trends, which could affect the future availability of students
- Economic trends, including the past and current employment rates for graduates from different disciplines
- Trends in labor force participation for men and women, which could affect the demand for different disciplines
- Projections of the demand for different types of skills in the future
- Past and current distributions of full- and part-time students across disciplines in that university
- Numbers and types of students attending other post-secondary institutions in the state or province

Many universities, for example, have started surveying their graduates to find out whether and where they are employed, whether they are satisfied with their university experience, and how they would want the university to change to improve the education and services it provided to them. Some of this feedback is relevant to assessing the need for particular programs, as well as identifying possible strategic issues that could inform strategic planning.

These strategic planning activities use some of the same sources of information that we will discuss later in conducting needs assessments. In fact, strategic planning, to the extent that it offers an opportunity to consider and redirect an organization's energies, *must* be done so that the needs/demands/wants/preferences of stakeholders are considered. Organizations that serve clients must consider their clients' feedback/input to develop appropriate strategic objectives; otherwise, there may be a lack of fit between the organization's directions and the expectations of those in its environment. Later in this chapter we will summarize the St. Columba Collaboration Project in Newark, New Jersey as an example of a strategic planning process that relied on client input to align the organization with neighborhood needs (Elliot, Quinless, and Parietti, 2000).

Frequently, needs assessments are conducted as programs are being developed. In the performance management cycle depicted in Figure 6.1, the strategic planning part of the cycle is intended to yield, among other products, strategic objectives for the organization as a whole. Often, these are general goal statements, which are intended as guides for more specific objectives that frame program and policy development.

Developing programs is informed not only by the strategic objectives, but also by sources of information that can shape the actual structure, operations, and intended outcomes. A key part of developing programs is to identify the client population(s) who are expected to benefit from or be affected by program outcomes, and design programs that are efficient and effective in resolving the problems or (more generally) achieving intended objectives with respect to those clients.

Needs assessments at this stage in the performance management cycle are typically focused on questions that will yield data that can be used to design specific programs, or amend the design or direction of existing programs. Later in this chapter, we will consider an example of a narrowly focused needs assessment, which is intended to provide input into designing a daycare program for the employees of one public sector agency.

In addition, program evaluations can incorporate components that focus on client/stakeholder needs. For example, if an evaluation includes a survey of existing program recipients, the survey instrument can solicit client experiences and assessments of their interactions with the existing program, as well as their perceptions of ways in which the program could be modified to better meet their needs. The information on gaps between the services received from the current program and perceived needs could be one source of data used to modify the design and/or the implementation of the program.

● WHAT ARE NEEDS?

Theorists and researchers in the last century have offered models of human needs that are intended to explain and predict human behaviour. One example of this kind of work is Charlotte Towle's book *Common Human Needs,* originally published in 1945. Towle was a professional social worker and argued that human beings have basic and higher needs and that, unless basic needs are satisfied, human beings will generally not realize their potential. She claimed that we often disregard the inter-related nature of people's needs and the fact that basic dependency needs must be met before people can realize their opportunities for independence (Towle & Younghusband, 1973).

Perhaps the most important contribution to our understanding of human needs is the work of Abraham Maslow. He, and others who have followed, claim that it is possible to identify and rank human needs, and that human beings will tend to behave in ways that satisfy more basic needs before they try to satisfy higher needs.

Table 6.1 summarizes Maslow's basic classification of needs and offers examples of specific needs in each of five main categories. These categories of needs are hypothesized to motivate behavior and to do so in a hierarchical manner. Persons with unmet physiological and safety needs, for example, would behave in ways that would satisfy their physiological needs first.

Maslow assumed that these needs were universal, that is, all human beings would be motivated by each level of needs successively. He did

Table 6.1	Maslow's Hierarchy of Needs
Need	*Examples*
Physiological needs	Air, food, water, sleep
Safety needs	Stability, consistency
Belonging/love needs	Group acceptance, friendship
Esteem needs	Respect, recognition
Self-actualization needs	Knowledge, peace, self-fulfillment

acknowledge that persons operating at one level could "jump" to other levels out of sequence, but his intention was to frame a theory of human motivation that would capture broad patterns of human behavior.

Much has been written about Maslow's theory, and it maintains its appeal to the present day. A cursory search of websites reveals a continuing interest in his work among a wide range of individuals and groups. The publication of *Maslow on Management* (Maslow, 1998), originally published in 1965 as *Eupsychian Management* (Maslow, 1965), in which Maslow details his views on organizational management, is one indication of the currency of his work.

For our purposes, the main impact of Maslow's work has been to create the sense that human needs exist, can be organized in relation to each other, and, with appropriate instruments, can be measured. Indeed, the existence of "needs assessments" as a distinct field in program evaluation owes much to Maslow's work.

We can, however, look at needs from another point of view. In our society, we develop and choose among policies using a variety of political processes. The process of making choices among competing policies entails choosing to support some values and priorities and not others. David Easton (1965) has described politics in Western societies as the authoritative allocation of values. What he means is that making political decisions and then implementing them as policies and programs is a process that chooses among differing viewpoints on what we *ought* or *should* do, in short, among different values.

Stating needs can be thought of as one way of expressing our values and our support for particular policies and programs. The *need* for a national policy on social housing, for example, is an expression of concerns about the homeless in cities and a statement that we *ought* to ameliorate this problem.

Regardless of whether we believe that there are intrinsic human needs or not, translating needs (or demands or wants) into policies and programs entails political choices. If politics is about allocating values and resources to support some values and not others, then needs assessments can be seen as one type of input into political processes. Assessing needs, reporting or documenting such assessments, and communicating the results of assessments will be scrutinized by a range of interested parties, some of whom will be supportive and others of whom will work to prevent resources being allocated to those purposes. The defensibility of needs assessments becomes an important issue in these circumstances.

Needs are relative. In our society, our conception of needs has changed over the last 50 years. Currently, we have policies and programs that support persons and families who are unable to earn a living. Although the levels of support and the conditions to become eligible for social assistance vary from one jurisdiction to another, we have generally agreed, as a society, that some basic level of income is a human need.

But, if one looks back in time to the Great Depression when there was massive unemployment and poverty, there was no general political will, at least initially, to provide income to those who had no other means of support. The experience of the 1930s in North America was an important stimulus to creating components of the social safety net in the post-World War II period.

Likewise, if we compare our society to others, our conception of publicly supported needs is different from others. In some cases, these differences can be attributed to a lack of resources to address problems, even where there is agreement that a need exists. In other cases, different values have resulted in radically different views of needs. In some countries, for example, public education for female children is not supported, in part because social and religious values militate against spending public resources for this purpose. We might disagree with such policies or question the morality of public policies that are apparently supported by values very different from our own but, in our own society, we also deeply disagree with each other on whether public resources should be used for some policies or programs. Public support for abortions is one example of an issue that continues to be contentious in our society.

● BENCHMARKING NEEDS: CRITERIA FOR ESTABLISHING THE EXISTENCE AND MAGNITUDE OF NEEDS

All needs assessments involve *comparisons*. These are not comparisons to determine whether a program was effective (we discussed those in Chapter 3) but instead are comparisons to measure the extent and types of needs. One strategy is to ascertain *current* levels and kinds of services being provided to a given **population**, and compare the programs and services to some set of **benchmarks** or reference points. The sources of such benchmarks include:

- Conceptions of human needs that are suggested by frameworks, models, or theories
- Moral or ethical values that indicate what human beings ought to do for each other or provide for each other
- Existing levels and kinds of services provided to other, comparable populations
- Service provider opinions/preferences
- Opinions/preferences of the clients (current and prospective) themselves

Theories, Models, or Frameworks as Benchmarks

The first benchmarking process refers in part to work done by researchers like Maslow. His hierarchy of needs appeals in part because it is embedded in a broader humanistic psychological theory about human nature, human personalities, and developmental processes. Maslow's work has an optimistic view of human potential and suggests that satisfying human needs will produce salutary outcomes.

This source of reference points can also refer to expert opinions or standards that have been established in some professions or service areas. In many areas of hospital-related care, for example, standards for waiting times for tests or treatments have been developed and are being used to assess the extent of deficiencies in service provision to specific client populations.

Theoretical models have also been used to estimate needs for programs and services. In highway maintenance programs, for example, formulae can be used to estimate desired intervals between various kinds of maintenance work. Given traffic volumes, weather conditions, snow and ice removal methods, surface and substructure types, age of roads, and previous maintenance, it is possible to estimate how often different types of maintenance and refurbishing should occur. An interval of 10 years for resurfacing would be one example of using this approach to estimate the need for highway maintenance activities.

In the field of drug and alcohol addiction treatment, it is costly and often very difficult to obtain valid estimates of the prevalence of problems and, hence, estimates of the need for services. One approach is to use theoretical models that estimate the prevalence of an addiction/health problem in a population and, hence, the need for services. An example of such an approach is the Ledermann model, which has been used to estimate alcohol consumption in populations (Dewit & Rush, 1996). The model assumes that alcohol consumption is log-normally distributed in any population: that is, most will consume very little, but there will be a small proportion who will consume a lot of alcohol, making for a skewed distribution of actual consumption rates overall (a logarithmic transformation of such a skewed distribution can produce a distribution that appears to be symmetrical and has properties akin to a normal distribution). Using available information on total consumption, making an assumption about the upper limit of consumption (no more than 1% of a given population consumes 1 liter of pure alcohol per day, for example) and an assumption that the variance in alcohol consumption is constant, it is possible to estimate the numbers of persons who are consuming different amounts of alcohol. Using that information, it is possible to estimate the numbers of persons who are addicted and, hence, the magnitude of the need for services in that population.

Moral or Ethical Values as Benchmarks

Moral or ethical values, informed by churches and other organizations, can be an important source for identifying needs. Serving the poor is an example of an important moral imperative that has motivated efforts to achieve greater social justice in our society by redistributing wealth from the relatively well-off to those in need. Moral and ethical values can be incorporated into the mission statements and strategic objectives of organizations, including political parties. In such situations, policies and programs will be affected by these objectives. If a church decides to operate a soup kitchen, for example, that program will be informed by values that point to a commitment to work with and for the poor/homeless in the community. Many churches operate services that are offered in partnership with other community agencies. Congregations will collect food around the Christmas season to donate to food banks for distribution to those who have either identified themselves as being in need, or have been identified by community agencies.

Comparisons Within or Among Jurisdictions as Benchmarks

Comparisons between services provided to a given population and those provided elsewhere are a common way of benchmarking needs. As an example, a community health center may be comparing its ratio of home care support workers per 100 clients to the ratios in agencies elsewhere in its state or province, to ascertain whether it is above or below its peers with respect to resourcing for these services. One variant on such comparisons is to conduct them over time within one agency. In our example of home care support workers, the agency may want to display the worker-to-client ratios for the past 5 years to show whether resources are increasing or decreasing relative to clients served. This time series could be one measure of performance—not focused on outcomes, but instead, focused on outputs (clients served).

Service Providers as Benchmarks

Service providers themselves can provide reference points that are useful in needs assessments. For services where qualitative differences among clients can be important in determining the level of need, service providers can lend their experience and familiarity with the clients to any needs assessment.

For example, a recent Erie County, New York, needs assessment of seriously emotionally disturbed youth (Kernan, Griswold, & Wagner, 2003) relies mostly on data collected through interviews with social workers, psychologists, and case managers working with the children. Although family members were also approached for information, in only 55% of cases were they available or would they consent to be interviewed, making the information from service providers even more central to the study.

Current or Prospective Clients as Benchmarks

Finally, some needs assessments offer current or prospective clients opportunities to rate or rank services that are currently provided or might be provided in the future. In effect, clients are expected to self-reference their needs and offer comparisons between what is and what should be. This approach reflects the recent trend toward emphasizing the empowerment of clients in needs assessments (Houston & Cowley, 2002; Lewis, Rudolph, & White, 2003). One example of this kind of needs assessment, used by the National Health Service in the United Kingdom, uses an open assessment tool called the "Field of Words" (FOW). The FOW is essentially a list of adjectives and words referring to certain topics of interest in health promotion, such as "good," "problems," "emotional support," and "violence," for the topic of "Relationships." Clients are asked to circle as many or as few words as they feel apply to their situation. The purpose of the FOW assessment tool is to trigger dialogue with clients and enable them to recognize and discuss self-identified needs (Houston & Cowley, 2002).

The advantage of asking current or future clients about their needs is that each person's circumstances will, in effect, be taken into account in the responses. However, substantial challenges exist around the reliability and validity of such data. We will discuss data sources and needs assessment methodologies later in this chapter.

● STEPS IN CONDUCTING NEEDS ASSESSMENTS

McKillip (1998) has suggested a series of steps that are included in most needs assessments. The following eight steps are adapted from his discussion:

1. Become familiar with the political context

2. Identify the users and uses of the needs assessment

3. Identify the target population(s) who will be or are currently being served

4. Inventory the existing services that are available to the target population, in part to identify gaps

5. Identify needs, using complementary strategies for collecting and recording data

6. Prepare a document that integrates evidence, benchmarks, conclusions, and recommendations

7. Communicate the results of the needs assessment. This involves marketing the work that has been done and maximizing its impacts on the political/resource allocation processes

8. Implement the recommendations from the needs assessment

Each of these steps is integrated with the others. In a given situation, a needs assessment may proceed iteratively, that is, steps will be reviewed or even repeated as the project evolves. Alternatively, steps may be combined, as was the case with the Newark, New Jersey, needs assessment summarized below as an example that applies these steps. An important consideration in a needs assessment is the level of stakeholder collaboration that will be included as part of the project. Some or all of the steps may include engagement with various stakeholders such as clients, the organization's leaders, government officials, related service providers, and others. Since it will be necessary to arrive at a reasonable level of consensus at the end of the assessment, careful thought must be put into who will be involved in the assessment.

Below, we will expand on some of the key issues involved within the steps of a needs assessment, and then further in the chapter we will use the steps to outline a needs assessment conducted by a coalition of community groups in a Latino neighborhood in Newark, New Jersey.

Become Familiar With the Political Context

Needs assessments, because they are usually intended to create information that feeds into resource allocation decisions, can be contentious. Proponents of expanding or creating a service based on a needs assessment can be challenged on several grounds, for example:

1. The provision of the service is wrong or a poor use of resources because the objectives of the service are different from or directly challenge strongly held values or existing commitments.

2. The needs assessment itself is flawed and the information produced is biased, inaccurate, or incomplete.

Disagreements based on differences in values are fundamentally political and usually are resolved through political decision-making processes. Convincing proponents of different or competing values of the "rightness" of one's own values/objectives is usually difficult. In practice, many political decisions are made by trading values: If you support program A (which I value), I will support your program B (which I do not necessarily value, but am willing to support if it is instrumental in my realizing program A).

Methodological challenges can, when anticipated, at least be addressed proactively (even if they cannot all be resolved). This chapter suggests areas where persons conducting needs assessments need to pay attention to possible methodological problems. Perhaps the most important consideration is an awareness of the incentives that various stakeholders have in providing information. For example, service providers typically will be interested in preserving existing services and acquiring more resources for expanded or new services.

Existing or prospective clients will generally be interested in improved services as well. Usually, the services they consume do not entail paying substantial user fees, so there is a general bias toward wanting more services than would be the case if they bore the full marginal costs associated with increasing services.

Other stakeholders (service providers who might be offering complementary or perhaps competing services) may or may not have an a priori tendency to want services expanded. Their views can be very useful as a way to triangulate the views of clients and service providers at the focus of a needs assessment.

Identify the Users and Uses of the Needs Assessment

The distinction between formative and summative evaluations is useful to describe the intentions of stakeholders who are involved in needs assessments. An example of a formative needs assessment would be one commissioned by a service provider to determine whether the existing services are getting to all those who can be identified as prospective clients. Service providers would be able to use the results to identify gaps in the services and perhaps make the case for more resources to augment existing programs. There is no threat to the scope or the funding for the existing program(s) in this kind of needs assessment.

An example of a summative needs assessment might be one commissioned by a major funder to determine whether the program(s) in an agency are still meeting a need. Possible outcomes of the needs assessment might include whether those programs might be downsized or terminated to free up resources for higher priority needs.

Examples of prospective users of a needs assessment include: service providers, funders, elected officials, board members, current and prospective clients, and the general public. Typically, there are several stakeholder groups who are interested in a needs assessment, and it is essential that a written agreement describing the terms of reference for the study be negotiated. Key to preparing the terms of reference is determining what the main questions are. Although most needs assessments focus on the gaps between what programs and services are currently available and what the need is, other questions are usually included. Examples would include asking current clients about their levels of satisfaction with existing services, or their willingness to pay for current or additional services.

Knowing who the key users are will influence how the reporting process unfolds. Engaging stakeholders in the study as it is happening is one way to increase their buy-in for the recommendations in the final report.

Identify the Target Population(s)
Who Will Be or Are Currently Being Served

There are two important dimensions to identifying target populations: (a) the geographic scope of a planned needs assessment and (b) the socio-demographic characteristics of the target population in those geographic areas. The more precise the designation of the geography and characteristics of people that are relevant, the easier it is to focus the methodologies of the study (sampling and data collection methods).

Among the socio-demographic attributes of current and prospective clients are: age, gender, ethnicity, first language, literacy level, education, occupation, income, and place of residence. These can be used as indirect measures of need and can be used to cross-classify reported needs to zero in on the sub-populations where reported needs are greatest.

In addition to identifying the target populations, it is important to identify any comparison populations that could be used to benchmark needs. For example, a needs assessment for job training services in a community where a sawmill is the main employer might include comparisons to other resource-dependent communities that have already established such programs, to gauge how many clients are served in these other communities, in relation to the populations. The idea would be to establish a benchmark that suggested an appropriate scale for a job training program based on the ratio of population to clients served in other, similar, communities.

Inventory Existing Services and Identify Potential Gaps

Assuming that the clients of the needs assessment have been identified, and the target population has also been specified, an important step in conducting a needs assessment is to develop an inventory of the services currently being offered in a given geographic area. McKillip (1998) suggests that once a target population in a geographic area has been identified, it is useful to contact existing service providers and find out:

- Who is providing services to the target population
- What services are actually being provided, including types, availability, and costs
- Who the clients are, that is, what their demographic characteristics are and their geographic locations in relation to the service providers

McKillip (1998) illustrates displays of service inventory information in a matrix format, with relevant client demographic characteristics across the top of the matrix, the types of services that are provided to that client population down the side of the matrix, and the names of the provider agencies in the cells of the matrix. Table 6.2 illustrates part of a matrix that might be constructed for a needs assessment that is focused on services for older people in a given population.

Table 6.2	Services Inventory for Elderly Persons in a Population			
	Relevant Client Characteristics			
Services Provided	*Over 65 years of age*	*Person living alone*	*Woman over 65 living alone*	*Person with short-term physical disabilities*
Meals delivered to the home	Agency A	Agency A	Agency A	Agency A
Light housekeeping				Agency B
Home care nursing				Agency C

During the process of developing a service inventory, service providers can also be asked, perhaps in interviews, to offer estimates of the extent to which existing clients' needs are being met by the services provided to them (Kernan et al., 2003). Using service provider input in this manner has the advantage of acquiring data from persons or agencies who are knowledgeable about current clients, past clients, and the patterns of service requests/demands/needs the providers observe. On the other hand, service providers clearly have a stake in the outcome of needs assessments and may have an incentive to bias their estimates of the adequacy of existing service to existing clients.

A closely related issue is using service provider input to estimate the numbers of additional clients with needs. Interviews or possibly a survey can be useful to obtain estimates of the numbers of clients with unmet needs. One useful way to corroborate provider estimates is to request evidence of waiting lists or other indicators of underserved clients.

Identify Needs, Using Complementary Strategies for Collecting and Recording Data

Data Sources and Data Collection Methods

Needs assessments can be conducted with combinations of **primary** (collected specifically for the needs assessment) and **secondary** (already existing) sources of data. Methods may involve quantitative and/or qualitative data collection, such as:

- Literature reviews
  - Similar studies
  - Demographic statistics
  - Government reports
- Surveys (by mail, phone, or in person)
- Focus groups
- Interviews (structured or open-ended)
- Direct observation

Literature reviews may uncover "meta-needs assessments," which rely entirely on synthesizing data from previously completed needs assessments on a particular topic. Gaber (2000) argues that a meta-needs assessment can be a very useful substitute for an in-depth needs assessment when resources are insufficient to conduct one. Although a discussion of this approach is beyond the scope of this chapter, Gaber (2000) provides a thorough summary of a meta-needs assessment conducted for the Nebraska Department of Social Services, including a clear outline of the steps taken and the strengths and weaknesses of the approach.

In our later examples of needs assessments, we will review several of the data collection methodologies most frequently used in needs assessments.

Demographic Data

Census information or similar data from government agencies can be used to estimate the occurrence of individual or group characteristics associated with known needs for services. For example, demographic information on the age and gender distributions in a region might be used to roughly gauge the need for services for the elderly. The assumption that age is strongly associated with the need for services that target older persons can be corroborated by developing demographic profiles of existing clients for services for the elderly where such services are provided, and then comparing client profiles to the population demographics (Soriano, 1995). Estimates of need can be constructed by comparing total numbers of prospective clients in the population to the numbers served by existing providers.

Surveying Current and Prospective Clients. Surveys have become a principal means of gathering data in needs assessments, and surveys of current or future clients are often used to estimate unmet needs. As was indicated earlier in this chapter, program evaluations can be combined with needs assessments in that clients surveyed to ascertain their experiences and levels of satisfaction with existing services can also be queried about gaps in the services in relation to their perceived needs.

One type of survey is focused on a **random** or **representative sample** from a population and questions are posed that get at uses of existing services, as well as respondent estimates of the adequacy of services vis-à-vis their needs. Respondent estimates of their use of existing services can be used to develop demographic profiles of current service users. Survey-based data on who uses services can be used in conjunction with census or other population-based characteristics to develop estimates of the total possible usage of services in other populations. For example, if 10% of the respondents to a population survey of senior citizens in a region indicated that they have used Meals-on-Wheels services in the past year, and if our **sample** is representative of the regional population, an estimate of total possible usage of Meals-on-Wheels for the population in that region can be calculated by constructing a **confidence interval** around the sample proportion of users, and multiplying the lower and upper limits of the confidence interval by the total population of seniors in the region. In our example, if 10% of a survey sample of 500 seniors indicated they have used Meals-on-Wheels in the past year, the 95% confidence interval for the population proportion of users is between .0737 and .1260. In other words, we can be 95% sure that the true percentage of seniors using Meals-on-Wheels in the population is between 7.37% and 12.6%.

Suppose we wanted to use this information to estimate the need for Meals-on-Wheels in another region. If we knew the population of seniors in that region (e.g., 10,000), we could estimate that the number of seniors who would use Meals-on-Wheels is between 737 and 1,260 persons. This approach to estimating need is similar to the Marden approach for estimating the number of individuals in a population who are at risk of having alcohol-related problems (Dewit & Rush, 1996).

A second use of population surveys is to obtain more direct estimates of the need for particular services. Table 6.3 displays a section from a population needs assessment that was conducted in Airdrie, Alberta, Canada (Gubbels, Hornick, & Nicoll, 1989).

The services listed in Table 6.3 were among those that reflected a broad concern that Airdrie, as a satellite community of Calgary, did not enjoy the range of community services that other similar-sized communities in Alberta enjoyed. By asking respondents to rate the importance of the services, the analysts obtained one measure of the importance of these services.

A more direct measure was then obtained by asking respondents to indicate whether they know of anyone who *needed* the services in the past year. Table 6.4 excerpts a part of that section of the questionnaire.

Survey respondents could pick any or all of the services, based on their perception of the need for each. The surveyors, by combining information

Table 6.3   Ratings of the Importance That Services be Made Available to Airdrie Residents

We would like your opinion on how important you feel it is for the following services to be made available to Airdrie residents. (Please circle the number which is closest to your opinion.)

		Very Unimportant				Very Important
Please give your opinion for the following:						
5.1	Counseling for people with alcohol abuse problems	1	2	3	4	5
5.2	Counseling for people with drug abuse problems	1	2	3	4	5
5.3	Counseling for unplanned pregnancies	1	2	3	4	5
5.4	Family planning and birth control information	1	2	3	4	5
5.5	Counseling for family problems	1	2	3	4	5
5.6	Counseling for people with emotional problems	1	2	3	4	5
5.7	Family life education	1	2	3	4	5
5.8	Job search assistance for people seeking employment	1	2	3	4	5
5.9	Full-time child care services	1	2	3	4	5
5.10	Before and after school care	1	2	3	4	5
5.11	Drop-in child care	1	2	3	4	5

Table 6.4   Need for the Services in Airdrie

Now, we would like to find out if, in the past year, you have known someone, in Airdrie, who needed any of the following services. Beside each item, please indicate if you know of someone who needed the service. (Circle the appropriate number.)

		No	Yes
Please give your opinion for the following:			
6.1	Counseling for people with alcohol abuse problems	1	2
6.2	Counseling for people with drug abuse problems	1	2
6.3	Counseling for unplanned pregnancies	1	2
6.4	Family planning and birth control information	1	2
6.5	Counseling for family problems	1	2
6.6	Counseling for people with emotional problems	1	2
6.7	Job search assistance for people seeking employment	1	2
6.8	Full-time child care services	1	2
6.9	Before and after school care	1	2
6.10	Drop-in child care	1	2

Table 6.5   Ranking of Options for Child Care in a Public Agency

There are a number of child care options which the Division could adopt. Please list your 1st, 2nd, and 3rd choices in the boxes on the right.

1 Greater flexibility in scheduling

2 More opportunities for part-time work and job-sharing

                                        1st Choice

3 An information and referral service to help
   locate daycare in my community

                                        2nd Choice

4 On- or near-site daycare center

                                        3rd Choice

5 None of these options

6 Other: (specify) _____

on the importance of each service with responses indicating the need, were able to develop a profile of services that were both important and for which there was a perceived need.

One concern with surveys that ask for ratings of the need for services is that there is no constraint or tradeoffs among the number of important services that can be identified. Respondents to the Airdrie Needs Assessment Survey could have indicated that *all* the services were important and were needed. Often, analysts are faced with survey outcomes that suggest that needs are great and that differences among needed services are small.

An alternative is to ask survey respondents to rank the importance of services, forcing them to prioritize. Although ranking techniques are limited in their use (most respondents will not rank more than about six choices), in situations where analysts want information on a limited number of alternatives, ranking is more valid than rating the choices.

Table 6.5 excerpts one question from a survey of possible users of an on-site daycare facility in a public agency in British Columbia, Canada. Survey respondents were asked to rank alternative ways of meeting their child care needs. Although the need for on- or near-site daycare was the main purpose of the survey, respondents were given the option of selecting and ranking other options.

Later in the same survey, respondents were asked about their willingness to pay for on-site daycare. Combining data on perceived needs with data on perceived willingness to pay served as a rough indicator of the effective demand for services.

Qualitative Methods in a Needs Assessment. Chapter 5 provides details on conducting qualitative research in evaluations. In that chapter, the Most Significant Change approach (Sigsgaard, 2002) was introduced as a way to use qualitative methods to construct a performance measurement/assessment approach for projects in developing countries. Here, we will briefly summarize an example where qualitative methods were used to conduct a needs assessment in a neighborhood of Johannesburg, South Africa (Lewis et al., 2003).

The needs assessment of health promotion needs in the Hillbrow neighborhood of Johannesburg (Lewis et al., 2003) offers an example of an approach to needs assessment called "rapid appraisal." It also illustrates how **triangulation** can be used in this kind of project.

The researchers set out to conduct a needs assessment that would focus on consulting with and involving the community itself, in an attempt to make the conclusions as relevant as possible to local needs. They used a method called "rapid appraisal," designed to "gain insights into a community's own perspective on its major needs, then to translate these into action and, finally, to establish an on-going relationship between service providers and local communities" (Lewis et al., 2003, p. 23). Problems they encountered when using this approach included issues of reliability and validity, which the researchers attempted to address through triangulation, using a four-step methodology.

Step one involved a review of the available written records concerning the neighborhood. These had been produced by institutions outside the community itself and were incomplete and questionable in their accuracy. Step two made an attempt to flesh out this background information with a series of semi-structured interviews with a small number of key stakeholders who worked and were in positions of influence in the neighborhood. One issue which emerged from these interviews was the lack of community engagement with youth and women, so step three involved two focus group discussions, one with area youth (14 participants) and the other with area women (12 participants). The main intent of these discussions was to get local people's views on some of the issues raised in steps one and two. These focus groups were carefully facilitated to allow participants to direct the discussion and to focus on the issues that were of greatest importance to them.

The fourth step was designed to create an opportunity for stakeholders and members of the neighborhood to consider the information gathered and the issues raised in steps one to three, and to attempt to reach agreements about possible courses of action. A community workshop was held with over 80 participants, made up not only of community representatives, but of service providers and decision makers. Emphasis was placed on ensuring that all participants felt completely involved in the discussions and that divergent views were fully exposed. To start, key messages from the earlier stages of the needs assessment were presented by representatives of the focus groups, followed by a break down of the larger group into smaller groups to discuss the issues. Each small group focused on one predominant theme that emerged from the overall assessment, such as crime, Hillbrow's physical environment, or cultural offerings. This smaller group work helped participants to interact and to understand each other's viewpoints. In the final stage of the workshop, each small group reported their key conclusions back to the whole group, and together, participants worked to develop an action plan.

The authors of the study report that the "rapid appraisal" methodology outlined above allowed them not only to gain many different perspectives on the unmet health and social needs of Hillbrow, but to further the involvement and positive interaction of community members and build partnerships for future action.

Sampling Procedures. In the examples of needs assessments used in this chapter, several references have been made to the assumption that samples were representative, that is, statistically representative of the client populations from which they were drawn. Generally, a key point of being able to defend the methodology of a needs assessment is a **sampling procedure** that is defensible. Of course, sampling is an issue that arises in most data collection activities, and as such, spans all areas of program evaluation, rather than being unique to needs assessments. However, as the topic has arisen several times throughout this chapter, we discuss the central considerations of sampling here.

Ideally, sampling for needs assessment surveys should be *random*. That means that any respondent has an equal chance of being selected, and no respondents or groups of respondents have been excluded from the sampling process. Selecting a **random sample** requires us to be able to enumerate the population and, using one of several methods (e.g., random number tables, computer software), pick our respondents. Where we cannot fully list all cases in a population, or where it is not practical to do so, it may be possible to obtain a list of all members of the population and draw a **systematic sample**. As an example, a directory of property owners that lists all residential owners could be used if the target population was all resident owners in that community. Typically, in systematic sampling, an estimate of the total population size is divided by the desired sample size to obtain a **skip factor** that is used to count through the list, picking cases that coincide with the skip interval. By using a random starting point in the first skip interval, it is possible to approximate a random sample.

One concern with systematic samples is that if the population listing is organized so that the order in which cases appear corresponds to a key characteristic, then two different passes through the population listing that began at different points in the first skip interval would produce two samples with different characteristics. If client files were organized by the date they first approached a social service agency, for example, then two different passes through the files would produce different average times the samples have been clients.

There are several other random sampling methods that result in samples designed for specific comparisons. **Stratified samples** are typically drawn by dividing a population into strata (men and women, for example) and then randomly sampling from each stratum. In populations where one group is dominant, but the analyst wants to obtain sufficient cases from all groups to conduct statistically defensible comparisons, stratified sampling will yield samples that are representative of each group or stratum. **Proportionate stratified samples** are ones where the proportion of cases sampled from each stratum is the same as the relative proportions of the strata in the population. If women are 25% of a population, a proportionate sample would be 25% women. A **disproportionate stratified sampling** method is sometimes used where an important group is relatively small. For example, if a needs assessment for community health services were being conducted in a region that had 5% Aboriginal residents, a disproportionate stratified sample might select more Aboriginal residents than the 5% in the population warranted to permit statistically valid comparisons between Aboriginal and non-Aboriginal health needs.

Sample Sizes. The cost of a needs assessment survey will vary with the size of the sample, so it is useful at the outset to have a general idea of how much precision is desired in any generalizations from the sample back to the population. Generally, the larger the sample, the more precision.

Existing methods for determining sample sizes are awkward in that they force us to make assumptions that can be quite artificial in needs assessments. To determine sample size (assuming we are going to use a random sample), we need to know:

1. How much error we are willing to tolerate when we generalize from the sample to the population.

2. What the population proportion of some key feature of our cases is (or is estimated to be), so we can use that to pick a sample size.

Typically, when we conduct a needs assessment, we are interested in a wide variety of possible generalizations from the sample to the population. We may have decided to conduct a survey and are now interested in estimating the sample size that we require to accurately estimate the perceived needs for the services in the population. The methodology of determining sample sizes requires that we assume some population proportion of need in advance of actually conducting the survey, and use that to estimate our required sample size. In effect, we have to zero in on one service, "estimate" the perceived need in the population for that service in advance of conducting the survey, and construct our sample size with respect to that estimate.

It is as if we were going to conduct a needs assessment of a range of possible community-based health services and, to get a sample size, we needed to come up with an estimate of the proportion of the population that needs home care nursing services *before* we survey our sample. Although important, home care nursing services would be only one of many possible service options for which we would be conducting our needs assessment.

Table 6.6 displays a typical sample size table (Soriano, 1995). Across the top are the expected proportions of responses to one key item in the needs assessment survey (the proportion needing home care nursing services, for example), and down the left side, the percentages of error when we generalize from a given sample back to the population.

Suppose we "guesstimate" that 5% of the population would indicate a need for home care nursing services. That would put us in the first column of the table. Now, suppose we wanted to be able to estimate the *actual* (as opposed to the "guesstimated") proportion of persons indicating a need for home care nursing to within ±2%. We would need a random sample of 479 cases.

There is one additional factor that is implicit in Table 6.6. In addition to specifying our desired level of precision in estimating the population proportion of persons needing home care (±2%), we must recognize that *all* of Table 6.6 is based on the assumption that we are willing to accept a *95%* **level of confidence** in our generalizations to the population. That means that even though we might, for example, conduct a needs assessment and estimate that the *actual* population percentage of persons needing home care nursing is 7%, with a possible error of ±2% *either way,* we would be able to say, with 95% confidence, that in the population the percentage of persons needing home care is between 5% and 9%.

Another way to look at this situation is to say that we are only *95% confident* that our estimating process has captured the true population proportion of persons needing home care nursing. What that implies is that if we were to do 100 needs assessments in a given community, using sample sizes of 479 each time, in 5 of those needs assessments our estimation procedure

| Table 6.6 | Sample Sizes for a 95% Level of Confidence Depending on Population Proportions Expected to Give a Particular Answer and Acceptable Sampling Error |

Acceptable Sampling Error (%)	Proportion of Population Expected to Give Particular Answer					
	5/95	*10/90*	*20/80*	*30/70*	*40/60*	*50/50*
1	1,900	3,600	6,400	8,400	9,600	10,000
2	479	900	1,600	2,100	2,400	2,500
3	211	400	711	933	1,066	1,100
4	119	225	400	525	600	625
5	76[a]	144	256	336	370	400
6	—	100	178	233	267	277
7	—	73	131	171	192	204
8	—	—	100	131	150	156
9	—	—	79	104	117	123
10	—	—	—	84	96	100

SOURCE: Soriano, F. I. (1995).

a. Samples smaller than this would be too small for analysis.

would *not* capture the true population proportion, even though our samples were random each time. Unfortunately, we do not know *which* of those needs assessments will produce the misleading results.

Clearly, estimating sample sizes involves assumptions that are quite restrictive and, perhaps, not based on much information. But, to be able to defend the findings and conclusions from a needs assessment, the sampling methodology must be transparent and consistent with accepted practices.

If you look carefully at Table 6.6, you will see that for any given level of **sampling error**, as the expected population proportion that gives a particular answer moves towards 50% (say, a positive response to a survey question about the need for home care nursing), the required sample size increases. So, an evaluator conducting a needs assessment can avoid having to make "guesstimates" in advance of the survey by *assuming* that the population responses will be 50/50. That is the most conservative assumption and is eminently defensible. However, it also requires much larger sample sizes for all levels of acceptable sampling error.

Validity Issues

In Chapter 3, we discuss the *validity* of measures and define validity as the extent to which a measure does a "good job" of measuring a particular construct. In the example below, surveying prospective bus riders yields a *biased* measure of the construct "actual transit ridership." Validity is about bias—if a measure is biased, we say that it is not valid (or, in our example, needs to be adjusted if the degree of bias can be estimated).

In a community in northwestern Pennsylvania, the local Public Transit Commission was interested in expanding the bus routes to attract more ridership. There were several areas of the community that were not served by existing routes, so the commission hired a transit planner on contract to estimate the costs and revenues that would result from a number of expansion options (Poister, 1977).

Among the methodologies selected by the planner was a household survey that targeted the areas of the city that were currently not served by public bus routes. Questions in the telephone survey asked respondents to estimate their own usage of public buses if they ran through their neighborhood:

Now, turning to your own situation, if a city bus were to run through your neighborhood, say, within three blocks of your house, how many times a week would you ride the bus?

_____ Less than once per week

_____ Once per week

_____ 2–3 times per week

_____ 3–4 times per week

_____ More than 4 times per week

_____ Would not ride the bus

_____ Don't know

_____ No response

Survey results indicated that nearly 30% of respondents would become regular users of the buses (3 or more times per week). When the sample proportion of regular users was generalized to the population, expansion of the bus system looked feasible. The increased ridership would generate sufficient revenue to more than meet the revenues to costs target ratio.

But the transit planner, who had done other studies of this kind, did not recommend the bus routes be expanded. In his experience, a 30% potential ridership would translate into an *actual* ridership of closer to 5%, which was insufficient to meet the revenues-to-costs ratio.

The transit planner was willing to use the survey results, but was also aware that they were seriously biased in favor of more transit ridership. His experience allowed him to discount the bias to a more realistic figure, but another person might not have been aware of this problem, resulting in a service provision decision that would not have been cost-effective.

It is important to keep in mind that needs assessments are subject to many of the threats to validity that we discuss in Chapter 4. Those of us conducting needs assessments must do our best to control for these elements of bias. Recently, Calsyn, Kelemen, Jones, and Winter (2001) published an interesting study of one common element of response bias in needs assessments—over-claiming of agency awareness. In needs assessments, a respondent's awareness of a particular agency is often used as a measure of their use of, and therefore need for, that agency. However, for reasons such as age or a desire to appear well-informed, subjects often claim awareness of agencies that they do not, in fact, have any knowledge of. The study by Calsyn et al. (2001) concluded that one of the best ways of discouraging such response bias is to warn respondents ahead of time that the list of agencies being used in the needs assessment contains the names of fictitious as well as real agencies. This warning tends to make respondents more cautious about their answers and produces more accurate estimates of agency awareness.

Prepare a Document That Integrates Evidence, Benchmarks, Conclusions, and Recommendations

Typically, needs assessment reports are layered, that is, constructed to make it possible for users to obtain varying levels of detail as they use the report. A typical needs assessment report would have the following major sections:

An executive summary that is usually two to three pages in length and is focused mainly on the recommendations from the study. Typically, the executive summary is intended for elected officials who do not have the time to read the whole document, and it is constructed after the full report has been completed.

An introductory section that states the purposes of the needs assessment, including the key questions or issues that have prompted the study. It would be in this section that suspected needs gaps would be identified, for example.

A methods and data sources section that describes the ways needs are being measured, and how information from different sources (both qualitative and quantitative) is being used to address the key questions or issues that motivate the study.

A findings section that succinctly summarizes the findings that are relevant to the questions and issues in the study. Here, the style of presentation is important, given the audiences for the study. Typically, relying on visual displays of information (graphs and charts) is superior to tables. Even bivariate relationships can be displayed graphically, in preference to cross-tabulations.

A conclusions section relates the findings back to the questions and issues that underlie the needs assessment. In this section, the study team is interpreting the findings and summarizing the evidence that will be relevant for interested stakeholders.

A recommendations section that offers advice to stakeholders based on the study. Recommendations are based on evidence in the study, and they offer ways of connecting the findings and the conclusions of the study to policy options. There are several styles of writing recommendations. For example, recommendations can be written and for each one, the advantages and disadvantages of implementing it are offered. Alternatively, for each recommendation, a rationale is offered, based on the evidence in the report. Since the recommendations are the principal part of the executive summary, they need to be written in plain language.

Appendices that offer stakeholders more detail on methods used, data sources, and analyses. Appendices permit a more detailed layer to the report for stakeholders who want to see these details.

The process of drafting a report is iterative. Typically, as a first draft is completed, it is made available to the primary users of the report. The draft can be reviewed, errors corrected, and the recommendations discussed. Engaging the users even as the analysis proceeds will familiarize them with the study and foster a "no surprises" report. Patton (1997) makes stakeholder involvement in evaluations an essential building block for utilization.

Communicate the Results of the Needs Assessment

Usually, needs assessments are done in a manner that encourages their eventual use. Valid research is necessary but not sufficient for using the study to make or modify policies or programs. Effective communication of results and recommendations entails communicating about the project as it is happening.

Building a support coalition for a needs assessment makes it possible to engage stakeholders in communicating the results and recommendations from the study. In the Newark, New Jersey, needs assessment for a poor, ethnically diverse neighborhood (Elliot et al., 2000), a coalition of neighborhood organizations called the Collaboration decided to conduct the needs assessment for their own use. As the report was being prepared, members of the Collaboration were active in communicating the results of the study.

Communication styles and content will depend on the audience. Although a written report is generally an expected part of a needs assessment, it is usually necessary to prepare and deliver oral presentations to stakeholders, and even make key findings and recommendations available in the public media. Because needs assessments almost always intersect with political processes, appropriate communications will be important in determining how the report is actually used.

Timing of a needs assessment project is important. Most public and non-profit organizations have an annual budget cycle that is fixed so needs assessments have to be completed in time to be an input in that cycle.

Implement the Recommendations of the Needs Assessment

Success in implementing recommendations depends on several factors. First, the extent to which key stakeholders have been engaged in the needs assessment process all along is likely to affect their willingness to treat the recommendations as a priority. Because needs assessments are about making claims on resources that are usually allocated via political processes, having policy makers onside at the outset enhances the likelihood that recommendations will be implemented.

Second, the scale and scope of recommendations also affect implementation. Needs assessments that recommend marginal adjustments to existing programs and services will typically be easier to implement than recommendations that point to new programs or greatly expanded programs. Much of political decision making is incremental, so needs assessments that recognize that fact will typically be less difficult to accommodate.

Third, needs assessments that engage stakeholders in coalition-building for additional resources have an advantage. In the St. Columba Collaboration needs assessment, described in detail below, members of the coalition that sponsored and conducted the needs assessment also actively pursued private sector funding for some of the initiatives recommended in the report (Elliot et al., 2000).

● THE ST. COLUMBA COLLABORATION PROJECT: NEEDS ASSESSMENT IN A NEWARK NEIGHBORHOOD

Using the steps of a typical needs assessment summarized in the previous section of this chapter, we show how the St. Columba Collaboration in Newark, New Jersey designed and conducted a needs assessment that identified gaps in the programs offered within the community. The Collaboration is a coalition that includes the St. Columba church (the Church), the St. Columba elementary school (the School), and the St. Columba Neighborhood Club (the Club) (Elliot, Quinless, & Parietti, 2000).

1. Familiarize yourself with the political context.

The St. Columba neighborhood is one of the poorest areas in Newark. It has a long history of community activism and has relied on voluntary organizations to contribute substantially to the well-being of the residents. In the early 1990s, the Collaboration was created to improve program and service delivery and to increase the resources available for coordinated social and political actions. In 1995–96, the Collaboration undertook a major community-based needs assessment.

New Jersey State policies in the mid-1990s reflected the political view that persons on public assistance had to take more responsibility for their own economic well-being. Legislation was passed that included WorkFirst New Jersey, which required mothers who had relied on public assistance to work once their youngest child was 12 weeks old. Health-related reforms included legislated Medicaid care for all public assistance recipients and a state-wide program called KidCare, which was intended to provide support for daycare for persons with lower incomes (Elliot et al., 2000).

Because these policy and program changes affected many of the residents of St. Columba, the Collaboration decided to conduct a grassroots needs assessment to see how the changes in programs and services would impact residents and to identify the key problems and issues that residents wanted to see remedied. The political context made it unlikely that there would be substantial new funds for programs and any funds that were available would need to be justified by evidence of need.

2. Identify the users and the uses for the needs assessment.

The principal user of the needs assessment was the Collaboration itself. The needs assessment was conducted as a part of a strategic planning process that was intended to mobilize community and governmental support for programs to fill the gaps that were identified. Elliot et al. (2000, p. 214) describe the purposes of the needs assessment:

A. To identify the perceived strengths, weaknesses, and opportunities in the neighborhood

B. To identify the self-reported demographics of local residents and Collaboration clientele (Church, Club, and School users)

C. To identify service priorities and needs of residents and clientele

D. To identify directions for financially sound strategic planning, effective community organizing, and efficient internal communications

3. Identify the target population(s) who will be or are currently being served.

The St. Columba neighborhood comprises three Census Tracts in south Newark. All three tracts have 35% or more residents below the poverty line. The total population in 1996 was 7,300 persons. Most of the residents are Black (64%), with another 34% Latino. The Latino population is concentrated on Census Tract 67 and this area is the part of the neighborhood where the Collaboration is most active.

Most residents have not completed high school, and the neighborhood is dependent on welfare for large numbers of residents. Elliot, Quinless, and Parietti (2000) summarize the income picture this way:

cash payments from Public Assistance account for 8% of all income, as compared to 3% for the city as a whole, and Social Security contributes 10% to Neighborhood incomes, compared to 6% for all of Newark. Only 46% of the population are in the workforce, as compared to 60% for the whole city. (p. 216)

Overall, the St. Columba neighborhood is poor. Nearly all residents are minorities, are relatively poorly educated, and more likely to be unemployed.

4. Inventory the existing services that are available to the target population, in part to identify gaps.

The Club is a principal provider of services to the neighborhood. Established in 1975, it offers social and welfare programs, "including a Family

Life Program, an HIV/AIDs program, youth groups, seniors groups, Healthy Mothers/Healthy Babies and a local Headstart program" (Elliot, Quinless, & Parietti, p. 213).

Given the changes in the state's policies in the mid-1990s, it was not clear how these programs would be affected—whether there would be an increased need for the services, or whether there would be a need for additional programs, or both.

5. Identify needs, using complementary strategies for measuring and recording data.

The St. Columba Collaboration collected data using four complementary strategies. They divided their data collection strategies into "outside" and "inside" views of services, neighborhood problems, and needs.

The "outside" view was based on demographic data, obtained from several sources. The City of Newark had applied to the federal government in 1995 for assistance for its poorer neighborhoods, and the three Census Tracts figured prominently in that application. Demographic data describe a neighborhood that has a majority of women (54%) with 29% of those under 18 years of age. Seventy-five percent of families had incomes less than $20,000 with an average per capita income of $6,300. The unemployment rate, obtained from a 1993 Newark Metropolitan Needs Assessment, was the highest in the city (23%) (Elliot, Quinless, & Parietti, 2000). The demographics of the neighborhood consistently suggest a community that is poor and has a high unemployment rate.

The "inside" view included surveys as a principal means of getting data on needs. Two different surveys were distributed. One was a door-to-door survey (n = 295) that was distributed by trained volunteers, graduate students, and Club staff. "This survey sought residents' perceptions of services needed, health status, neighborhood problems and social services needed" (p. 216).

A second survey (n = 471) was distributed to users of existing services and programs provided by the Collaboration. The survey was written in both English and Spanish and covered the same topics as did the residential survey.

A total of 10 focus groups, using the same questions, were held with persons who were providing programs and services. Teachers from the elementary school (five focus groups) and Club staff (five focus groups) participated.

6. Prepare a document that integrates evidence, benchmarks, conclusions, and recommendations.

The user survey indicated that residents who use Collaboration services are satisfied with the services they receive. Over 75% indicated that services are good or excellent. A key finding for that survey was the large number of persons (>40%) who wanted new programs for youths. Just under 40% wanted more crime prevention programs, and nearly 35% wanted job training and English-as-a-second-language (ESL) programs.

The residential survey highlighted health problems (e.g., asthma, diabetes, and high blood pressure) as well as community concerns. The principal problems in the neighborhood were: drug addiction (72% said that was a problem), crime (61%), youth violence (52%), and unemployment (38%).

The focus groups offered a somewhat different picture of neighborhood needs. Mental health care was identified as a need by over half of the focus group participants. Although unemployment, low paying jobs, and poor housing were all identified as serious problems, the focus groups generally offered a more optimistic picture of the neighborhood.

Elliot et al. (2000) summarize the findings from the needs assessment:

> In summary, the St. Columba Neighborhood needs assessment describes a poor, uneducated, minority population falling well below the state and federal norms for average income. It is a neighborhood in which violence, crime, unemployment, lack of education, and drugs are identified as problems. On an individual health status level, the problems of asthma, diabetes, hypertension, drug and alcohol addiction, and HIV/AIDS are commonplace. Women's and children's issues surfaced in relation to violence and mental health. The neighborhood is overwhelmed by a high rate of unemployment. Many residents, particularly women, are not equipped with the tools and skills necessary for employment (e.g., basic highschool education or general equivalency diploma and functional literacy skills). (p. 220)

Because the Collaboration conducted the needs assessment for its own uses, recommendations were acted on as they came out of the study process. This is possible where a needs assessment is done internally, but the more usual circumstance is that the needs assessment is "handed off" to organizations that are responsible for making policies. Collaboration with internal stakeholders and other agencies at various stages in the needs assessment can enhance teamwork and facilitate coordination of later program recommendations and implementation.

The needs assessment generated two kinds of recommendations. One type of recommendation was for a broadening of the Collaboration to include the School of Nursing in the University of Medicine and Dentistry of New Jersey. Broadening the Collaboration was intended to bring more staff into the health center and expand the coverage and services available. The School of Nursing in turn developed a partnership with the University Hospital to "provide infrastructure support to the health center, including on-site patient registration, financial counsel or support, clinical supplies, and a shuttle van service to route ill patients to the hospital for acute treatment" (p. 220).

Bringing the School of Nursing into the Collaboration also resulted in establishing a Women, Infants and Children Program at the same site as the existing Healthy Mothers/Healthy Babies program. Services were also expanded into the neighborhood with funding from private foundations.

A second partnership between the School of Nursing and the New Jersey Institute of Technology created an ESL program for Latina women. This program was intended to be the foundation for building employment skills.

The second kind of recommendation focused on specific clusters of programs based on the needs assessment. One cluster focused on violence, primarily youth violence in the neighborhood. Through a process of training trainers in alternative dispute resolution methods, these persons trained Club staff and health center personnel so that they were better able to deal with angry or violent clients.

A second cluster of programs focused on health issues. Existing programs for children were enhanced, including immunizations, and more screening and treatment for asthma, diabetes, and high blood pressure. For women, health services have been expanded, including a study to understand and intervene in depressive illnesses. Teen pregnancies were also included in programming—efforts to prevent teen pregnancies were highlighted.

7. Communicate the results of the needs assessment through marketing the work that has been done and maximizing its impacts on the political/resource allocation process.

Communicating the results of the needs assessment marked the transition from the study to planning for its implementation. Elliot et al. (2000) highlight the importance of communication in the entire needs assessment process:

Although part of a highly mobile, diverse immigrant neighborhood, the St. Columba Neighborhood residents, Club users, faculty, parishioners, and parents of school children recognized that their participation in the needs assessment was a valuable contribution to change, both at the local neighborhood level, and possibly the municipal and state policy levels. (p. 220)

At the completion of the assessment phase, which took 9 months, residents celebrated the collection and analysis of the data with a fiesta. This marked the commencement of the strategic planning phase of the process. This was also the beginning of the eighth stages of the needs assessment: *Implement the recommendations from the needs assessment.*

SUMMARY ●

Needs assessments can be conducted at several different points in the performance management cycle: a part of the strategic planning process; a part of program or policy design; or a part of evaluating the results of programs or policies that have been implemented. Needs are fundamentally about what we value in our society. If we can agree that a given service is a need (basic income support payments, for example), we will usually find ways through the political process to allocate resources to that need. But because needs reflect our values, there can be sharp disagreements over whether it is desirable to fund programs, even if a need has been demonstrated.

Because needs assessments can be contentious, it is important that they be conducted in ways that are methodologically defensible. There are a number of complementary ways to elicit information that can be used to estimate needs for programs or services. Among those, surveys have been used frequently to obtain information from current and prospective clients. Designing and implementing a needs assessment survey demands attention to sampling issues, as well as instrument design and administration. Bias in survey responses is a significant problem that can be mitigated by careful survey design and by triangulating survey results with other sources of information, including population demographics.

Qualitative methods play an important role in needs assessments. In fact, it is possible to conduct a needs assessment using only qualitative methods. The "rapid appraisal" approach to needs assessment is intended for developing nation contexts and is motivated by similar methodological concerns as discussed in Chapter 5 when the Most Significant Change approach was introduced as a qualitative approach to program performance assessments for projects in developing countries.

● DISCUSSION QUESTIONS

1. What are needs? How do we know that something is a need?

2. Why do all needs assessments involve comparisons?

3. What does it mean to benchmark needs?

4. Comment on the following statement, "If needs assessments are well-designed and well-executed, the findings and recommendations will speak for themselves—policy makers will pay attention to the results of a carefully constructed needs assessment."

5. Why is sampling important for needs assessments?

6. Why is validity important for needs assessments?

● APPENDIX: DESIGNING A NEEDS ASSESSMENT FOR A SMALL NONPROFIT ORGANIZATION

Needs Assessments in a Political World: A Case Study

The purpose of this case is to give you an opportunity to design a needs assessment, based on the situation described below. In this chapter, we outline eight steps in designing and implementing a needs assessment. Those steps were illustrated with an actual needs assessment in Newark, New Jersey. You can use the steps as you develop your design, but do not commit to doing anything that is not realistic, that is, beyond the means of the stakeholders in this case.

Your task in this case is to design a needs assessment for a nonprofit organization. There is not much money, the organization involved is small, and the needs assessment is being demanded by a key funder of the agency's programs. When you have read the case, follow the instructions at the end. Once you have developed your needs assessment design (we would suggest you work with one or two classmates to develop the design), discuss it with other class members.

The Program

A Meals-on-Wheels program in a community is currently being funded by the United Way and private donations. The United Way is under considerable budget pressures because the total donations have not kept up with the demand for funds. The Board of the United Way has recently adopted a policy that requests that the program managers of all funded agencies demonstrate the continuing relevance of their program in the community in order to receive funding.

The program manager of Meals-on-Wheels is very concerned that the needs assessment that she must conduct will be used to make future funding decisions, but does not feel she has a choice. She has limited resources to do any kind of needs assessment on her own. There is basically her time, the time of an office staff member, and the time of volunteers.

The Meals-on-Wheels program is intended to bring one hot meal a day to her clients—all of whom are elderly, single members of the community. Most have physical limitations that limit their ability to cook their own food, and some are experiencing memory loss and other problems that make it hard for them to remember when or even how to prepare regular meals. There are a total of 150 clients at this time in the program and that number has been fairly steady for the last several years. Volunteers, most of whom are elderly, pick up the meals from a catering company in the city, and are assigned a group of deliveries each day. The volunteers have to be able to drive, and because they have other commitments, most volunteers do not deliver meals every day.

In addition to making sure that the program clients get at least one hot meal a day, the volunteers can check to make sure that they have not fallen or otherwise injured themselves. If volunteers find a client in trouble, they decide whether to call 911 directly, or instead, call the Meals-on-Wheels office.

Most of the volunteers have been delivering meals for at least 3 years. Their continued commitment and enthusiasm are a key asset for the program. The program manager recognizes their importance to the program and does not want to do anything in the needs assessment that will jeopardize their support.

Your Role

The program manager approaches you and asks you to assist her with this project. She is clearly concerned that if the needs assessment is not done, her funding will be cut. She is also concerned that if the study does not show that her clients need the services, she and her volunteers who offer the program will still be vulnerable to cuts.

You are a freelance consultant, that is, you work on your own out of a home office and do not have access to the time and resources that a consultant in a larger firm would have. She can pay you for your work in the design; but any suggestions that you make to the program manager will have to be realistic, that is, cannot assume that large amounts of money are available.

Your Task

Working in a team of two to three persons, draft a design for a needs assessment that is focused on whether there is a continuing need for the Meals-on-Wheels program in the community. In your design, pay attention to the eight steps in conducting a needs assessment that were discussed in this chapter. Make your design realistic, that is, do not assume resources are just going to be available.

Outline your design in two to three pages. Discuss it with other teams in your class.

● REFERENCES

Calsyn, R. J., Kelemen, W. L., Jones, E. T., & Winter, J. P. (2001). Reducing overclaiming in needs assessment studies: An experimental comparison. *Evaluation Review, 25*(6), 583–604.

Dewit, D. J., & Rush, B. (1996). Assessing the need for substance abuse services: A critical review of needs assessment models. *Evaluation and Program Planning, 19*(1), 41–64.

Easton, D. (1965). *The political system.* New York: Alfred A. Knopf.

Elliot, N. L, Quinless, F. W., & Parietti, E. S. (2000). Assessment of a Newark neighborhood: Process and Outcomes. *Journal of Community Health Nursing, 17*(4), 211–224.

Gaber, J. (2000). Meta-needs assessment. *Evaluation and Program Planning, 23*(2), 139–147.

Gubbels, P., Hornick, J. P., & Nicoll, M. (1989). An Alberta community needs assessment survey. *Canadian Journal of Program Evaluation, 4*(1), 49–66.

Houston, A. M., & Cowley, S. (2002). An empowerment approach to needs assessment in health visiting practice. *Journal of Clinical Nursing, 11*(5), 640–650.

Kernan, J. B., Griswold, K. S., & Wagner, C. M. (2003). Seriously emotionally disturbed youth: A needs assessment. *Community Mental Health Journal, 39*(6), 475–486.

Lewis, H., Rudolph, M., & White, L. (2003). Rapid appraisal of the health promotion needs of the Hillbrow Community, South Africa. *International Journal of Healthcare Technology and Management, 5*(1/2), 20–33.

Maslow, A. H. (1965). *Eupsychian management: A journal.* Homewood, IL: Irwin and Dorsey Press.

Maslow, A. H. (1998). *Maslow on management.* New York: John Wiley & Sons.

McKillip, J. (1998). Need analysis: Process and techniques. In L. Bickman & D. J. Rog (Eds.), *Handbook of applied social research methods* (pp. 261–284). Thousand Oaks, CA: Sage.

Office of Management and Budget. (2002). *Instructions for the program assessment rating tool.* Retrieved August 4, 2004, from http://www.whitehouse.gov/omb/budget/fy2004/pma/Instructions.pdf.

Patton, M. Q. (1997). *Utilization-focused evaluation: The new century text* (3rd ed.). Thousand Oaks, CA: Sage.

Poister, T. (1977). *The Newcastle transit update.* Harrisburg: Pennsylvania Department of Transportation.

Sigsgaard, P. (2002). MCS approach: Monitoring without indicators. *Evaluation Journal of Australasia, 2*(1), 8–15.

Soriano, F. I. (1995). *Conducting needs assessments: A multidisciplinary approach.* Thousand Oaks, CA: Sage.

Towle, C. (1945). *Common human needs, an interpretation for staff in public assistance agencies.* Washington, DC: Social Security Board.

Towle, C., & Younghusband, E. L. (1973). *Common human needs* (new edition prepared by Eileen Younghusband, Ed.). London: Allen and Unwin.

Treasury Board of Canada Secretariat. (2001). *Evaluation policy.* Retrieved August 4, 2004, from http://www.tbs-sct.gc.ca/pubs_pol/dcgpubs/TBM_161/ep-pe_e.asp

Treasury Board of Canada Secretariat. (2003). *President of the Treasury Board announces expenditure and management reviews.* Retrieved August 5, 2004, from http://www.tbs-sct.gc.ca/media/nr-cp/2003/0529_e.asp

CHAPTER 7

CONCEPTS AND ISSUES IN ECONOMIC EVALUATION

Irene Huse
James C. McDavid
Laura R. L. Hawthorn

INTRODUCTION ●

Over the last two decades, the combination of greater demands for public services concurrent with increased fiscal pressures has created demands for evaluations to help optimize economic efficiency in the use of public resources. While the focus of program evaluation is typically on how well the actual results/outcomes compare with the objectives of a program, economic evaluations explicitly connect the costs of programs to their outcomes and can be used to help decision makers make program choices and allocate resources in public and nonprofit organizations.

Grounded in the principles of welfare economics (see Mishan, 1988), the aim of this type of economic analysis is to be able to compare various program options, including the status quo, and choose the option or options that maximize **net social value**. **Efficiency**, of course, is not the only consideration when making program decisions. In reality, the interrelationship between efficiency and social equity, as well as the influence of politics on public sector policy making, means that economic evaluations are one component of a complex decision-making process (Gramlich, 1981).

Dhiri and Brand (1999, p. 11) list the following key questions to be answered in an economic evaluation:

- What was the true cost of an intervention?
- Did the outcome(s) achieved justify the investment of resources?
- Was this the most efficient way of realizing the desired outcome(s) or could the same outcome(s) have been achieved at a lower cost through an alternative course of action?
- How should additional resources be spent?

Economic evaluations, typically conducted by persons trained in economic methods, systematically compare costs and outcomes, and in doing so, address the question of whether program alternatives caused the outcomes that were measured. Sound economic evaluations share some of the same attributes as sound program evaluations: attention to understanding actual causal linkages and being able to validly measure inputs and outcomes.

Why a Program Evaluator Needs to Know About Economic Evaluation

Budgetary and accountability trends indicate that economic evaluations will likely have an increasing role to play in program evaluations (Clyne & Edwards, 2002). While most program evaluators cannot expect to conduct an economic evaluation without a background in economics, it is important to have an understanding of how economic evaluation intersects with program evaluation, and how some of the terms, such as cost-effectiveness, can be defined differently within the two approaches to evaluation.

Sometimes, program evaluations can be done in a way that facilitates an economic evaluation of the outcomes of an intervention or program (Dhiri & Brand, 1999). Similarly, a program evaluator may review economic analyses that were done prior to the implementation of a program, to incorporate the same variables, if appropriate, so that future economic evaluations can use the data. It is important for evaluators to be able to use their skill and knowledge about evaluation methods to critically assess economic evaluations that are being used to justify program or policy decisions. Jefferson, Demicheli, and Vale (2002), for example, examined reviews of economic evaluations in the past decade in the health sector and concluded that "[T]he reviews found consistent evidence of serious methodological flaws in a significant number of economic evaluations" (p. 2809). Neumann, Stone, Chapman, et al. (2000) came to similar conclusions in their analysis of cost-utility analyses in the health sector from 1976 to 1997; and Dhiri and Brand (1999) note that studies of cost-effectiveness of crime interventions "have typically been hampered by incomplete data on the cost of inputs and the lack of clear specification of outputs and outcomes" (p. 8). Levin and McEwan (2001) argue that educational evaluators should become familiar with **cost-effectiveness analysis** to better prepare them for making judgments when applying their craft.

In an article examining economic evaluation in the health sector, Kernick (1998, p. 1664) lists the following as questions to be asked when critically assessing an economic study:

- Are all the relevant alternative treatments considered?
- What is the viewpoint or perspective of the study, for example, society, the National Health Service, the purchasing authority, the general practitioner?
- Were the economic data collected alongside a trial, applied retrospectively to an existing trial, or modeled?
- Are all costs measured and sources of costs credible?
- What outcome measures were used?
- Are the results generalizable, particularly to your practice?

A program evaluator should be able to appreciate the relevance of these questions, in that they also speak to the validity of typical program evaluations. The list underscores the need for critical appraisal when incorporating economic evaluation information or studies.

This chapter is intended to help program evaluators (a) become knowledgeable users of economic evaluations; (b) be able to see how these evaluations relate to nonfinancial ones; and (c) be able to discern potential weaknesses in the validity of specific studies.

Connecting Economic Evaluation With Program Evaluation: Program Technologies and Outcome Attribution

High- and Low-Probability Program Technologies

As mentioned in previous chapters, program technologies that are relatively certain tend to yield outcomes that are more easily measurable and countable. An example of a high-probability program technology would be a program that is focused on maintaining highways within a region to some set of audited standards. If we assume that the level of maintenance is generally constant over time, that is, the program passes an annual audit based on minimum standards of maintenance completeness and quality, then we could construct a ratio of the cost per lane-kilometer of highway maintained in that region. This measure of cost-effectiveness could be compared over time, allowing us to detect changes in cost-effectiveness. Because the program technology is high probability, we can be reasonably sure that, when we compare program costs to program outcomes, the costs (program inputs) have indeed been responsible for producing the observed outcome. Our measure of cost-effectiveness is a valid measure of program accomplishment.

In our highway maintenance example, we could use the cost per lane-kilometer to assess the effects of program-related interventions or even changes in the program environment. Suppose that highway maintenance was contracted out, as has been done in many jurisdictions. If we compared the cost per lane-kilometer before and after outsourcing the service, we could estimate the change in cost-effectiveness due to that change in service delivery. Likewise, if winter weather in the region was severe for several years, we would expect that to affect the cost-effectiveness of the service.

An example of a program with a low-probability technology would be a program to encourage adults to exercise. Although it may be easy to measure the level of change in people's exercise habits, it is difficult to predict what the effect of the program will be on the level of exercise. As well, it would be challenging to make the causal link between the program and the actual changes in level of exercise, especially without a randomized controlled study or at least a quasi-experimental study with a control group (more fully described in Chapter 3).

The Attribution Issue

While the issue of whether a program or treatment is actually the cause of the outcome(s) is often assumed in economic evaluations, causality is actually subject to the same challenges as it is in regular program evaluations.

Recall the York crime prevention program, which had an objective of reducing burglaries committed in the community. In Chapter 4 we considered some of the potential difficulties of developing valid measures of the construct "burglaries committed," and ended up using burglaries reported to the police. One way of assessing the cost-effectiveness of that program would be to calculate the ratio of cost per one percent reduction in burglaries in the city. Creating this ratio *presupposes* that we have successfully dealt with the attribution problem—the reductions in reported burglaries are due to program efforts and, hence, can be connected to program costs.

Levin and McEwan (2001), in their discussion of cost-effectiveness analysis, point out the importance of establishing causality before linking costs with observed outcomes. They introduce and summarize the same threats to internal validity that we discuss in Chapter 3. They point out that experiments are the strongest research designs for establishing causality:

> Randomized experiments provide an extremely useful guard against threats to internal validity such as group nonequivalence. In this sense, they are the preferred method for estimating the causal relationship between a specific . . . alternative and measures of effectiveness. (pp. 124–125)

If we are reasonably confident that we have resolved the attribution problem, then our ratio—cost per 1% reduction in burglaries—would be a summary measure of the cost-effectiveness of that program. Its usefulness depends in part on whether reducing burglaries is the only program objective and in part on how thoroughly we have estimated the costs of the program.

In the remainder of this chapter, we further examine the background of **cost-effectiveness analysis** (CEA), **cost-utility analysis** (CUA), and **cost-benefit analysis** (CBA), the key steps in conducting these types of analyses, and then use an example of a cost-effectiveness study to illustrate the basic steps needed to perform an economic analysis. In the example we include the issues to keep in mind to be able to read such a study critically, and we then review some of the controversy and limitations surrounding the use of CEA, CUA, and CBA in the summary of this chapter.

THREE TYPES OF ECONOMIC EVALUATION ●

Cost-benefit analysis and its two variants, cost-effectiveness analysis and cost-utility analysis, are the three main types of economic evaluation applicable to public sector program evaluation. With all of these types of analyses, the *costs* of the programs or potential programs are monetized but, for each, the *benefits* are quantified differently. With CBA, both the costs *and* the resulting benefits to society are monetized to determine whether there is a net social benefit. However, with CEA the costs are weighted against a single non-monetized outcome such as "lives saved," and with CUA they are weighted against a nonmonetized multiple outcome indicator such as "quality-adjusted life-years."

In all cases, it is important that the effectiveness units (or outcome units), whether monetized or not, are consistent with the intended outcomes that have been predicted for the program as part of the performance management cycle. As well, with all three types of economic evaluation, it is important to establish a baseline of inputs and outcomes that demonstrate the costs and benefits of either (a) the no-program situation, or (b) the current costs and benefits of the existing program before modifications are made. In other words, what we are trying to determine is the *additional,* or *incremental* costs and benefits/outcomes of a new or modified program or program alternatives, so we know the differences that have been achieved. Because of external influences, it is usually not simple to separate out the additional outcomes from those that would have occurred in the absence of a new or modified program.

The Choice of Economic Evaluation Method

Analysts may choose to use CEA or CUA when the key benefit(s) are difficult or controversial to monetize. Or, since CBA entails the monetization of *all* social benefits, CEA or CUA are sometimes used where just *some* of the important benefits are difficult to monetize. For example, a cost-effectiveness study might examine a seniors' falls reduction program on the basis of *cost per fall prevented.* It would be difficult to monetize all the benefits of a prevented fall because, aside from averted medical costs, the costs would also have to include other variables such as the incremental value of labor and voluntary work of the persons whose falls were prevented, as well as the value of the time gained by friends and family who might otherwise be called on for caregiving duty.

While CEA captures the main benefit in one outcome, such as number of children vaccinated, with CUA several (typically, two) outcomes are combined into one measurement unit. The most common outcome unit for CUA is **quality-adjusted life-years** (QALY) gained, which combines the number of additional years of life with subjective ratings of quality of life expected in those years, to create a standardized unit for analysis that can be used to compare across various programs for health or medical interventions. The quality of life ratings are standardized in that the value normally ranges between 1 for perfect health and 0 (zero) for death. CUA, then, is most commonly used in the health sector, where it is both (a) important to capture the benefit of extra years lived and the quality of life in those extra years lived, and (b) difficult to monetize all of the social benefits of a treatment or program.

CBA is most often used in determining whether a particular program will add value to the economic welfare of a society, as compared to alternative programs or the status quo. With CEA and CUA, specific outcomes, such as lives saved by implementing a smoking cessation program, have already been established as desirable, and the question is not *whether* to initiate or expand a particular program or project, but *how* to most efficiently expend resources to attain the desired level of outcomes (or, conversely, how to increase effectiveness while maintaining current levels of expenditure). CBA is most applicable for high-probability technologies that require capital investment, such as constructing highways, building bridges, or developing recreational areas or electricity production facilities. In these sorts of cases it is not as difficult to monetize the social benefits, as the benefits are mostly tangible.

In their discussion of priority setting in healthcare, Mitton and Donaldson (2003) make the distinction between **allocative efficiency** and **technical efficiency** to highlight the differences between CBA, CEA, and CUA in economic evaluations:

> Cost-benefit analysis (CBA) . . . informs questions about allocative efficiency, because the technique is used to determine which goals are worth achieving. . . . In terms of opportunity cost, the question is raised as to whether the benefit to be obtained from one service is more or less than the benefit obtained from alternate uses of the resources. Cost-effectiveness analysis (CEA) can be used to address questions of technical efficiency. That is, given that a particular goal is to be achieved for a given fixed budget, CEA can provide a response to the question, "What is the best way to obtain that goal?" Cost-utility analysis (CUA) is a form of economic evaluation that can be used to determine either technical or allocative efficiency. As CUA often does involve comparisons across domains, it implies some form of shifting resources from one patient group to another. (p. 100)

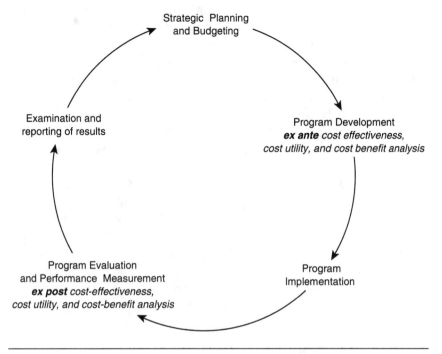

Figure 7.1   Economic Evaluation in the Performance Management Cycle

These three types of analysis are discussed in more detail further in the chapter.

ECONOMIC EVALUATION IN ●
THE PERFORMANCE MANAGEMENT CYCLE

Figure 7.1 shows the two main points in the performance management cycle where economic evaluations can be conducted. Earlier in the cycle, CBA or its variants can occur as programs are being developed. These are usually *ex ante* **analyses**—occurring before program implementation. *Ex ante* analyses typically use existing theoretical models or experience to predict how costs and benefits will distribute themselves over time, if the program is implemented and works as the program logic indicates it will. For example, Smith and Widiatmoko (1998) addressed the cost-effectiveness of a hypothetical home assessment and modification program for falls reduction among the elderly. They used a computer-based "decision analytic model . . . developed to simulate the potential costs and outcomes of the intervention" (p. 436), using data largely from Australian published studies.

Analyses can also be conducted *ex post* (after the program has been implemented or after completion). An *ex post* **analysis** is based on the *after-the-fact,* rather than hypothesized costs and benefits accruing to a program. As with typical program evaluation, higher-probability program technologies, such as the building of a stretch of highway, produce outcomes that are easier to measure and attribute to the program and, hence, easier to value in dollar terms. Conversely, low-probability technologies, such as programs to facilitate employment of social assistance recipients, will tend to produce outcomes that are a mix of tangibles and intangibles, many of which may be qualitative rather than countable.

Because many public sector outcomes, such as those in the health and education sectors, are often difficult to monetize, and because it is common that a decision has already been made to pursue a particular outcome (and the question is how to best achieve the chosen outcome), *cost-effectiveness* and *cost-utility* evaluations are done more commonly than *cost-benefit* evaluations.

● HISTORICAL DEVELOPMENTS IN ECONOMIC EVALUATION

Cost-benefit analysis and its variants have a long history, particularly in the United States where they began in the early 1800s with a federal treasury report on the costs and benefits of water projects. The use of CBA grew in the 1930s when it was seen as a tool to help decide how to best spend public funds during Roosevelt's "New Deal," when large-scale job-creation programs were used to stimulate the depression-era economy. From the 1930s through the 1950s various forms of CBA were applied to water resource projects, such as flood control, navigation, irrigation, electric power, watershed treatment, and to the related areas of recreation, fish, and wildlife (Poister, 1978). From this foundation, CBA began to be applied to other public investments in the 1960s and 1970s.

The use of CBA increased in both Canada and the United States during the 1970s, with increasing pressures to determine "value-for-money" in public expenditures. Cost-effectiveness based on operational costs has been adapted by the public sector auditing profession as **value-for-money auditing.** Value-for-money auditing was introduced in Canada in 1978 by then Auditor General J. J. Macdonell (Canadian Comprehensive Auditing Foundation, 1985) as a way to broaden the purview of auditors from their traditional focus on financial accountability to inclusion of the relationships between resources and results. Unlike cost-effectiveness analysis, value-for-money audits typically use a mix of qualitative methodologies to construct an understanding of the **economy** (were the inputs to a program purchased economically?), **efficiency** (what are the relationships between inputs and outputs?), and **effectiveness** (have the managers of the program implemented procedures that allow them to tell whether the program was effective?)

By the 1980s and 1990s there were increasing federal requirements, particularly in the U.S., for CBA for large public projects. Regulations applying to the environment, or to health and safety, often were first subjected to a cost-benefit analysis. Examples are President Reagan's Executive Order 12291 in 1981 and, later, President Clinton's Executive Order 12866, which "require agencies to prepare Regulatory Impact Analysis (RIA) for all major federal regulations" (Hahn & Dudley, 2004, p. 4). Aside from the assessment of regulatory issues, economic evaluations are increasingly being used in almost all areas of public expenditure, including transportation, education, pollution control, and protection of endangered species (Fuguitt & Wilcox, 1999). Cost-benefit analyses are included as part of decision-making processes, and sometimes their contributions are to highlight costs or benefits that may not have been properly understood or to more clearly identify the winners and losers of a policy.

Evaluators need to be aware that many purported "cost-effectiveness" studies are not actually cost-effective analyses in the economists' sense, but program evaluations with calculations of operational cost inputs. "Cost-effectiveness" is often used in government reports, more related to transparency and financial audit than to economic evaluation or program evaluation. In British Columbia, Canada, for example, "cost-effectiveness" was built into the accountability framework developed by the Auditor General and the Deputy Ministers' Council (1996), yet it is mentioned in terms of government efficiency and accountability, without the actual expectation that cost-effectiveness evaluations would be conducted within the ministries. Rather, government is expected to be "clear about its objectives and targets, the strategies it will employ to meet its objectives, the full costs of these strategies, and its actual results" (p. 33). This information is to be gleaned from "information required for managing at the program level" (p. 33).

Distinguishing Operational Costs From Social Costs: Key to Developing Economic Evaluation Methods

There is a significant difference between the *operational* costs of programs and the *social* costs of programs. While operational costs have been, and continue to be, based on government expenditures only, social costs can include intangibles and externalities, such as loss of leisure time, loss of air quality, and loss of recreational opportunities. The National Center for Education Statistics (1999) provides a comparison of two types of operational-cost CEA in the education sector. One approach is an "accounting model" based on operating costs specified in the traditional accounting system, and the other is the "resource cost model approach," which relates costs to service delivery of specified outcomes. However, neither of these approaches uses welfare economics theory to arrive at a social value of costs and benefits. Instead, both approaches are based on market costs gathered from accounting records. As with CBA, welfare economics theory is the basis for social-value CEA, where costs are based on "opportunity costs," which is the value of the next best use of the resources to society, and benefits are based on "willingness to pay," which is the aggregate subjective social value of the program or policy. When government auditors audit programs for efficiency, or even for effectiveness, they are basing their judgments on the budgetary costs for given outcomes, not on *social* costs and benefits.

● COST-EFFECTIVENESS ANALYSIS

Cost-effectiveness analysis combines information about program costs, measured in dollars, with tallies of one specific outcome, measured in **units of effectiveness** to create a ratio of *cost per unit of outcome achieved*. What we end up with, then, is cost per unit of effectiveness criteria, such as cost per life saved, cost per person returning to employment, or cost per case of illness prevented. In many cases the *size* of the program or project makes a difference, so it is important to look at not only the *average* cost per unit of effectiveness, but the *marginal* cost-effectiveness ratio. This ratio shows the marginal cost of each *additional* unit of outcome achieved, which can then be compared to the alternative marginal cost-effectiveness ratios, to determine the most effective choice (Clyne & Edwards, 2002).

Cost-effectiveness analysis facilitates comparisons across program alternatives that share the same objective, as well as comparison with the status quo, in order to determine the least-cost method of achieving an outcome. Equally, it can be used to determine the best way to maximize the number of a given outcome for a predetermined budgetary outlay. CEA is commonly used in the social services and the health sector, but is becoming important in the education sector and is also used in the crime prevention and transportation sectors (Levin & McEwan, 2001).

A general problem with cost-effectiveness ratios is the tendency to create ratios based on what is measurable. Hospitals routinely calculate costs per patient-day and use this as a measure of their performance (in fact, in many jurisdictions, funders expect hospitals to use this measure). But "patient-days" does not adjust for quality and by itself is more likely to be a measure of technical efficiency (cost per unit of output).

Cost-effectiveness ratios are powerful tools when they are used summatively, that is, to make judgments about the future of programs. Public sector and nonprofit managers are increasingly expected to create a "bottom line" analogous to the bottom line their private sector counterparts use, and cost-effectiveness measures help to fulfill this expectation. However, it is necessary when doing cost-effectiveness analysis to remember that the most effective approaches are not necessarily the least costly ones (Royse, Thyer, Padgett, & Logan, 2001) and that, when reaching conclusions about the value of a program, the quality of its outcomes is at least as important as their quantity and cost. For this reason, some evaluators have argued that cost-effectiveness (and cost-benefit) analysis should only be used in conjunction with other forms of program evaluation that focus more on whether stated outcomes have been achieved and on the quality of those outcomes (Krause, 1996).

Steps for Cost-Effectiveness Analysis

Aside from the monetization of benefits, the main steps of a CEA (and CUA) parallel those of CBA. We outline these steps in this section and expand on the monetization of benefits in the CBA section that occurs later in the chapter.

The following are the nine major steps common to economic evaluation. The square bracketed components are applicable only to CBA (adapted from Boardman, Greenberg, Vining, & Weimer, 2001, p. 7):

1. Specify the set of alternative projects

2. Decide whose benefits and costs count (standing)

3. Catalogue the costs and benefits and select measurement indicators (units)

4. Predict the impacts quantitatively over the life of the project

5. Monetize (attach dollar values) to all costs [and benefits]

6. Discount costs [and benefits] to obtain present values

7. Compare costs with outcomes or [compute the net present value (NPV)] of each alternative

8. Perform sensitivity analysis

9. Make a recommendation based on the comparisons and sensitivity analysis

Specify the Set of Alternative Projects

This step may seem deceptively simple. Perhaps the simplest example is the case where we are trying to determine the cost-effectiveness of a new program as compared to the status quo, such as the introduction of an innovative preschool program to enhance the reading ability of aboriginal children. In that case we would compare the cost-effectiveness of the project versus the status quo. But specifying alternatives is often more complex. As we shall see in the example further in the chapter, even in the case where we are studying a home hazard intervention program to prevent falls among the elderly, we are actually looking at three alternatives: intervention for seniors who have a previous history of a fall within the past year, intervention for seniors without a history of falls, and no intervention.

Cost-effectiveness analyses will entail choosing from among one or more interventions that all have a common outcome, such as number of falls prevented or units of increased reading ability. However, in the case of cost-utility analyses we can be comparing alternative programs or interventions that have a more diverse range of outcomes, as long as the outcomes can be converted to common units such as QALY. An example might be a health authority who wishes to compare investment in a program that treats patients with HIV/AIDS to a program that treats hepatitis. The program that resulted in the lowest cost per QALY or, similarly, the program with the highest number of QALY for a given expenditure, would be given higher consideration in terms of cost-utility.

Although a cost-benefit analysis can be done to compare virtually any combination of public projects since the costs and outcomes are all converted to dollar values and then are boiled down to the **net present value** of each potential option, it is most commonly done to determine the overall feasibility of a single project. For example, a state government may arrange for a cost-benefit analysis of investment in a highway project, and within that analysis may compare various alternative approaches to the project, such as the location of the road, the number of lanes of highway, and number of overpasses to offer high-speed intersections with other roads.

The key point to keep in mind when specifying the alternatives to be compared is that as more alternatives are added to the mix, the analysis becomes more challenging and more resource-intensive. It is important to carefully choose the alternatives, or the analysis may become bogged down later.

Decide Whose Benefits and Costs Count (Standing)

In this step, the different perspectives of who might be affected by a program must be taken into consideration. Public programs involve costs and benefits within a society, and the analyst must determine whose costs will be counted and whose benefits will count, so as to facilitate the calculations—deciding who "counts" in the calculations is often a political or jurisdictional issue. For example, in a cost-benefit analysis of a highway project, if a state receives funding from the federal government only if it goes ahead with the project, and the calculations are to be based on only state taxpayers as having "standing," then the federal funds are considered a benefit. However, if all the funds are originating from the state government, then they are not considered a benefit because if they are not spent on the project they will be spent on something else in the state—the "benefit" of spending money to build the highway would be offset by the costs incurred by not spending the money on alternative projects. We refer to offsetting benefits and costs within a jurisdiction as **pecuniary benefits** (or **costs**). In the same example, we would need to decide who has "standing" as far as counting and monetizing the number of lives saved by the new highway. Will only travelers from the state count, or all travelers, including international tourists?

Catalogue the Costs and Benefits and Select Measurement Indicators (Units)

Theoretically, all inputs (costs) and all outputs or outcomes (benefits) of each option should be listed. This includes both **tangible costs** (that we can usually enumerate because information is available) and **intangible costs** (which by their nature are difficult or impossible to express in dollar terms) and, in the case of cost-benefit analysis, both tangible and intangible social benefits (Kee, 1994). In the case of cost-effectiveness analysis or cost-utility analysis, the benefits will not need to be monetized, but they must be numerically countable.

The theory of welfare economics indicates that all the social costs of the project alternatives should be included (and the social benefits in the case of cost-benefit analysis). That means that unintended costs and benefits would be included as part of the analysis. In reality this is extremely challenging, and often only the key inputs and outputs or outcomes are included. Typically, operational and support costs that are obtainable from accounting records and capital costs that can be obtained from market prices are the ones that will be used for the analyses. Cataloguing the benefits in a cost-benefit analysis can be the most challenging, because items such as lives saved, traveling time saved, burglaries prevented, additional years of education attained, or additional disability-free years on the job must be somehow calculated and, later, monetized. Again, particularly in an *ex ante* analysis, it is important to make sure that the specified outcomes include the stated outcomes of the project. In an *ex post* analysis, unintended outcomes should also be included. When defining the variables and their related indicators, an evaluator should also be considering which variables and indicators were included in similar economic evaluations, so that later evaluators can compare the various studies.

Dhiri and Brand (1999), in a U.K. Home Office report, explain how units and monetization methods in cost-effectiveness analyses and cost-benefit analyses should be standardized so as to facilitate long-term comparability across various projects of a Crime Reduction Programme (CRP):

> Each CRP intervention will be subject to a cost-effectiveness analysis. Their evaluations need to be conducted to a sufficient standard to ensure that the cost-effectiveness information they produce can support strategic decisions. Projects must make a significant contribution to the evidence base that will enable comparative cost-effectiveness to be calculated and/or strategies to be adopted for improving cost-effectiveness. (p. 7)

In an innovative approach, evidence will be analyzed at two levels: Evaluators will conduct cost-effectiveness analyses of individual intervention projects, and the evidence from the various evaluations will be pooled by the Home Office (Research, Development and Statistics Directorate), who will monetize the outcomes in a standardized way, and conduct cost-benefit analyses (Dhiri & Brand, 1999).

Many public sector and nonprofit organizations have accounting systems that report some or most of the **operating costs** of a program, but often omit the **support costs**. Operating costs include expenditures for items that pay for the operation of the program: salaries and wages, supplies, and communications costs (fax, telephone, couriers) would be examples of operating costs.

Examples of support costs would be: depreciation on equipment used, heating, lighting, building rental, insurance, administration (human resources, budgeting, accounting, and information system support would be examples), and other costs that often do not appear as program expenditures. Other costs like fringe benefits, capital expenditures, or debt retirement, for example, may not be included, but generally should be.

There is a second set of issues regarding program costs. Estimating accounting costs may not provide a complete measure of the **economic costs** of the program. Economic costs typically are estimated by asking what the **indirect** as well as the **direct costs** are of spending money for a program. Kee (1994) points out that *direct costs* typically are costs associated with the program itself and will often include what we have categorized as operating and support costs. *Indirect costs* are costs that accrue that are usually not intended (Kee, 1994). A program that plants trees in logged areas will eventually eliminate opportunities for hunters who use the clear-cut areas to stalk game; that would be an example of an indirect cost.

Predict the Impacts Quantitatively Over the Life of the Project

Once we have identified the measures that will be included, we need to estimate (or, in the case of an *ex post* analysis, we can calculate) the quantitative amounts of these measures for the timeframe of the analysis. For example, if we have decided that "reported robberies" over the 2-year time span of a project will be one of the measures, the next step would entail determining the number of reported robberies over that time—in effect, defining a set of operations that would result in data being collected for analysis. Or, if we have determined that one of the indicators of cost would be volunteer hours of a given program, then at this step we would quantify the number of hours of volunteer labor provided over the given time span. A critical point here is that we will need to determine the quantities *against the base case that would have occurred had there not been an intervention.* For example, when we count up the robberies we need to determine the difference between the robberies that occurred and the robberies that *would* have occurred even if there had been no intervention. We are looking for the incremental difference that has occurred because of the program and to determine that difference, we would need to have a research design that facilitates estimating incremental outcomes.

Monetize (Attach Dollar Values) to All Costs [and Benefits]

This step involves attaching the dollar values to all the quantified input indicators and, in the case of a cost-benefit analysis, the quantified outputs or outcomes indicators. It can be the most controversial part of an economic evaluation, particularly when social benefits are to be calculated. Nonmarket values such as environmental impacts, human lives, and improved education levels can be difficult to monetize. Economists measure the value of social benefit indicators by determining the public's "willingness to pay" for these benefits. There can be a difference between "social value" and "price" where the price of a benefit is less than the price that a stakeholder group would be willing to pay. Some services, in fact, are offered to the public "free," such as some immunization programs, yet the value to the public is obviously greater than zero. These values can be determined in many ways: surveys, secondary data and, where possible, extrapolations from market values.

When measuring costs, economists measure "opportunity costs," which are not necessarily the same as market costs. **Opportunity costs** are the expense that is equivalent to the next-best economic activity that would be forgone if the project or program proceeds. For example, sometimes the value of an indicator such as an hour of labor can vary according to the local unemployment rate and the likelihood of whether the laborers would have been able to find alternate work if the project did not exist. Taxes are another important consideration when determining opportunity costs; projects have a variety of tax implications. Normally the values are listed in a table that shows the dollar values of costs and benefits in each year of the program, when it is a multi-year project.

Discount Costs [and Benefits] to Obtain Present Values

Discounting benefits and costs that accrue over time is a key part of conducting a benefit-cost analysis. The present value of the costs of a project is the sum of the discounted values of the costs in each of the years of the project. Similarly, the present value of the benefits of a program is the sum of the discounted values of the benefits from each of the years of the project. Usually, most costs occur at the beginning of a project, and the benefits occur over time.

One way to understand discounting is to remember two simple principles:

- A cost owed in the future is less onerous than a cost owed now.
- A benefit received in the future is less valuable than a benefit received now.

With respect to discounting costs, the rationale for the above principle is that in situations where people and organizations behave rationally (from an economic perspective), future costs can be partially offset by setting aside money now, investing it, and using the proceeds of the investment to help defray the cost when it is actually due. The magnitude of the return on the investment is affected by the **discount rate** (the interest rate) and is compounded annually until the payment is due.

Similarly, if you are to receive a benefit sometime in the future, you are unable to access that money now for investment purposes. In effect, you forgo the returns on the benefit until the money is in your hands. Another way to look at discounting benefits is to ask: What amount of money in the present time, if invested at the prevailing discount (interest) rate, would yield the amount that is due to you in the future?

Discounted benefits or costs are calculated with formulae that keep track of *when* benefits or costs occur, *what* the prevailing interest rate is, and, hence, what the present value of each benefit or cost is. The formulae for calculating the present value of costs and benefits divide the expected cost or benefit at some point in the future (t) by the compounded discount rate.

$$C_o = \frac{C_t}{(1+i)^t} \text{ or } B_o = \frac{B_t}{(1+i)^t}$$

where

C_t = cost accruing at time t

C_o = present value of the cost

t = number of years in the future when the cost is expected

i = the prevailing interest rate

    (the discount rate)

B_t = benefit at time t

B_o = present value of the benefit

As an example, suppose that a cost is expected to accrue 3 years from the present. Assume the prevailing rate of interest is 10%, and the expected amount of the cost in year 3 is $100. The present value of that cost is:

$$C_o = \frac{\$100}{(1 + 0.10)^3} = \$75.10$$

Thus, $100 3 years from now is equal to $75.10 today.

Clearly, the rate at which costs and benefits are discounted is a key determinant of the overall bottom line for programs or projects. For example, if a cost-benefit analysis is being conducted in advance of constructing a natural gas pipeline, the fact that the costs will be substantially front-end loaded in the project will mean that the overall net present value will depend on how steeply the **benefit stream** is discounted into the future. For example, if the discount rate is relatively high (15%), the value of future benefits will be sharply reduced compared to a discount rate of 5%. By manipulating discount rates, it is possible to come up with a bottom line that appears to be favorable, that is, where the benefits exceed the costs.

With respect to public sector programs or projects, economists have debated whether the discount rate ought to be comparable to the prevailing interest rate used by private companies when they invest in projects or, instead, a lower discount rate reflecting less risk for public sector projects. Generally, most analysts believe that the discount rate for public sector projects should be less than for the private sector, in part because the risk that government-financed programs or projects will "fail" is less (governments are less likely to go bankrupt). Second, governments generally have the capacity to borrow at lower rates than the private sector and, finally, government projects or programs can be viewed as long-term investments that are made because the private sector typically has a shorter time horizon and may not have incentives to invest in areas where at least some of the benefits cannot be captured by market pricing.

Historically, the typical range for discount rates has been between 5% and 15% if **nominal prices** are used and between 3% and 8% if **real prices** are used. Real prices have been adjusted to remove the effect of inflation over time.

Compute the Net Present Value (NPV) of Each Alternative

Once discounted benefit and cost streams have been calculated for the project or program, the totals of benefits and costs can be compared. This information, comparing the net present value of the costs and the net present value of the benefits, is basically a tabular presentation of the above calculations, laid out in a way that is easily read by the decision makers.

Perform Sensitivity Analysis

Because we cannot make perfect predictions about variables such as interest rates, utilization rates of a program, the success rates of a medical treatment, density of traffic on a new highway, or other outcomes, the evaluator will need to do **sensitivity analyses** to show the range of possibilities for some key variables. It is particularly common to do a sensitivity analysis using a range of discount rates, since changes in interest rates can have a large effect on the viability of some projects. Normally, results of sensitivity analyses are presented in tables so that the decision makers can easily make comparisons across several possibilities.

A second component of the sensitivity analysis is to explain the *distribution* of the costs and benefits that will occur if the project or program goes ahead. Particularly for cost-benefit analysis, where the net present value has been basically boiled down to a single value for society, it is important to show whether any particular groups will be disproportionately affected. For example, a cost-benefit analysis may have shown that there will be an overall net benefit to society if the government invests in a program to flood a valley to create a hydro-electric project, but it will be critical to clearly show which stakeholders will benefit overall (for example, the users of the electricity who may pay lower prices) and which groups may lose (for example, aboriginal people who live in the valley and depend on the current waterways for food and transportation).

Make a Recommendation Based on the NPV and Sensitivity Analysis

Theoretically, the economist's recommendation is based on the project or program that has the highest NPV, or the option that achieves the given effectiveness level at the lowest price (or, alternatively, the highest effectiveness level for a given price). However, it is important to remember that the economic analysis is only one part of the decision-making process. The policy maker is given the evaluation information, and then must take into consideration equity issues and the political, legal, and moral climate. The policy maker will also need to consider the various scenarios offered in the sensitivity analysis. Fuguitt and Wilcox (1999), while acknowledging that professional judgment is inevitably a part of the process of an economic evaluation, offer the following suggestions for the analyst's role:

> The analyst's responsibility is to conduct the analysis with professional objectivity and minimize inappropriate subjective influences. Specifically, in performing the analysis, the analyst must (1) minimize any biases or otherwise misleading influences reflected in the measurements of individuals' subjective preferences, (2) explicitly identify any value judgments embodied in the analysis and make transparent their implications for the outcome and (3) where analyst or decision-maker discretion is possible, choose an approach that minimizes one's subjective influence and demonstrates the sensitivity of the analysis to alternative (subjective) choices. (p. 18)

Thus, the reader of an economic evaluation report should expect to find assumptions, value judgments, technical choices, and possible errors that were made during the evaluation, and even subjective biases that may have affected the outcome of the analysis (Fuguitt & Wilcox, 1999).

● COST-UTILITY ANALYSIS

Even where numerical information on effectiveness is available, programs typically consist of multiple components, as well as multiple outcomes. Because cost-effectiveness analysis generally implies the creation of ratios between costs and single outcomes, it is necessary to either (judgmentally) choose one key outcome, or weight and combine them (where more than one is selected). **Cost-utility analysis** is a variation of cost-effectiveness analysis, and its most common form is in the health sector, with the use of QALY gained as an outcome. It is used for comparing the health and economic consequences of a wide variety of medical and health-related interventions. The method combines both the years of additional life as an outcome and the subjective value of those years, in cases where quality of life is affected. Once the cost of each type of intervention is calculated, it is possible to create a ratio of "cost per quality-adjusted life-year." It is recommended for the following reasons:

It captures gains from both prolongation and quality of life in a single measure, it incorporates the value that people place on different health outcomes or their preferences for particular health states, and it provides a convenient means of comparing analyses of diverse interventions and conditions. (Neumann et al., 2000, p. 965)

Although cost-utility analysis is one way that multiple objectives can be weighted and combined, the process is resource intensive and time consuming. The subjective valuations of the quality of life of various health outcomes must be determined to create the QALY index, and researchers have developed several methods to do this. Three common methods are the *health rating method,* the *time trade-off method,* and the *standard gamble method.* With the *health rating method* "analysts derive a *health rating* from questionnaires or interviews with health experts, potential subjects of treatment, members of society in general, or else on the basis of their own expertise" (Boardman et al., 2001, p. 446). The time trade-off method involves gathering subjective ratings of various combinations of length of life and the quality of life in that time, whereas the standard gamble method involves

Table 7.1   Example of a QALY League Table

Cost per Quality-Adjusted Life-Year (QALY) of Competing Therapies—Some Tentative Estimates

Intervention	Cost per QALY (£, 1990 prices)
General practioner's advice to stop smoking	270
Antihypertensive therapy	940
Pacemaker insertion	1,100
Hip replacement	1,180
Valve replacement for aortic stenosis	1,410
Coronary artery bypass graft	2,090
Kidney transplant	4,710
Breast screening	5,780
Heart transplant	7,840
Hospital hemodialysis	21,970

having the subjects make choices in decision-tree scenarios with various timings of life, death, or survival with impaired health (Boardman et al., 2001). The results of this research facilitate the development of QALY league tables such as the one shown in Table 7.1 (from Kernick, 1998, p. 149), which can be used to determine the best use of health care resources.

The QALY approach has had broad application in health policy analyses, but has not been used extensively in other program or policy areas (Levin & McEwan, 2001). The difficulties of working through scenarios where several outcomes (which may be valued differently by different stakeholders) are ranked or weighted are formidable. Sensitivity analysis of the results should include examining a range of the following: costs, preference weights, estimates of effectiveness, and the discount rate (Neumann et al., 2000).

COST-BENEFIT ANALYSIS •

Conceptually, cost-benefit analysis is straightforward and goes a step beyond cost-effectiveness analysis in its monetization of the benefits. Its main premise is that by converting costs and benefits that accrue to a project or program over time, compensating for the impact of time, adding all benefits and costs together, and comparing total benefits to costs, an elegant "bottom line" for programs or projects can be constructed.

Ex ante cost-benefit analyses typically forecast streams of costs and benefits for the duration of a project or program, and apply valuation and discounting methods to arrive at an estimate of the **net present value** of the project. *Ex post* benefit-cost analyses proceed in a similar manner except that the evaluator must measure the *actual* benefits and costs that accrue over time. An important issue for both *ex ante* and *ex post* analyses is the problem of attributing costs and benefits *to the program,* as opposed to other factors. In *ex ante* cost-benefit analyses it is necessary to assume that the program will actually produce the forecasted costs and benefits. Likewise, for *ex post* analyses, the observed outcomes need to be linked to the program (the attribution issue) and they also need to be converted to a common monetary measure. Yee (1994) calls this attribution issue the **multiple-causation problem**, and it is not unique to benefit-cost analyses, as we have seen in Chapter 3.

As was mentioned earlier in this chapter, some portions of both benefits and costs can be intangible, which usually means that it is difficult, if not impossible, to measure them defensibly. In some situations, their magnitudes can be estimated qualitatively (more or less than their tangible counterparts), and their impact on the bottom line judgmentally assessed as the analysis concludes.

Key Concepts in Cost-Benefit Analysis

Evaluations of programs using cost-benefit techniques must distinguish between *real* and *pecuniary* benefits and costs. **Real benefits** include improved lives, more trees planted, safer highways, increased earnings, time saved, or other program outcomes that are *net gains* to society (or to the jurisdiction for which the cost-benefit analysis is being conducted). In other words, these gains are not offset by losses due to the program elsewhere in the society or jurisdiction.

Correspondingly, **pecuniary benefits** are transfers that cancel out from a societal perspective. Subsidizing the operation of a car manufacturer in a state produces benefits for the employees and the community in which they reside. But these benefits are offset by the costs to state taxpayers which, when cumulated, equal the value of the subsidy to the company.

Similar distinctions hold for **real costs** and **pecuniary costs**. An example of pecuniary costs would be a decision by a local government to change the level of residential solid waste collection service from yard pickup to curb side pickup. From a typical homeowner's perspective, there is a cost—each household now has to carry its garbage cans to the curb. But, each homeowner also realizes a benefit in that the municipality can now reduce crew sizes on the routes, or reduce the number of crews needed to service all residents, producing a saving for residents as taxpayers.

Estimating Willingness-to-Pay

Measuring intangible benefits is sometimes approached by constructing "willingness to pay" scenarios for program or project recipients. Basically, analysts use either survey research methods, focus groups, or simulations to obtain estimates of how much, if anything, persons would be willing to pay for a projected benefit (say, reducing their travel time to work by 20 minutes if a new highway was built) or how much they would be willing to pay to avoid an anticipated cost (say, the increased noise and pollution associated with living near a new highway).

Constructing willingness-to-pay scenarios is a useful way to extend the reach of cost-benefit analysis, but can be costly to do well. If survey methods are used, representative samples of sufficient size to generalize to a population of affected recipients must be drawn. As well, instruments need to be constructed so that scenarios are not biased or have not omitted a key benefit or cost (Boardman et al., 2001).

Internal Rate of Return

An alternative approach to using the **net present benefit** to determine whether a program or project should proceed is to calculate the **internal rate of return** for a given program. Using this approach means that, once the projected benefits and costs have been calculated, the discount rate can be manipulated so that it yields a net present benefit of zero (or a **benefit-cost ratio** of one). If the "zero-benefit" discount rate is higher than the rate at which money can be borrowed for the project (higher than the opportunity cost of investing in the project or program), then it is reasonable to proceed with the project, since the actual discount rate will result in a positive net present benefit.

The Marginal Value of Money

Earlier in this chapter, pecuniary benefits and costs were introduced and defined as situations where transfers occur that offset each other. Benefits to the workers in a subsidized manufacturing company were assumed to be offset by the **marginal costs** to all tax payers in the state.

But, it is possible that the **marginal value** of income to the workers in such a situation will be greater than the marginal cost of the taxes needed to pay the subsidy. This would mean that the transfers are not pecuniary, but could produce a net benefit to society.

Analyzing such situations in benefit-cost analyses can be complex. It is not easy to estimate the marginal benefits or costs associated with income redistribution, although it is generally believed that the marginal value of money tends to be greater for those with lower incomes. Generally, analysts tend to assume that transfers are neutral with respect to the marginal value of money simply because estimates of the magnitudes of differences are difficult to construct and defend.

● COST-EFFECTIVENESS ANALYSIS EXAMPLE: A STUDY OF FALLS PREVENTION AMONG THE ELDERLY

In this section, we discuss an Australian study that examined a specific strategy for reducing the incidence of falls among the elderly. We examine the study in terms of the nine steps for conducting a cost-effectiveness analysis that were summarized earlier in the chapter.

As the baby boomers age and the proportion of elderly persons in North America continues to grow, and because falls among the elderly are common and often have devastating health impacts, it is becoming increasingly important to determine which types of programs most cost-effectively contribute to falls prevention. While many studies have examined the effectiveness of various strategies to reduce the incidence of falls among the elderly, very few have assessed the cost-effectiveness of these strategies (for an overview, see Kochera, 2002; Salkeld, Cumming, O'Neill, et al., 2000).

As noted earlier, in cost-effectiveness analysis the costs of a program are determined in monetary units but the effectiveness, rather than being monetized, is measured in number of units affected (in this case, number of falls prevented). The rationale for cost-effectiveness analysis is that it can be used to compare projects or programs that have a specific effectiveness target, such as prevented falls, without having to monetize the specified effect.

In some cases, cost-effectiveness is considered not only in terms of program cost per unit of effectiveness (e.g., per fall prevented), but also in terms of whether the costs of the specific form of intervention are less, on average, than the savings that result from the incremental reduction in health costs. One way to conclude that the program was successful is that the intervention costs were less than the savings per unit of outcome (falls prevented), and hence, the program paid for itself. In some ways, then, these studies are like cost-benefit analyses except that only the monetized value of the incremental reduction in health costs is considered as the social benefit.

Salkeld et al. (2000) carried out a randomized controlled trial in Australia to examine the cost-effectiveness of a home hazard reduction program designed to reduce falls among the elderly. In a study of 530 elderly people who had been recently discharged from a hospital stay, Salkeld et al. examined the costs of a home hazard reduction program, the incremental changes in rate of falls, and the incremental changes in participants' use of health care services. They calculated the changes in falls-related health care resource use[1] for each of the study participants, and the costs of the interventions for the treatment group[2] to determine an "incremental cost-effectiveness ratio." The study found that the intervention was only "cost-effective," that is, the reduction in costs for health care resource use was greater than the cost of the intervention, for those with a previous history of falls. Overall, the intervention resulted in a reduction in falls but with an accompanying increase in the health care costs of the treatment group. By applying a cost-effectiveness criterion that included all health care costs (regardless of whether these were caused by falls), the analysts took the perspective of the funders of all health care services, not just those related to falls.

Below, we discuss this study in terms of the nine steps of a typical cost-effectiveness analysis, and point out some of the key issues in reading such a study critically.

1. Specify the Set of Alternative Projects.

The "programs" under consideration were two approaches that would entail home hazard modifications for elderly persons, designed to reduce falls. Because the costs and benefits were based on the end results of the study, it would be considered an *ex post* analysis. Specifically, this study can be seen as comparing two variations of interventions: (a) in one case, only hospitalized seniors with a history of falling in the previous year would have home hazard intervention when returning home; and (b) in the other case, all seniors who had been hospitalized and would be returning home would qualify for the program. By also having a control group that does not receive the program and by looking at incremental cost-effectiveness, by definition the status quo is also under consideration. That is, we are able to compare the results of the program (number of falls and dollars of health resources used) with the results that would have occurred with no program.

Note that the experiment could have included seniors who had never been hospitalized. However, there are always resource limitations on research, and the researcher must use professional judgment in specifying the alternatives. As well, by limiting the experiment in this way, these researchers could more easily do random assignment of the individuals to the intervention group and the nonintervention group. By selecting participants only from seniors who had been hospitalized, though, the researchers did limit the generalizability (external validity) of the study. That is, care must be taken in generalizing the results to seniors in Australia who have not been previously hospitalized, or seniors in other countries, whether they have been hospitalized or not.

2. Decide Whose Benefits and Costs Count (standing).

The benefit, in this case, is the incremental reduction in number of falls for the elderly who received home hazard intervention. As well, the study looked at the changes in health care resource use for those who participated in the program. These health care costs were calculated in a way that shows that various stakeholders had standing: the taxpayers, the elderly, and the families/friends that would be impacted via their caretaker roles. In similar studies (e.g., Smith & Widiatmoko, 1998) only the taxpayers and the elderly were considered to have standing.

3. Catalogue the Impacts and Select Measurement Indicators (units).

The difference between the rate of falls in the intervention group as compared to the nonintervention group, over a 1-year period, is the measure of incremental reduction in falls. Measures of the costs were as follows:

- Days of hospitalization per person
- Per-person number of usages of other health care institutions
- Per-person hours of patient services provided in the home
- Per-person hours of informal health care provided by family and/or friends
- Average number of occupational therapist hours per intervention (assessment and supervision of the modifications)
- Home modifications per person

Baseline data, to be used for determining incremental changes in health care costs, were determined using per-person healthcare records based on all the above factors (for the 12 months previous to the intervention) except for the intervention (program) costs. Note that research costs are not included in the analysis.

The outputs are defined as the number of falls reduced by the program. One shortcoming in this approach, it should be noted, is that by viewing only the incremental reduction in falls as the output of the program, numerous other benefits, admittedly difficult to quantify, are ignored. For example, an avoided injurious fall also has the benefits of facilitating a higher social productivity level from the individual and preventing the increased fear of falling that occurs among many fallers (affecting their activity levels). Lost productivity and quality of life are not considered. So, by doing an examination of the cost-effectiveness of reducing falls, especially in terms of intervention costs and reduced health care costs (the approach most studies have taken), we are in fact undervaluing the outcomes of reduced falls. This should be kept in mind if, for example, a trial such as this one showed that the intervention costs are reasonably close to, but more than, the saved health care cost. It may still be worthwhile to implement one of the programs.

4. Determining the Impacts Quantitatively Over the Life of the Project.

The costs and impacts were quantified over the life of the project as follows:

The baseline costs were determined by having an interviewer go through a questionnaire with participants to determine their level of health care use, in various domains including home help, over the previous 12 months, as well as the number of times they had fallen. These data included number of days in hospitals, number of trips to clinics, number of visits to doctors, hours of unpaid help with housework, etc.

Quantitative *changes* in resource use, then, were determined by comparing these data to the intervention costs and the 1 year of health care costs following the intervention (as listed in step 3), once all was monetized. The follow-up data were collected using a method whereby participants are given a set of follow-up calendars that they or their caregivers filled in with information on the first four health care-related items on the above list. The calendars were mailed in monthly.

5. Monetize (attach dollar values).

The following monetization strategies were used to specifically attach dollar amounts to the health care costs:

- Hospitalization costs were calculated on a per-day rate based on average rates provided by local hospitals for each type of care level.

- Other health care institutional costs were calculated on per-use rate based on average hourly or daily rates provided by the government.
- Costs for patient services provided in the home were based on average hourly charges of the specific providers.
- Costs for informal health care provided by family and/or friends were based on average hourly wages for caregivers and housekeepers, depending on the type of care being given.[3]
- Occupational therapist costs were based on average number of hours per intervention times hourly costs.

Home modifications were based on the actual market price of the home modifications, including the modifications and the hourly wages of the installers.

6. Discount Costs and Benefits to Obtain Net Present Values.

Because some of the cost figures (e.g., daily hospital rates for particular levels of care) were not readily available for the current year, they were estimated based on historical costs, taking inflation into account. Overall, "all costs are counted for the base year 1997" (Salkeld et al., 2000, p. 266). Since intervention costs were borne at the beginning of the project, and all following resource costs occur within the year, no further discounting was done for those. However, since there is long-term use of the home hazard equipment, such as bathtub bars, these costs were annualized over 10 years (i.e., expecting a 10-year life for the equipment), so only 10% of the costs were actually counted within the first year. As part of the sensitivity analysis, costs of interventions and incremental health care costs were estimated over a 10-year period, adjusting certain variables (details below in sensitivity section), then discounting to determine NPV. Thus, the NPV of the two intervention alternatives, one for all seniors who had been hospitalized and one for only those with a history of falls, was compared with the status quo of zero cost for interventions and zero incremental changes in health care costs.

7. Compute the NPV of Each Alternative.

Since this study was actually a cost-effectiveness analysis (though a modified one that had similarities to a cost-benefit study), there was not a calculation of a net social value for the program. However, the study did look at the "incremental cost per fall prevented," using a single baseline year, and found that there was, when using median figures, a minor cost saving for those seniors in the study who had fallen in the 12 months before the intervention, but that overall there was an incremental cost per fall prevented of (AUD$4,986). That is, for each fall prevented, the combination of the intervention costs and the somewhat puzzling increased health care cost (although most was not related to falls) led to a cost of almost AUD$5,000 per prevented fall.

8. Perform Sensitivity Analysis.

Because this cost-effectiveness analysis was done via a 1-year controlled randomized study, much of the sensitivity analysis was done in terms of varying the ways of looking at the collected health care data. For example, it was important to eliminate outliers and to do comparisons using both means and medians of the two groups (intervention and control groups, and also splitting out those who had a history of falls from those who did not) to determine whether there were statistically significant differences between the groupings. In addition, savings, or costs, could have been examined over a 10-year period, with varying assumptions about the rate of increases in health care-related costs in relation to the intervention costs. Health care costs can sometimes expand more rapidly than the regular inflation rate.

Because all of the expenses of the intervention occurred at the beginning of the program, and the health care "savings" occur over time, if doing a 10-year estimation it might have been worth looking at a high, low, and medium discount rate that would reflect a reasonable range of possibilities.

A distribution analysis could more explicitly consider some of the benefits of a prevented fall that accrue to two of the groups with standing: the elderly and their friends and family. The costs of the intervention to taxpayers, then, might be more balanced out by the benefits to these two groups, and society might better see the value of supporting the intervention program.

SUMMARY OF THE FALLS ● PREVENTION EXPERIMENT

This example gives some insight into the breadth of factors that must be taken into account in conducting a cost-effectiveness analysis, as well as issues to consider when critically reading and using the analysis. Although each economic evaluation will have unique features, the steps that have been summarized for the falls prevention evaluation exemplify the issues that need to be addressed. In addition, core issues for program evaluation—measurement (validity and reliability of measures) and research design (internal and external validity)—are also embedded in economic evaluations. Economic evaluations build on the same platform that supports defensible program evaluations.

One of the key limitations of a cost-effectiveness analysis, as compared with a cost-benefit analysis, is that important benefits that are difficult to monetize are disregarded; only one outcome is considered. This can become misleading when, as in this example, the costs of an intervention are then combined with the changes in health care resource use, to see whether the intervention is "cost-effective." In this study, the benefits that accompany an elderly person's prevented fall, such as maintained personal independence and productivity, a sense of confidence of movement, and the ability to live at home rather than in an institution, have significant social value. However, these benefits were not monetized, and are therefore not counted as part of the benefits. The benefit, a "prevented fall," may not seem worth the intervention if the other benefits are not explicitly taken into consideration.

● STRENGTHS AND LIMITATIONS OF ECONOMIC EVALUATION

Cost-effectiveness analysis, cost-utility analysis, and cost-benefit analysis have a strong appeal for evaluators and decision makers who desire precision. In addition, cost-benefit analyses, when well-conducted, offer a relatively complete consideration of intended as well as unintended costs and benefits. All three methods of economic evaluation create ratios of resources to outcomes, either monetizing them (CBA) or constructing cost-effectiveness ratios (CEA and CUA).

Historically, one "promise" of program evaluations was to be able to offer decision makers information that would support resource allocation and reallocation decisions (Mueller-Clemm & Barnes, 1997). Economic evaluations, particularly ones that explicitly compare program or project alternatives, can offer decision makers "bottom line" information that suggests the most rational choice, if the analyst's conclusions are translated into policy or program decisions.

There is considerable evidence from fields like health care that economic evaluations have become an essential part of the process whereby programs, policies, and technologies are assessed. Growing costs, growing demands for services, and resource scarcities collectively support analytical techniques that promise ways of determining the best use of new or even existing funds.

Strengths of Economic Analysis

Economic analysis works best in situations where the intended logic of a program or a project is high probability. That means that if resources are invested, we can be reasonably sure that the intended outcomes will materialize. Examples include flood control projects, highway construction projects, and natural gas pipelines. If we can forecast the benefits of a project, given a projected investment of resources, then we are in a position to conduct *ex ante* analyses—offering decision makers information that indicates whether a project has a NPV greater than zero, for example. *Ex ante* analyses can be conducted at the program planning stage of the performance management cycle, strengthening the process of translating strategic objectives into well-considered, implemented programs.

Methods of economic analysis demand that analysts and other stakeholders identify the assumptions that underlie an evaluation. Because CEA, CUA, and CBA all focus on comparing inputs and outcomes (monetized or not), it is necessary for analysts to wrestle with issues like standing (from whose perspective is the evaluation being done?); what to include as costs and benefits; how to measure the costs and benefits; how to discount to present values; and how to rank program or project alternatives. Competent consumers of an economic evaluation can discern what assumptions are explicit or implicit in the analysis—this makes it possible to identify possible biases or ways that values have been introduced into the process.

It is important to keep in mind that even though economic evaluations offer precision, their execution depends on the exercise of professional judgment. A competent economic evaluator will rely on her experience to navigate the steps in the process. As we explain in Chapter 12, every evaluation entails professional judgments—their extent and nature will vary from one evaluation to the next, but professional judgments are woven into the fabric of each evaluation.

Limitations of Economic Evaluation

The precision that is inherent in the methods of economic evaluation depends on the quality and completeness of the data and on the accuracy of the assumptions that undergird the analysis. For programs where key costs and outcomes cannot be quantified or monetized, the validity of any ratios comparing costs to benefits can be weakened. For many social programs, where program technologies are low probability, it is more questionable to forecast outcomes in advance of implementing the program, which reduces the value of *ex ante* economic evaluations.

Even *ex post* evaluations may end up relying on methodologies that introduce a substantial amount of uncertainty in the findings. For example, a cost-benefit analysis of a mobile radio system (MRDS) that was piloted in the Vancouver, British Columbia, police department (McRae & McDavid, 1988) relied on questionnaires administered to police officers to assess the likelihood that having an in-car computer terminal (with information access to a Canada-wide arrest warrant database) was instrumental in making an arrest. Over a period of 3 months, a total of 1,200 questionnaires were administered for the arrests logged in that period. Having identified which arrests the officers attributed to the MRDS system, estimates of the incremental effects of the system were calculated.

Because one of the frequent arrest types was for outstanding parking tickets in the city, the city realized a substantial benefit from the installation of the in-car terminals. This benefit was monetized and forecasted for the duration of the project and became a key part of the overall net present benefit calculation supporting the expansion of the system to the rest of the fleet. Clearly, relying on officer assessments of incrementality introduces the possibility of bias in estimating benefits.

In any economic analysis, it is possible that questionable assumptions are made. In some cases, manipulation of key variables affects the entire economic evaluation, biasing it so that the results are not credible. Manipulating discount rates is one example of this kind of problem. Fuguitt and Wilcox (1999) offer an example of institutional capture that illustrates what can happen when discount rates are manipulated to achieve a political objective:

> President Nixon ordered the use of a relatively high discount rate for analyses of most federal projects; thus, some relatively efficient projects were determined to be inefficient. Many authors relate this requirement to Nixon's promise to reduce public expenditures. . . . Moreover, in response to Western states' interests, federal water projects were excluded from this stipulation; analyses for these projects were allowed to use a specified low discount rate. . . . By instituting different discount rates for two separate categories of federal projects, Nixon effectively shaped all federal cost-benefit analyses produced during his administration. (p. 20)

A second example of how economic evaluations can be affected by political factors is offered to suggest that political involvement can be salutary. In 1990, the Oregon Health Services Commission developed a ranking of all medical procedures funded by the federal Medicaid program. The measure of effectiveness that was used was obtained by multiplying the expected improvement in a patient's life by the number of years over which the expected benefit obtained (Pinkerton, Johnson-Masotti, Derse, & Layde, 2002). One purpose of this ranking was to be able to ration Medicaid funds— ostensibly funding the procedures that were more highly ranked.

Problems with the data emerged, and it became clear that using the original ranking of procedures to decide which procedures would be funded would not be acceptable to many stakeholders, including the federal government. By 1993, the original list of procedures was modified to focus on a 5-year survival horizon (instead of a lifetime survival time frame). But more significantly, ties on the list of ranked procedures were broken by comparing average cost per treatment and alphabetical order of the procedure. Pinkerton et al. (2002) conclude that extra-rational factors prevented the use of a flawed analysis, "Thus, in the Oregon experience, the political process constrained the application of a seriously flawed cost-effectiveness methodology for the prioritization of health services" (p.78).

Reviews of economic evaluations suggest that even though the use of this approach is increasing, many studies that are done are flawed. Jefferson, Demicheli, and Vale (2002) conducted a meta-review of economic evaluations in health care. They summarized a wide range of reviews of existing studies and generally concluded that there are serious problems with the methods that are used. In one cluster of nine reviews, a total of 1,407 economic evaluations were assessed for the quality of their methods. Jefferson et al. (2002) summarize these reviews: "All reviews cast serious doubts on the validity of the conclusions reached by the economic evaluations assessed and all reviews propose stricter criteria for quality control" (p. 2810); and conclude: "We believe that urgent action should be taken to address the problem of poor methods in economic evaluations" (p. 2811).

Those words suggest that conducting economic evaluations is challenging. The public and nonprofit sectors are moving to embrace cost-effectiveness and other criteria that imply comparisons of program resources to results. This trend is part of a broader movement in the public sector to embrace values similar to those in the private sector—governments are expected to be accountable and to demonstrate value for money in the programs and services that are delivered. Appropriate uses of cost-benefit, cost-effectiveness, and cost-utility analyses are useful ways of supporting summative policy and program decisions. Like all analytical approaches, they can be applied well, or "stretched" into situations where the assumptions that are required are not realistic. Sound professional judgment is an important part of choosing whether to use economic analysis.

● SUMMARY

Economic evaluation is an approach to program and policy evaluation that relies on the principles of welfare economics—broadly speaking, that choices among programs and policies need to take into account their benefits and costs from a societal point of view. The three main methods of conduting economic evaluations are cost-benefit analysis, cost-effectiveness analysis, and cost-utility analysis.

Each of these three methods of doing economic analysis differs in the ways that it treats the costs and the outcomes of programs or policies. Cost-effectiveness analysis takes into account the economic costs of decision alternatives among programs or policies and compares these costs to a quantitative measure of the key intended outcome. Economic costs can be different from the kinds of costs recorded in budgets. Typically budgeted costs do not take into account the opportunity costs of inputs. Opportunity costs are the costs that would be incurred if the inputs, instead of being used to support a given program, were expended on their next best alternative use.

Cost-utility analysis compares the costs of program or policy alternatives to the utility expected from their implementation. Utility is usually estimated by taking two or more outcomes and constructing a measure that combines them. For example, health program alternatives are often compared by estimating the number of quality-adjusted life-years that the program will yield for a typical client.

Cost-benefit analysis compares the costs and benefits of program or policy alternatives, where all benefits and costs have been converted into some currency (dollars, for example). The challenges in estimating costs and benefits vary with the types of programs or policies that are being compared. The biggest challenge is in estimating intangible costs and benefits—they typically cannot be estimated directly and often require assumptions that might be challenged for their credibility.

All three approaches to economic analysis depend on being able to predict the actual outcomes for program or policy alternatives that are being compared. Where programs have a high probability of achieving their intended outcomes (e.g., dam construction), estimating future costs and benefits is relatively less challenging. But where we are comparing social program alternatives (e.g., job training programs), estimating the costs and the actual benefits is challenging. The validity of economic evaluations depends, in part, on the validity of the assumption that the intended outcomes for program or policy alternatives would actually be achieved, if implemented.

Economic evaluation is growing in importance as governments are increasingly expected to demonstrate value for money for their expenditures. This growing demand for economic evaluations, particularly in the health sector, is not necessarily being matched by the quality of the studies being done.

DISCUSSION QUESTIONS ●

1. What are the principal differences between cost-effectiveness analysis (CEA), cost-utility analysis (CUA), and cost-benefit analysis (CBA)? Give an example of each approach.

2. Why are research design and the attribution of outcomes important to CBA, CEA, and CUA?

3. What are opportunity costs? How do they differ from accounting or budgeted costs?

4. Why is a benefit (e.g., $100) that is realized in the future not as valuable as the same benefit accrued now?

5. Why is a cost that is incurred in the future not as onerous as the same cost incurred now?

6. What is the difference between nominal prices and real prices?

7. What is the difference between real and pecuniary benefits?

8. The fire chief in a coastal city that surrounds a harbor has long expressed his concerns to the city manager that the harbor is not adequately protected in case of a boat fire or a fire in a structure on the harbor. In a memorandum to the city manager, the chief offers two options:

Marine fire protection is currently provided on a contract basis with Seaspan Tugboats at an annual cost of $12,000. For the past 7 years, Seaspan has not been able to man their fire boat between the hours of midnight and 7 A.M. This represents a very serious deficiency in the city's fire defense plan in that most of the serious fires occur during this time frame.

Option 1: Maintain the present level of service with Seaspan which provides a manned fire boat 17 hours a day and recall off-duty personnel for 7 hours a day with a response time of approximately 60 to 90 minutes.

20-year total cost with annual increases of 5%: $397,000.

Option 2: Fire department operating a city-owned fire boat which could also be used for marine rescue, code enforcement, and monitoring and containment of water pollution on a 24-hour basis.

Capital cost	$175,000
Mooring facility	$10,000
Total maintenance over 20 years	$50,000
Total	**$235,000**

It is recommended that Option 2 be adopted for the following reasons:

1. The fire boat would be available for prompt response to fire and life rescue assignments on a 24-hour basis.

2. The cost saving projection over the 20-year period would be $162,000.

Do you agree with the fire chief's recommendation? Why or why not? Be specific in your response to this case.

● REFERENCES

Auditor General of British Columbia and Deputy Ministers' Council. (1996). *Enhancing accountability for performance: A framework and an implementation plan—Second joint report.* Victoria: Queen's Printer for British Columbia.

Boardman, A. E., Greenberg, D. H., Vining, A. R., & Weimer, D. L. (2001). *Cost-benefit analysis: Concepts and practice* (2nd ed.). Upper Saddle River, NJ: Prentice Hall.

Canadian Comprehensive Auditing Foundation. (1985). *Comprehensive auditing in Canada: The provincial legislative audit perspective.* Ottawa, ON: CCAF.

Clyne, G., & Edwards, R. (2002). Understanding economic evaluations: A guide for health and human services. *Canadian Journal of Program Evaluation, 17*(3), 1–23.

Dhiri, S., & Brand, S. (1999). *Analysis of costs and benefits: Guidance for evaluators.* London: Research, Development and Statistics Directorate, Home Office.

Fuguitt, D., & Wilcox, S. J. (1999). *Cost-benefit analysis for public sector decision makers.* Westport, CT: Quorum.

Gramlich, E. M. (1981). *Benefit-cost analysis of government programs.* Englewood Cliffs, NJ: Prentice Hall.

Hahn, R. W., & Dudley, P. (2004). *How well does the government do cost-benefit analysis?* (Working paper 04–01). Washington, DC: American Enterprise Institute, Brookings Center for Regulatory Studies.

Jefferson, T., Demicheli, V., & Vale, L. (2002). Quality of systematic reviews of economic evaluations in health care. *Journal of the American Medical Association, 287*(21), 2809–2812.

Kee, J. E. (1994). Benefit-cost analysis in program evaluation. In J. S. Wholey, H. P. Hatry, & K. E. Newcomer (Eds.). *Handbook of practical program evaluation* (pp 456–488). San Francisco: Jossey-Bass.

Kernick, D. P. (1998). Economic evaluation in health: A thumb nail sketch. *British Medical Journal, 316,* 1663–1665.

Kochera, A. (2002). *Falls among older persons and the role of the home: An analysis of cost, incidence, and potential savings from home modification.* Washington, DC: Public Policy Institute, American Association of Retired Persons.

Krause, D. R. (1996). *Effective program evaluation: An introduction.* Chicago: Nelson-Hall.

Levin, H. M., & McEwan, P. J. (Eds.). (2001). *Cost-effectiveness analysis: Methods and applications* (2nd ed.). Thousand Oaks, CA: Sage.

McRae, J. J., & McDavid, J. (1988). Computer-based technology in police work: A benefit-cost analysis of a mobile digital communications system. *Journal of Criminal Justice, 16*(1), 47–60.

Mishan, E. J. (1988). *Cost-benefit analysis: An informal introduction* (4th ed.). London: Unwin Hyman.

Mitton, C., & Donaldson, C. (2003). Tools of the trade: A comparative analysis of approaches to priority setting in healthcare. *Health Services Management Research, 16*(2), 96–105.

Muller-Clemm, W. J., & Barnes, M. P. (1997). A historical perspective on federal program evaluation in Canada. *Canadian Journal of Program Evaluation, 12*(1), 47–70.

National Center for Education Statistics. (1999). *Measuring resources in education: From accounting to the resource cost model approach* (Working paper no. 1999-16). Washington, DC: U.S. Department of Education.

Neumann, P. J., Stone, P. W, Chapman, R. H., Sandberg, E. A, & Chaim, M. B. (2000). The quality of reporting in published cost-utility analyses, 1976–1997. *Annals of Internal Medicine, 132*(12), 964–972.

Pinkerton, S. D., Johnson-Masotti, A. P., Derse, A., & Layde, P. M. (2002). Ethical issues in cost-effectiveness analysis. *Evaluation and Program Planning, 25,* 71–83.

Poister, T. H. (1978). *Public program analysis: Applied research methods.* Baltimore: University Park Press.

Royse, D. D., Thyer, B. A., Padgett, D. K., & Logan, T. K. (2001). *Program evaluation: An introduction* (3rd ed.). Belmont, CA: Brooks/Cole-Wadsworth Thompson Learning.

Salkeld, G., Cumming, R. G., O'Neill, E., et al. (2000). The cost effectiveness of a home hazard reduction program to reduce falls among older persons. *Australian and New Zealand Journal of Public Health, 24*(3), 265–271.

Smith, R. D., & Widiatmoko, D. (1998). The cost-effectiveness of home assessment and modification to reduce falls in the elderly. *Australian and New Zealand Journal of Public Health, 22*(4), 436–440.

● NOTES

1. These costs included hospitalization costs, other health care costs in institutional settings, other health care costs provided at home, and informal care costs.

2. These costs included the cost of the home modifications and the related occupational therapist expenses.

3. One might argue that this would be too high an amount, given that some caregivers would be unemployed or retired, yet there is value in the hours of work provided by a caregiver.

CHAPTER 8

PERFORMANCE MEASUREMENT AS AN APPROACH TO EVALUATION

● INTRODUCTION

Measuring the performance of programs, organizations, governments, and the people who work in them is becoming a universal expectation in the public and nonprofit sectors. In the last 20 years, there has been phenomenal growth in the attention and resources being devoted to **performance measurement**. This has been connected to a broad shift in expectations about the roles and responsibilities of public and nonprofit organizations, and their managers in particular. There appears to be a general consensus that pressures such as globalization of trade, public debt burdens, citizen dissatisfaction with public services, limited gains in public service efficiencies, and advances in information technology have led many governments to adopt a pattern of public sector reforms which includes measuring performance (see, for example, Borins, 1995; Gruening, 2001; Hood, 2000; Shand, 1996).

Increasingly, managers and organizations are expected to be accountable for achieving intended (and stated) outcomes. Emphasis on process—following the rules and compliance with authoritative directives—is being supplemented and, in some cases, supplanted by an emphasis on identifying, stating, and achieving objectives; planning and operating in a business-like manner; and, like businesses, being accountable for some surrogate for a "bottom line." For many public and nonprofit managers, evaluative criteria such as **value-for-money** and **cost-effectiveness** are intended to link resources to outcomes and to produce evidence that is analogous to private sector measures of success.

Managing for results entails being able to *measure* the extent to which intended results have been achieved. Performance measurement is generally seen to be a principal means of providing information that can be used to see whether and how well organizations and program managers have accomplished what they intended. Feedback via **performance measures** can be used for at least two broad purposes: making adjustments in the process that produces outcomes (**formative** uses) and reporting the actual results (outcomes) to stakeholders, including the public, elected officials, and governing boards (**summative** uses).

In Chapter 10, we examine the incentives for managers to develop and use performance measures, particularly as they assess the perceived benefits and costs that arise from combining formative and summative uses of performance measurement in the same system. Related to these issues is the important matter of how to sustain performance measurement systems in public and nonprofit organizations.

GROWTH OF PERFORMANCE MEASUREMENT •

Performance measurement is not new. Historically, it has been connected primarily with financial accountability—being able to summarize and report the ways that resources have been expended in a given period of time. The accounting profession emerged as the need for financial accountability grew—organizations became more complex and the regulation of ways that financial reporting occurred resulted in a need for expertise in assessing/ evaluating the completeness, honesty, and fairness of the "books" in organizations (Hopwood & Miller, 1994). In addition, because accounting emphasized a systematic description of the dollar values of resources that were expended in complex organizations, efforts to improve efficiency depended on being able to rely on the dollars expended for activities. Accounting provided a framework for calculating the inputs in efforts to estimate efficiency and productivity ratios.

While we have tended to situate the beginnings of performance measurement in the United States, in the 1960s, with the development of performance management systems such as programmed planned budgeting systems (PPBS) (Perrin, 1998), there is good evidence that performance measurement and reporting was well-developed in some American local governments early in the 20th century. Williams (2003) discusses the development of performance and productivity measurement in New York City, beginning as early as 1907 with the creation of the Bureau of Municipal Research. The bureau had a mandate to gather and report statistical data on the costs, outputs and some outcomes (infant mortality rates, for example) of municipal service delivery activities.

One of the innovations that the bureau instituted was an annual "Budget Exhibit"—a public display in city hall of the annual performance report that included:

> Facts and figures graphically displayed, intermingled with physical objects [that] informed the visitor of the city's activities—what had been and what was expected to be done with the taxpayer's money. (Sands & Lindars, 1912, p. 148, quoted in Williams, 2003)

Because the bureau adapted the accounting system to make it possible to calculate the cost of service activities, it was possible to produce information on the unit costs of services delivered. By comparing these figures over time or across administrative units, it was possible to track and improve efficiency. Reducing waste by improving productivity meant that taxpayers were getting better value for their dollars. Reporting publicly meant that New York administrative departments, the mayor, and the council were more accountable—valid and reliable information was available in a form that was intended to be accessible to the public.

By 1916, there were comparable bureaus of municipal research in 16 northeastern U.S. cities—each having a similar mandate as the New York bureau. This movement to measure and report local government performance extended to the first years of World War II (Williams, 2003). Understanding the successes and limitations of performance measurement (see Chapter 10) needs to reference the reasons why, in the public sector, results-based performance measurement originated in local governments. Because local governments before World War II generally included services where both costs and results were relatively easy to measure (services like roads and streets maintenance, sewer and water supply and distribution, and waste removal), it was relatively easy to calculate efficiency ratios—cost per lane mile of streets maintained, for example.

As concerns over public sector efficiency grew, measuring performance became an increasingly important government priority in the United States. In the 1960s the major movement was to develop and implement planned programmed budgeting systems (Perrin, 1998). PPBS were developed at the federal level in the U.S. as a way to link budgetary expenditures to program results. Government organizations implementing PPBS were expected to conceptualize clusters of activities as programs (programs did not need to coincide with organizational units or sub-units), each with identifiable objectives. Clusters of programs, conceptualized as open systems, contributed to organizational goals, and clusters of goals contributed to broader sectoral goals.

A key part of building PPBS was identifying performance measures for each program, and that in turn depended on specifying objectives that could be translated into measures. The overall intention of these systems was to be able to specify the costs of programs and link costs to results so that measures of efficiency and cost-effectiveness could be obtained and reported. Through a public reporting process wherein elected decision makers were principal clients for this information, advocates believed that the input-focused line-item budgeting process could be transformed to focus on the linkages between expenditures and results. The belief that data on efficiency and effectiveness would present decision makers with information to improve efficiency and effectiveness was a key part of the rationale underlying PPBS. The logic of this argument had a major impact on the ways that American (and later Canadian) jurisdictions conceptualized their budgeting processes. Although PPBS encountered what would eventually be insurmountable implementation problems—time and resources limitations; difficulties in defining objectives and measuring results; lack of management information capacity; and lack of accounting system capacity to generate meaningful program costs—the basic idea of relating governmental costs to results has endured.

Implementation problems resulted in a general abandonment of PPBS by the early 1970s in the United States (Perrin, 1998). Canada persisted with its own version at the federal level until the late 1970s (Savoie, 1990). Lack of success with PPBS did not, however, result in the abandonment of systems that focused on results. Successors to PPBS included zero-based budgeting (ZBB) which was similar to PPBS in linking costs to results but insisted that program budgets be build from a zero base rather than the previous year's base; and management by objectives (MBO) which emphasized the importance of stating clear objectives and making them the focus of organizational activities (Perrin, 1998). These alternatives did not prove to be any more durable than PPBS, but the core idea—it is desirable to focus on results—survived and has since flourished as a central feature of contemporary performance measurement systems.

Government Deficits and the Transformation of Public Expectations for Governments

A key feature of the fiscal environment in which governments operated in the 1970s and early 1980s was large and persistent operating deficits. Indeed, one of the appeals of ZBB was its emphasis on deconstructing organizational budgets and demanding a rationale for the full amount, and not just the incremental increase or decrease. The scale of government activities consistently outstripped revenues and the combination of deficits and inflation produced an environment wherein analysts and public officials were looking for ways to balance budgets. In Britain, the election of Margaret Thatcher in 1979 was a turning point in that country. Her government systematically restructured the way that the public sector operated. It emphasized reductions in expenditures and, hence, the scope and scale of government activities. It introduced competition in the production and delivery of services, and it generally articulated a view that emphasized the importance of diminishing the role of government in society and creating incentives and opportunities for the expansion of the private sector. Public accountability for results was also a key feature of the restructuring of government.

Taxpayers in the United States were making it increasingly clear that they were no longer willing to finance the growing demands for resources in the public sector. Proposition 13 in California in 1978 was a taxpayer initiative that effectively capped local government tax rates in that state. Similar initiatives spread rapidly across the United States so that during the late 1970s, 23 states experienced local or state legislation aimed at limiting government expenditures (Danziger & Ring, 1982).

This taxpayers' revolt was accompanied by a different vision of government. In the United States, Osborne and Gaebler's (1992) book *Reinventing Government* articulated 10 principles that they saw exemplified in successful government organizations, and amounted to a different philosophy of government:

1. Government should steer rather than row, creating room for alternatives to the public sector delivery of services.

2. Government should empower citizens to participate in ownership and control of their public services.

3. Competition among service deliverers is beneficial, creating incentives for efficiency and enhancing accountability.

4. Governments need to be driven by a mission, not by rules.

5. Funding should be tied to measured outcomes rather than inputs, and performance information should be used to improve results.

6. Meet the needs of customers rather than focusing on interest groups and the needs of the bureaucracy.

7. Foster enterprise in the public sector, encouraging generation of funds, rather than just spending.

8. Focus on anticipating and preventing problems and issues rather than remediating them (relatedly, strategic planning is essential to drive the framework for managing performance).

9. Use a participatory and decentralized management approach, building on teamwork and encouraging innovation.

10. Use market mechanisms to achieve public purposes.

The combination of criticisms of existing ways that governments operated, including government deficits, and an emerging unwillingness to finance higher levels of public sector expenditures contributed to a transformation of public sector management thinking and practice. What emerged was a management approach that took up the challenges of downsizing, reduction of "red tape," and more emphasis on efficiency and effectiveness. The focus on managing for results which had threaded its way through public sector innovations and reforms from the early years of the 20th century was now combined with the fiscal imperative, and a broad belief that private sector business practices needed to be emulated in the public sector.

The **new public management** (NPM), which has emerged as the principal framework within which contemporary administrative reforms have been articulated, has become a central feature of the administrative and governance landscape in the United States, Canada, and most Western democracies (Shand, 1996). Hood (1991) traces its origins to the mid-1970s when governments were wrestling with deficits and looking for ideas that offered the promise of limiting government growth and introducing market-like discipline into the ways that services were delivered. Although NPM has been and continues to be controversial (Denhardt & Denhardt, 2003; Hood, 1991; Savoie, 1995), it now dominates contemporary thinking about the design, implementation, and operation of government programs and services.

Similar to the 10 principles articulated by Osborne and Gaebler (1992), NPM centers around a core set of components

> providing high-quality services that citizens value; increasing the autonomy of public managers, particularly from central agency controls; measuring and rewarding organizations and individuals on the basis of whether they meet performance targets; making available the resources that managers need to perform well; and, appreciative of the virtues of competition, maintaining an open-minded attitude about which public purposes should be performed by the private sector, rather than the public sector. (Borins, 1995, p. 122)

This package clearly expresses a focus on results and offers an ideal for government reform. This approach has been widely adopted in the United States, Canada, and other Western countries and accounts for the current and widespread emphasis on performance measurement in the public sector. In the United States, a concurrent driver of the reforms that focused on improving the efficiency and effectiveness of government operations has been the Governmental Accounting and Standards Board (GASB). Created in 1984 "to establish standards of financial accounting and reporting for state and local government entities" (Governmental Accounting Standards Board, 2002, p. 1), the GASB has increasingly championed user-friendly performance reporting as a component of financial reporting. It has provided information on governing-for-results practices and standards and has supported nonfinancial performance reporting of Service Efforts and Accomplishments (SEA) (Burnaby & Fountain, 1994).

Performance measurement and reporting in the United States has been legislated in most states and in the federal government. Congress passed the

Government Performance and Results Act (GPRA) in 1993 to mandate results-based management in U.S. federal departments and agencies. Starting with pilot implementations, departments and agencies are now expected to develop and implement performance measures and reporting capabilities government-wide, linked to 5-year strategic plans. Requirements of the GPRA have survived the transition to a new administration and implementation continues. Although implementation issues have arisen, performance measurement and reporting continue to be centerpieces of federal accountability requirements (Feller, 2002).

In Canada, although there is no comparable legislation to the GPRA, there are Treasury Board expectations, at the federal level through the *Management Accountability Framework,* that government departments identify performance measures and report on their success in achieving stated objectives, as part of the budgetary process (Treasury Board of Canada Secretariat, 2003). More than half of Canadian provinces have legislated performance measurement and reporting requirements (Manitoba Office of the Auditor General, 2000).

More recently, the nonprofit sector in the United States and Canada has been expected to emulate the changes that have occurred in the public sector. New public management precepts have become a part of a general movement toward contractual relationships between funders and service providers (Eikenberry & Kluver, 2004; Scott, 2003). Measuring and reporting performance is a key part of this process.

● METAPHORS THAT SUPPORT AND SUSTAIN PERFORMANCE MEASUREMENT

Measuring the performance of public and nonprofit programs and organizations is rooted in metaphors that shape our perceptions and beliefs about what is possible and desirable as we conceptualize government and nonprofit activities. These metaphors serve as "theories" or models that guide management change efforts (Morgan, 1997) and suggest a rationale for pursuing performance measurement. In this chapter we summarize three metaphors that have important influences on how government organizations and programs are "seen": as a business, as an open system, and as a machine. These metaphors have widespread appeal, being rooted in our everyday experiences. Applying them makes it possible for various stakeholders to better comprehend the meaning of the key features of designing, implementing, and using performance measurement systems.

Government as a Business

A metaphor that has come to dominate our thinking about government and, increasingly, nonprofit organizations is the belief that government is a business. This metaphor, which has guided efforts to infuse business practices into governments in North America, the U.K., and Australasia (Pollitt, 1998), emphasizes the importance of clearly stated objectives, programs that are planned and managed to achieve those objectives, efficiency (including the positive effects of competition, privatization, contracting out), and, ultimately, attention to the bottom line. Performance measures are a key part of a management philosophy that emphasizes results and encourages managers to manage for results.

It is clear that the new public management embraces many of these same values. But business thinking in and for the public sector predates NPM, being traceable in part to the Reform Movement in the United States. With respect to state and local governments, the Reform Movement emerged as a response to the widespread concerns with political corruption and machine politics in American state and local governments around the turn of the 20th century. In Woodrow Wilson's (1887) article, "The Study of Administration," he exemplified the efforts of reformers who wanted to introduce political and organizational changes that would eliminate the perceived ills of U.S. public sector governance:

> Bureaucracy can exist only where the whole service of the state is removed from the common political life of the people, its chiefs as well as its rank and file. Its motives, its objects, its policy, its standards, must be bureaucratic. (p. 217)

A key part of this movement was its emphasis on business-like practices for government organizations. Indeed, the creation of New York's Bureau of Municipal Research was a part of this transformation of local government. Performance reporting was intended to provide information to key decision makers (the mayor) that would result in improved efficiency.

Organizations as Open Systems

In Chapter 2, we introduced and discussed the **open-systems metaphor** and its effects on the way we see programs. "Open systems" has become a dominant way that managers and analysts have come to view programs and organizations (Morgan, 1997) and has exerted a major influence on the way we structure program evaluations.

The key originating source of the open-systems metaphor is the biological metaphor (Bertalanffy, 1968). Gareth Morgan (1997) introduces the biological metaphor by pointing out that it is perhaps the dominant way that organizations are now seen. Looking first at the biological domain, organisms interact with their environments as open systems, and have structures that perform functions that, in turn, contribute to a goal of homeostasis (the ability to maintain a steady state in relation to fluctuations in the environment) (Morgan, 1997).

Biological organisms, to maintain themselves, operate within certain parameters. For example, warm-blooded animals have species-specific ranges of normal body temperature. Fluctuations above or below the normal range indicate that the organism is "not well," and if the temperature is not corrected, permanent damage or even death will result.

Although we generally do not explicitly assert that organizations are organisms, the biological/open-systems metaphor exerts an important influence on the way we think about measuring performance. In a report that was a part of the development of the performance measurement system in British Columbia, Canada (Auditor General of British Columbia & Deputy Ministers' Council, 1996), the State of Oregon's exemplary efforts to create state-wide performance measures and **benchmarks** is summarized this way:

> The State of Oregon has generally been recognized as one of the leading jurisdictions in reporting state-wide accountability information. It has defined a wide range of benchmarks to use as indicators of the progress that the state has had in achieving its strategic vision. *Just as blood pressure, cholesterol levels and other such indicators serve as signs of a patient's health,* benchmarks serve as signs of Oregon's vision of well-being in terms of family stability, early childhood development, kindergarten to grade 12 student achievement, air and water quality, housing affordability, crime, employment and per capita income. (Auditor General of British Columbia & Deputy Ministers' Council, 1996, p. 70, italics added)

It is appealing to compare the process of indicating a person's health to the process of indicating the "health" of a government, a public organization, an economy, or even a society. We generally agree that blood pressure is a valid measure of our physical health. We have well-established theories backed by much evidence that departures from the accepted ratios of diastolic to systolic pressure, as measured by blood pressure cuffs, result in health problems. Our research and experience with normal and abnormal blood pressures have established widely accepted benchmarks for this performance measure.

Using this metaphor as a basis for measuring public sector performance suggests that we also have a good understanding of the cause and effect linkages in programs and even whole organizations such that a performance measure or a combination of them will indicate "how well" the organization is doing. Finding the right performance measures, then, will be a powerful shorthand way to monitor and assess complex organizations.

Organizations as Machines

There is a third metaphor, that of organizations as machines, that has emerged as part of the intuitive and visual basis for performance measurement. Claiming that a suite of performance measures is to an organization as the instruments on a dashboard are to an automobile suggests that the appropriate performance measures can provide users with "at a glance" readings which are similar in functionality to those obtained with the instruments and gauges in cars or airplanes. In France, for example, the Tableau de Bord was developed early in the 20th century, originally as a "dashboard" system of indicators used by managers to monitor the progress of the business (Epstein & Manzoni, 1998). Similarly, the balanced scorecard approach (Kaplan & Norton, 1996) provides for cascading performance indicators nested within a framework of four interrelated perspectives: financial, customer, internal business process, and learning and growth.

If a suite of performance measures is a dashboard, then a public organization is a machine that consists of complex but understandable systems and subsystems which are linked and can be monitored in valid and reliable ways. By measuring analogues to indicators like compass direction (congruence with strategic objectives), managers and other stakeholders are expected to be able to "fly" or "drive" their organizations successfully.

The business, open-systems/biological, and even the machine metaphors combine to produce a strong intuitive rationale when we design, implement, and assess performance measures for public and nonprofit programs and organizations. Morgan (1997) states:

> [W]e see how the insights of different metaphors can contribute to a rich understanding of the situations with which we are trying to deal, suggesting their own favored methods of tackling the issues at hand. In approaching the same situation in different ways they extend insight and suggest actions that may not have been possible before. (pp. 352–353)

If these metaphors or combinations of them facilitate a successful mapping of key features of the metaphor onto the organizations or programs for which performance measures are being constructed, then the metaphors have done their job. Suitably constructed measures will track progress toward achieving objectives/intended outcomes, at what cost, and will provide that information to managers, who can use it formatively, and to executives and other stakeholders, who can use it summatively.

● COMPARING PROGRAM EVALUATION AND PERFORMANCE MEASUREMENT

In the first several chapters of this book, we suggested that basic program evaluation tools were also useful for performance measurement. Logic models (Chapter 2) can be used to construct models of programs or organizations and in so doing, identify key constructs that are included in cause and effect relationships which predict intended outcomes. Research designs (Chapter 3) focus our attention on the attribution question and guide analysts and managers in their efforts to interpret and report performance measurement results. Measurement (Chapter 4) outlines criteria that can guide the process of translating constructs (in logic models, for example) into measures for which data can be collected. Together, these three chapters focus on knowledge and skills that can be adapted by managers and evaluators who are involved in designing, implementing, or assessing performance measurement.

Because core program evaluation knowledge and skills can be adapted to design and implement performance measurement systems, it is clear that there is substantial overlap between these two important evaluation approaches. Persons trained as program evaluators usually have core skills that are needed to design and implement performance measurement systems (Scheirer & Newcomer, 2001).

The growth of performance measurement and its implementation in settings where resources are constrained or even diminished has prompted some managers to question the value of program evaluation—treating it as an "expensive luxury" (Scheirer & Newcomer, 2001). In contrast, some analysts have emphasized the importance of both approaches to evaluation, pointing out that they are mutually reinforcing (Wholey, 2001).

In this textbook the two approaches are presented as complementary evaluation strategies. Both program evaluation and performance measurement are a part of the performance management cycle that was introduced in Chapter 1. In that cycle, they are both intended to be a part of the feedback loop that reports, assesses, and attributes outcomes of policies and programs.

Table 8.1   Comparisons Between Program Evaluation and Performance Measurement

Program Evaluation	Performance Measurement
1. Episodic	Ongoing
2. Issue-specific	Designed and built with more general issues in mind. Once implemented, performance measurement systems are generally suitable for the broad issues/questions that were anticipated in the design
3. Measures are usually customized for each program evaluation	Measures are developed and data are usually gathered through routinized processes for performance measurement
4. Attribution of observed outcomes is usually a key question	Attribution is generally assumed
5. Targeted resources are needed for each program evaluation	Because it is ongoing, resources are usually a part of the program or organizational infrastructure
6. Program evaluators are not usually program managers	Program managers are usually expected to play a key role in developing performance measures and reporting performance information
7. The intended purposes of a program evaluation are usually negotiated up front	The uses of the information can evolve over time to reflect changing information needs and priorities

Although some analysts have argued that performance measurement can serve many of the same purposes that program evaluations have served (Mayne, 2001), there are important differences between program evaluation and performance measurement. This section of the chapter offers comparisons between evaluation and performance measurement on a number of important criteria. By contrasting performance measurement with program evaluation, it is possible to offer an extended definition of performance measurement as an approach to evaluation.

Table 8.1 summarizes key distinctions between program evaluation and performance measurement. Some of these distinctions have been noted by analysts who discuss the relationships between the two approaches to evaluation (Scheirer & Newcomer, 2001; General Accounting Office, 1998). Each of the comparisons in the table is discussed more fully in the subsections that follow.

1. Program evaluations are episodic, whereas performance measurement is ongoing.

Typically, program evaluations are circumscribed by time. As was indicated in Chapter 1, a program evaluation is a process that has a starting point, which can be driven by ad hoc information needs or by organizational policies governing periodic evaluations of all programs. Developing the terms of reference for a program evaluation typically marks the beginning of the process, and reporting the evaluation findings, conclusions, and recommendations typically marks the end point of a given program evaluation.

Performance measurement systems are designed and implemented with the intention of providing regular and continuing monitoring capacity and information for program and organizational purposes. Once implemented, they usually become part of the information infrastructure in an organization. Current information technologies make it possible to establish databases and update them routinely. As long as the data are valid, reliable, and complete, they can be used by managers and other stakeholders at their desktops and to generate periodic reports.

2. Program evaluations are issue-specific, whereas performance measurement systems are designed with more general issues in mind.

Program evaluations are usually developed to answer questions that emerge from stakeholder interests in a program at one point in time. The clients of the evaluation are identified, terms of reference for an evaluation are developed and tailored to that project, and resources are usually mobilized to do the work and report the results.

Some organizations have an ongoing infrastructure that supports program evaluation and, when evaluations are required, use their own people and other resources to do the work. Even where infrastructure exists and a regular cycle is used to evaluate programs, there is almost always a stage in the process wherein the terms of reference are negotiated by key stakeholders in the evaluation process—these terms of reference are at least partially individualized and situation-specific.

Contrast this with performance measurement systems that are intended to be ongoing information-gathering and dissemination mechanisms. Typically, developing performance measures entails agreement in advance on what general questions or issues will drive the system, and, hence, what is important to measure. Examples of general questions might include: *What are the year-over-year trends in key outcomes?* and *How do these trends conform to annual or multi-year targets?* Key **constructs** in a program and/or an organizational logic model can be identified, measures developed, and processes developed to collect, store, analyze, and report the data. Once a system is in place, it typically functions as a part of the organization's information infrastructure, and unless its proponents go back and make changes, the measures, the data structure, and the reporting structure stand.

3. For program evaluations, measures are at least partially customized for each evaluation, whereas for performance measurement, measures are developed and data are gathered through routinized processes.

Since the terms of reference are usually specific to each program evaluation, the evaluation issues and the data needed to address each issue are also individualized. The measures and the comparisons needed to answer key questions typically require a mixture of **primary** and **secondary data** sources. Primary data (and the instrumentation to collect them) will reflect issues and questions that that evaluation must address. Secondary data that already exist in an organization can be adapted to an evaluation, but it is rare for an evaluation to rely entirely on preexisting data.

Performance measurement systems tend to rely heavily on existing sources of data, and the procedures for collecting those data will typically be built into organizational routines. Program managers often have a role in the data collection process, and program-level data may be aggregated upwards to construct organizational performance measures. Even where primary data are being collected for a performance measurement system, procedures for doing so are usually routinized, permitting periodic comparisons of the actual performance results and, usually, comparisons between actual and targeted results.

For example, the Workers' Compensation Board (WCB) in British Columbia, Canada, regularly collects client satisfaction data as part of its monthly performance measurement process. Each time a survey occurs, a random sample of about 200 clients is drawn and a private polling company (under contract to the WCB) is hired to administer a pre-set survey constructed so that satisfaction ratings with WCB service can be scaled from 1 to 10 and compared over time and across administrative regions. These are posted around the headquarters building in Richmond, British Columbia so that employees can see how client satisfaction is tracking over time. The targets for satisfaction ratings are typically an average rating of 9.0 out of a possible 10. Visually, the actual averages can be compared to the targets.

4. For program evaluations, research designs and the comparisons they entail are intended as ways to get at the attribution issue, whereas for performance measurement systems, attribution is generally assumed.

The history of program evaluation is replete with discussions of ways to design evaluations to make it possible to isolate and measure the actual outcomes of a program (Scriven, in progress, unpublished). In Chapter 3, we introduced and discussed the importance of **internal validity** in evaluation research designs. Although there is by no means universal agreement on the centrality of internal validity as a criterion for defensible research designs, the practice of program evaluation has generally emphasized the importance of sorting out the incremental effects of programs and being able to make statements about the extent to which observed outcomes are actually attributable to the program, as opposed to other, environmental causes.

The attribution problem can be illustrated succinctly with Figure 8.1, which has been adapted from an experimental design model for evaluating European Union expenditure programs (Nagarajan & Vanheukelen, 1997). A key question for program evaluators is whether the observed outcomes (in this figure, the creation of 75 new job placements) is due to the training program in question. The purpose of the comparison group is to "calibrate" the observed outcome of the program, by offering evidence of what *would have happened without the program*. This is sometimes called the **counterfactual condition**. For the comparison group, 50 found new jobs during the same time frame that the program group found 75 new jobs. Because both groups found jobs, the *incremental* outcome of the program was 25 jobs. In different words, 25 new jobs can be attributed to the program—the other 50 can be attributed to factors in the environment of the program.

Clearly, there are issues of the feasibility of creating and using comparison groups. One of the major criticisms of experimental or even quasi-experimental research designs is that they are simply not practical because comparison groups are not available. Program evaluators often find themselves in situations where options for comparisons are quite limited, and their own professional judgment becomes a part of the process of determining whether the program was effective.

But even in these situations, most program evaluators are aware that the attribution problem affects their confidence in the interpretation of the patterns they see in the data. That they cannot offer firm evidence that bears on

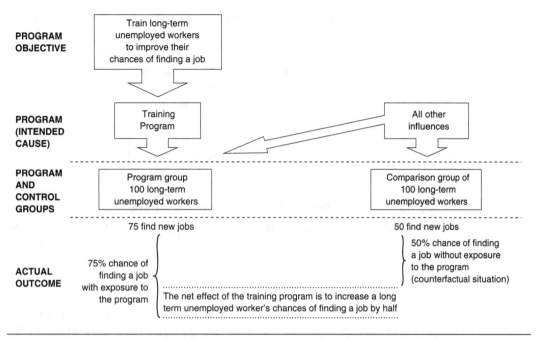

Figure 8.1 How an Internally Valid Research Design Resolves the Attribution Problem

this issue in a given evaluation means that conclusions must be tempered by the knowledge that observed outcomes may or may not be attributable to the program. Conclusions about program effectiveness can best be viewed as being tentative—a reduction in the uncertainty surrounding this core evaluation question in such situations.

In performance measurement systems, it is rare to build in the capacity to conduct comparisons that sort out outcome attribution (Scheirer & Newcomer, 2001). Typically, when performance measures are being developed, attention is paid to their **construct validity** (*are the measures valid indicators of constructs in the program logics?*). This is often a substantial challenge, given reliance on existing data sources, wherein measures must be adapted to "fit" the constructs that are important for the performance measurement system.

Recall the earlier mention of the Oregon benchmarking and performance measurement system (Auditor General of British Columbia & Deputy Ministers' Council, 1996). An organismic metaphor was invoked to provide a rationale for state-wide performance indicators. If we can use blood pressure as a measure of bodily well-being, and transposing the metaphor to the government of the State of Oregon is valid, then we might assert that using a measure like affordability of housing indicates the socio-economic well-being of Oregon citizens.

For persons developing performance measures, one challenge is to construct measures that are not only valid indicators of constructs (measure what they are intended to measure), but also give us an indication of *what the program or programs is/are actually doing,* under a variety of conditions.

Returning to the example of measuring blood pressure, because we know how blood pressure is **causally connected** to the circulatory and other systems in our bodies, we can be reasonably confident that measuring blood pressure will yield valid and useful information about our circulatory system and, less directly, about other systems in our body.

But if we think of programs, and of program outcomes in particular, an open-systems approach usually situates outcomes in the environment of programs. Thus, programs are intended to *produce* outcomes, and performance measures are intended to indicate whether and to what extent intended outcomes have been achieved.

Because outcomes occur in the environment of programs, other factors can and usually do affect outcome measures, besides the program. Unlike blood pressure, which indicates the performance of systems *within* our bodies, performance measures that focus on outcomes indicate the behavior of variables *outside* programs themselves. How well performance measures indicate what the program (or organization) actually accomplished depends in part on the types of measures chosen. Output measures, for example, are typically "closer" to the program process than outcomes and are often viewed as being more closely linked to what the program actually accomplished.

In Chapter 2 we introduced program technologies as a factor that helps to explain the likelihood that a well-implemented program will deliver its intended outcomes. For some programs (highways maintenance programs, for example), the **core technologies** are **high-probability**, meaning that resources will reliably be converted into results. Low-probability program technologies (a drug rehabilitation program, for example) are much less likely to directly produce intended outcomes because the "state of the art" of rehabilitating illegal drug users is more like a craft or even an art rather than an engineering science. The core technology of a program affects the likelihood that program outcomes can be "tracked" back to the program. Other things being equal (sometimes called the ***ceteris paribus*** assumption), observed program outcomes from high-probability programs will be more likely attributable to the program than outcomes from low-probability programs. In other words, attribution will be a more important problem for **low-probability program technologies**.

5. Targeted resources are needed for each program evaluation, whereas for performance measurement, because it is ongoing, resources are a part of the program or organizational infrastructure.

Program evaluations can be designed, conducted, and reported by contracted consultants, in-house staff, or both. Even where evaluations are conducted in-house, each evaluation typically includes a budget for primary data collection and other activities that are unique to that study.

Availability of resources for developing, implementing, and reporting from performance measurement systems can vary considerably but, typically, managers are expected to play a key role in the process. In organizations where budgets for evaluation-related activities have been reduced, managers often are expected to take on tasks associated with performance measurement as part of their work. Managerial involvement in developing performance measures makes good sense, since program managers are in a position to offer input, including pointing to measures that do a good job of capturing the work they are doing (outputs) and the results that are intended (outcomes).

6. For program evaluations, evaluators are usually not program managers, whereas for performance measurement, managers are usually key players in developing and reporting performance results.

Program evaluations in government organizations are typically conducted with the advice of a steering committee, which may include program managers among the stakeholders represented on such a committee. In nonprofit organizations, particularly smaller agencies, it is more likely that program managers will play a key role in conducting program evaluations.

Where program managers are involved in developing and implementing performance measures, the purposes for which the information from such a system is used can affect how managers see the incentives for participating. In the field of program evaluation, there is considerable discussion about the pros and cons of managerial involvement. Both Austin (1982) and Love (1991) have argued that managerial involvement is essential to ensuring that evaluations are "owned" by those who are in the best position to use them. Wildavsky (1979) has questioned whether evaluation and management can ever be joined in organizations. We look at this issue in more detail in Chapter 11.

7. The intended purposes of a given program evaluation are usually negotiated up front, whereas for performance measurement, the uses of the information will evolve over time to reflect changing information needs and priorities.

Program evaluations can be formative or summative in intent, but typically the purposes are negotiated as evaluations are planned. Terms of reference are usually established by the steering committee that is overseeing the program evaluation, and the process and product is expected to conform to the terms of reference.

Performance measurement systems can also be used formatively or summatively, but because they are ongoing processes in dynamic organizational environments, it is typical for the uses of information produced by the system to evolve as organizational information needs and priorities change. Because modern information technologies are generally flexible, it is often possible to re-focus a performance measurement system as information needs evolve. As well, typical performance measurement systems will include measures that are deemed to be less useful over time, and managers or other stakeholders will add measures as their relevance increases.

The Province of Alberta in Canada, for example, had been measuring provincial performance since 1995, and each year reports performance results for a set of province-wide measures. Each year, some of the measures are revised for future reporting opportunities (Alberta Finance, 2004).

● SUMMARY

Performance measurement in the United States began in local governments at the turn of the 20th century where it was possible to measure both costs and outputs of local government services. Since that time, it has been a part of different governmental reforms that extend up to the present day. In previous reform movements (principally in the 1960s and 1970s), although each government reform package did not realize its intended results, performance measurement survived. Most recently, the new public management reform movement, which is now a part of government reforms in the United States, Canada, Europe, Australasia, and increasingly, the rest of the world, has made performance measurement a central feature of a broad expectation that governments be accountable for their results.

Performance measurement and program evaluation are complementary ways of acquiring and analyzing information that is intended to reduce the uncertainty around program and policy decisions. They both rely on a common core of methodologies that is discussed in Chapters 2, 3, and 4. The differences between these two approaches to evaluating programs reflect a growing trend toward integrating evaluation databases into the information infrastructure of organizations. Managers are increasingly adopting a "just in time" stance with respect to acquiring, analyzing, and reporting evaluation information.

Although performance measurement can be a cost-effective alternative to program evaluation research, particularly where the goal is to describe patterns of actual program and/or organizational results ("what happened?" questions), it does not allow the user to directly address questions of why observed program results occurred.

DISCUSSION QUESTIONS ●

1. Why did performance measurement have its origins in local governments in the United States?

2. One of the metaphors that has provided a rationale for performance measurement in government is that government is a business. In what ways is government business-like?

3. What are some of the differences between governments and businesses? How would those differences affect performance measurement?

4. Assume that you are an advisor to a public-sector organization, with several hundred employees, that delivers social service programs. At present, the organization does not have the capability to do program evaluations or measure program performance. Suppose that you are asked to recommend developing either program evaluation capabilities (putting resources into developing the capability of conducting program evaluations) or performance measurement capability (putting resources into measuring key outputs and outcomes for programs), as a first step in developing evaluation capacity. Which one would you recommend developing first? Why?

5. What are the key performance measures for a driver of an automobile in a city? In other words, what would you want to know about the driver's performance to be able to decide whether he or she was doing a good job of driving the vehicle? Where would the data come from for each of your measures?

● REFERENCES

Alberta Finance. (2004). *Government of Alberta annual report 2003–04: Measuring up.* Edmonton, Alberta: Government of Alberta.

Auditor General of British Columbia and Deputy Ministers' Council. (1996). *Enhancing accountability for performance: A framework and an implementation plan—Second joint report.* Victoria: Queen's Printer for British Columbia.

Austin, M. J. (1982). *Evaluating your agency's programs.* Beverly Hills, CA: Sage.

Bertalanffy, L. V. (1968). *General system theory: Foundations, development, applications* (Rev. ed.). New York: G. Braziller.

Borins, S. (1995). The New Public Management is here to stay. *Canadian Public Administration/Administration Publique Du Canada, 38*(1), 122–132.

Burnaby, P. A., & Fountain, J. R., Jr. (1994). Service efforts and accomplishments: Its time has come. *Government Accountants Journal, 43*(3), 43–53.

Danziger, J. N., & Ring, P. S. (1982). Fiscal limitations: A selective review of recent research. *Public Administration Review, 42*(1), 47–55.

Denhardt, J. V., & Denhardt, R. B. (2003). *The new public service: Serving, not steering.* Armonk, NY: M. E. Sharpe.

Eikenberry, A. M., & Kluver, J. D. (2004). The marketization of the nonprofit sector: Civil society at risk? *Public Administration Review, 64*(2), 132–140.

Epstein, M., & Manzoni, J.-F. (1998). Implementing corporate strategy: From tableaux de bord to balanced scorecards. *European Management Journal, 16*(2), 190–203.

Feller, I. (2002). Performance measurement redux. *American Journal of Evaluation, 23*(4), 435–452.

Governmental Accounting Standards Board. (2002). *Facts about GASB.* Retrieved July 23, 2004, from http://www.gasb.org/facts/index.html

Gruening, G. (2001). Origin and theoretical basis of new public management. *International Public Management Journal, 4*(1), 1–25.

Hood, C. (1991). A public management for all seasons? *Public Administration, 69*(1), 3–19.

Hood, C. (2000). Paradoxes of public-sector managerialism, old public management and public service bargains. *International Public Management Journal, 3*(1), 1–22.

Hopwood, A. G., & Miller, P. (Eds.). (1994). *Accounting as social and institutional practice.* Cambridge, UK: Cambridge University Press.

Kaplan, R. S., & Norton, D. P. (1996). *The balanced scorecard: Translating strategy into action.* Boston: Harvard Business School Press.

Love, A. J. (1991). *Internal evaluation: Building organizations from within.* Newbury Park, CA: Sage.

Manitoba Office of the Auditor General. (2000). *Inter-jurisdictional comparison on trends and leading practices in business planning and performance measurement.* Winnipeg, Manitoba: Queen's Printer.

Mayne, J. (2001). Addressing attribution through contribution analysis: Using performance measures sensibly. *Canadian Journal of Program Evaluation, 16*(1), 1–24.

Morgan, G. (1997). *Images of organization* (2nd ed.). Thousand Oaks, CA: Sage.

Nagarajan, N., & Vanheukenen, M. (1997). *Evaluating EU expenditure programmes: A guide.* Luxembourg: Office for Official Publications of the European Communities.

Osborne, D., & Gaebler, T. (1992). *Reinventing government: How the entrepreneurial spirit is transforming the public sector.* Reading, MA: Addison-Wesley.

Perrin, B. (1998). Effective use and misuse of performance measurement. *American Journal of Evaluation, 19*(3), 367–379.

Pollitt, C. (1998). Managerialism revisited. In B. G. Peters & D. J. Savoie (Eds.), *Taking stock: Assessing public sector reforms* (pp. xi, 417). Montreal: McGill-Queen's University Press.

Sands, H. R., & Lindars, F. W. (1912). Efficiency in budget making. *Annals of the American Academy of Political and Social Science, XLI,* 138–150.

Savoie, D. J. (1990). Reforming the expenditure budget process: The Canadian experience. *Public Budgeting and Finance, 10*(3), 63–78.

Savoie, D. J. (1995). What is wrong with the New Public Management. *Canadian Public Administration-Administration Publique Du Canada, 38*(1), 112–121.

Scheirer, M. A., & Newcomer, K. (2001). Opportunities for program evaluators to facilitate performance-based management. *Evaluation and Program Planning, 24*(1), 63–71.

Scott, K. (2003). *Funding matters: The impact of Canada's new funding regime on nonprofit and voluntary organizations, summary report.* Ottawa, ON: Canadian Council on Social Development.

Scriven, M. (In progress, unpublished). *Causation.* New Zealand: University of Auckland.

Shand, D. A. (1996). *Performance auditing and the modernisation of government.* Paris: Organisation for Economic Co-operation and Development.

Treasury Board of Canada Secretariat. (2003). *Management accountability framework.* Retrieved July 20, 2004, from http://www.tbs-sct.gc.ca/maf-crg/dwnld/maf-crg_e.pdf

General Accounting Office. (1998). *Performance measurement and evaluation: Definitions and relationships* (GAO/GGD-98-26).Washington, DC: United States GAO.

Wholey, J. S. (2001). Managing for results: Roles for evaluators in a new management era. *American Journal of Evaluation, 22*(3), 343–347.

Wildavsky, A. B. (1979). *Speaking truth to power: The art and craft of policy analysis.* Boston: Little Brown.

Williams, D. W. (2003). Measuring government in the early twentieth century. *Public Administration Review, 63*(6), 643–659.

Wilson, W. (1887). The study of administration. *Political Science Quarterly, 2*(2), 197–222.

CHAPTER *9*

DESIGN AND IMPLEMENTATION OF PERFORMANCE MEASUREMENT SYSTEMS

INTRODUCTION ●

The process of designing and implementing performance measurement systems uses core knowledge and skills that are also a part of designing, implementing, and reporting program evaluations. In Chapter 8, we pointed out that program evaluation and performance measurement share core attributes including logic modeling and measurement. In addition, understanding research designs and how the four kinds of validity affect the quality of program evaluations is a valuable set of tools for appreciating the strengths and limitations of performance measurement systems.

In Chapter 1, we outlined steps that comprise a typical program evaluation. In this chapter we will do the same for performance measurement systems, understanding that for each situation, there will be unique circumstances that can result in differences between the checklist below and the process that is appropriate for a given context. Each of the 12 steps in designing and implementing a performance measurement system is elaborated to clarify issues and possible problems. We distinguish designing and implementing performance measurement systems from the uses for such systems. System usage is a critical topic on its own, enough so that we discuss uses and related issues in Chapter 10.

Given the ubiquity of performance measurement in the public sector and, increasingly, in the nonprofit sector, the *sustainability* of performance measurement systems is an important issue. Although the steps discussed in this chapter are intended to outline a process that optimizes the chances that a performance measurement system will be successfully implemented and sustained, the conditions under which actual organizations undertake this change are often less than ideal. In the Summary, we identify the steps that are most critical if organizations want sustainable performance measurement systems that can contribute to managerial and organizational efforts to improve efficiency, effectiveness, and accountability.

KEY STEPS IN DESIGNING AND IMPLEMENTING ● A PERFORMANCE MEASUREMENT SYSTEM

Table 9.1 summarizes the key steps in designing and implementing a performance measurement system. Each of these steps can be viewed as a guideline—no single performance measurement development and implementation process will conform to all of them. In some cases, the process may diverge from the sequence of steps. Again, this could be due to local factors. Each of the steps in Table 9.1 is discussed more fully in the following sections.

Table 9.1   Key Steps in Designing and Implementing a Performance Measurement System

1. Identify the organizational champions of this change.

2. Understand what a performance measurement system can and cannot do, and why it is needed.

3. Establish multi-channel ways of communicating that facilitate top down, bottom up, and horizontal sharing of information, problem identification, and problem solving.

4. Clarify the expectations for the uses of the performance information that will be created.

5. Identify the resources available for developing, implementing, maintaining, and renewing the performance measurement system.

6. Take the time to understand the organizational history around similar initiatives.

7. Develop logic models for the programs or lines of business for which performance measures are being developed.

8. Identify additional constructs that are intended to represent performance for aggregations of programs or the whole organization.

9. Involve prospective users in reviewing the logic models and constructs in the proposed performance measurement system.

10. Measure the key constructs in the performance measurement system.

11. Record, analyze, interpret, and report the performance data.

12. Regularly review feedback from users and, if needed, make changes to the performance measurement system.

Identify the Organizational Champions of This Change

The introduction of performance measurement, particularly measuring outcomes, is an important change to both the organization's way of doing business and to its culture (de Lancer Julnes & Holzer, 2001). Unlike program evaluations, performance measurement systems are ongoing, and it is therefore important that there be organizational leaders who are champions of this change, to provide continuing support to the process, from its inception onward. In many cases, the recent emphasis on outcome measurement is a significant departure from existing practices of tracking program inputs (money, human resources), program activities, and program outputs (work done). Most managers have measured/recorded inputs, processes, and even outputs for some time so the challenge in outcome-focused performance measurement is in specifying the expected outcomes (stating clear objectives for programs, lines of business, or organizations) and facilitating organizational commitment to the process of measuring and reporting actual outcomes.

Including outcomes, performance measurement commits organizations to comparing their actual results to stated objectives. In many jurisdictions, objectives are parsed into annual targets and actual outcomes are compared to the targets for that year. Thus, the performance measurement information commonly serves multiple purposes, such as enhancing managerial decision making, encouraging organizational alignment, and promoting transparency and accountability.

As we said in Chapter 8, the new public management emphasizes the importance of freeing managers from "red tape" so that they can more efficiently and effectively use the resources that are available (Norman & Gregory, 2003). The Government Performance and Results Act (GPRA) in the United States creates an option for managers to request managerial flexibility waivers from the Office of Management and Budget (OMB). These waivers are intended to recognize possible tradeoffs between adhering to rule-focused procedures and getting performance results (Kravchuk & Schack, 1996). Managerial flexibility coupled with measures for intended outcomes is expected to improve their operations. Later in this chapter we consider the incentives for managers to become involved in developing and using performance measures.

Because performance measurement systems are ongoing, it is important that the champions of this change support the process, from its inception onward. The nature of performance measures is that they create information—a key resource in public and nonprofit organizations. Information can reduce uncertainty with respect to questions it is intended to answer, but the process of building performance measurement into the organization's business can significantly increase uncertainty for managers. Understandably, some will resist this change, and if leadership commitment is not sustained, the transition to performance measurement as a part of managing programs will wane with time (de Waal, 2003).

A results-oriented approach to managing has implications for public sector accountability. In many jurisdictions, public organizations are expected to operate in ways that conform to process-focused notions of accountability. In Canada, for example, the Westminster parliamentary tradition makes the minister who heads up a government department accountable for all that happens in his or her domain. The adversarial nature of politics and the tendency of the media to emphasize mistakes that become public usually biases managerial behavior toward a rule-focused process wherein only "safe" decisions are made (Propper & Wilson, 2003).

In most government settings, leadership at two levels is required. Clearly the senior appointed officials in a ministry or department must actively support the process of constructing a performance measurement system. But it is equally important that the political leadership be supportive of the development, implementation and use of performance information. The ultimate users of performance information that is publicly reported are intended to be the elected officials to whom the reports are addressed.

In British Columbia, Canada, for example, the Budget Transparency and Accountability Act (2001) specifies that annual performance reports are to be tabled in the legislative assembly. The goal is to have standing committees of the legislature review these reports and use them as they scrutinize ministry operations and future budgets. Each year, the public reports are tabled in June and are based on the actual results for the fiscal year ending March 31. Strategically, the reports should figure in the budgetary process for the following year, which begins in the fall. If producing and publishing these performance reports is not coupled with scrutiny of the reports by legislators, then a key reason for committing resources to public accountability is undermined.

The initial organizational commitment, which typically includes designing the system, can produce "results" that are visible (a website with the performance measurement framework, for example). But implementing and working with the system over 3 to 5 years is a better indicator of its sustainability, and for this having organizational champions is critical.

Understand What Performance Measurement Systems Can and Cannot Do

There are limitations to what performance measurement systems can do, yet in some jurisdictions performance measurement has been treated as a cost-effective substitute for program evaluation (Martin & Kettner, 1996). Public sector downsizing has diminished resources committed to program evaluations, and managers have been expected to initiate performance measurement instead (McDavid, 2001).

Performance measurement can be a powerful tool in managing programs. If the measures are valid and the information is timely, emerging trends can identify possible problems (a negative feedback mechanism) as well as possible successes. But performance measures *describe* what is going on—they do not explain *why* it is happening (Newcomer, 1997).

Recall the distinction between intended outcomes and actual outcomes (introduced in Chapter 1). Programs are designed to produce specified outcomes and one way to judge the success of a program is to see whether the intended outcomes have actually occurred. If the actual outcomes match the intended outcomes, we might be prepared to conclude that the program was indeed effective.

However, we *cannot* conclude that the outcomes are due to the program, unless we have additional information that supports the assumption that other factors in the environment could not have caused the observed outcomes. Getting that information is at the core of what program evaluation is about and it is essential that those using performance measurement information understand this distinction. As Martin and Kettner (1996) comment when discussing the cause-and-effect relationship that many people mistakenly understand to be implied in performance measurement information: "Educating stakeholders about what outcome performance measures really are, and what they are not, is an important—and little discussed—problem associated with their use by human service programs" (p. 56).

Establishing the causal link between observed outcomes and the program that was intended to produce them is the **attribution** problem. Recently, some analysts have explicitly addressed this problem for performance measurement. Mayne (2001) offers six strategies intended to reduce the uncertainty about whether the observed outcomes can be attributed to the program. Briefly, his suggestions include: develop an [intended] results chain; assess the existing research/evidence that supports the results chain; assess alternative explanations for the observed results; assemble the performance story; seek out additional evidence, if necessary; and revise and strengthen the performance story. Several of his suggestions are common to both program evaluation and performance measurement as we have outlined them in this book. His final (seventh) suggestion is that if the performance story is not sufficient to address the attribution question, do a program evaluation. This suggestion supports a key theme of this book— that performance measurement and program evaluation are complementary and each offers ways to reduce uncertainty for managers and other stakeholders in public and nonprofit organizations.

Establish Multi-Channel Ways of Communicating That Facilitate Top-down, Bottom-up, and Horizontal Sharing of Information, Problem Identification and Problem Solving

The process of designing and implementing a performance measurement system is a major change to an organization. Assuming a commitment to creating performance information that actually gets used to manage program performance, this process is similar to other major organizational changes. Kotter (1995) enumerates eight steps or guideposts for organizational changes and key among them is communicating during the intended change process.

It is quite common for public sector organizations to begin a performance measurement system informally. Managers who are keen to develop information that they can use formatively will take the lead in developing measures and procedures for gathering and using the data. This bottom-up process is one that encourages a sense of ownership of the system. In British Columbia, this managerial-driven process spanned a period from 1995 to 2000. Some departments made more progress than others, in part because some department heads were more supportive of this process than others. Because they were driven by internal performance management needs, the systems that developed were adapted to local needs.

To support this evolutionary process, in British Columbia, the Treasury Board Staff (a central agency responsible for budget analysis) hosted an informal network of government practitioners with an interest in performance measurement and performance management. The Performance Measurement Resource Team held monthly meetings with speakers from ministries who provided information on their problems and solutions. Attendance and contributions were voluntary. Information sharing was the principal purpose of the team.

When legislation was passed in 2000 (Legislative Assembly of British Columbia, 2000) mandating performance measurement and public reporting government-wide, the stakes changed dramatically. Performance measurement systems that had been intended for formative uses were now exposed to the requirement that performance results would be published in an annual report. This top-down directive to report performance for summative purposes needed to be meshed with the bottom-up cultures that had been developed in some ministries.

Departments responded to this challenge in different ways. Some of those that had existing performance measurement systems confronted the challenge of melding the formative and summative thrusts of the (new) system by communicating. For example, one department responsible for the publicly funded college and university system in the province conducted a series of formal and informal workshops and meetings with executives and senior and middle managers in attendance. Over a period of a year, using an iterative process, the department was able to develop a general understanding of how the new, externally focused performance measurement system would look, what the new system would do, and how it would connect with the internal performance management system that managers were keen to sustain.

More generally, public organizations that undertake designing and implementing performance measurement systems that are intended to be used internally must include the intended users (Kravchuk & Schack, 1996), the organizational leaders of this initiative, and the methodologists (Thor, 2000). Top-down communications can serve to clarify direction, offer a framework and timelines for the process, clarify what resources will be available, and affirm the importance of this initiative. Bottom-up communications can question or seek clarification of definitions, timelines, resources, and direction. Horizontal communications can provide examples, share problem solutions, and offer informal support.

The communications process outlined here exemplifies the culture that needs to emerge in the organization if performance management is to take hold. Key to developing a performance management culture is treating information as a resource, being willing to "speak truth to power" (Wildavsky, 1979), and not treating performance information as a political weapon. Kravchuk and Schack (1996) suggest that the most appropriate metaphor to build a performance culture is the learning organization. This construct was introduced by Senge (1990) and continues to be a goal for public organizations that have committed to performance measurement as part of a broader performance management framework.

Clarify the Expectations for the Organizational Uses of the Performance Information That Is Created

Normally, developing performance measures is intended, in part, to improve program performance by providing managers with information they can use to make adjustments to program processes. The promise of these uses is often viewed by managers as an asset and an incentive to get involved in constructing a system that provides feedback and allows them to monitor performance.

To attract the buy-in that is essential for successful design and implementation of performance measurement systems, we believe that performance measurement should be used *primarily* for internal performance management. Public reporting can be a part of the process of using performance measurement data, but should not be the primary reason for developing a performance measurement system. A good performance measurement system should support improvements to programs and/or the organization. It should help to identify areas where activities are most effective in producing outcomes and areas where improvement could be made (de Waal, 2003). Designing and implementing a performance measurement system primarily for public accountability will usually create a disconnect between the activities associated with measuring and reporting performance and the "real" work that goes on in the organization.

If the staff who are collecting data do not believe in the usefulness of the system, the performance measurement system will be flawed. The staff may not collect data consistently or follow processes that support the system. Further, if the staff do not see the value in the process but are required to take time away from their activities to participate in the process, the management will have staff morale and productivity issues to deal with that stem from the process.

Balancing the incentives for participants in this process is one of the significant challenges for the leaders of the organization. As we mentioned earlier, developing and then using a performance measurement system creates uncertainty for those whose programs are being assessed. They will want to know how the information that is produced will affect them, both positively and negatively. It is essential that leaders of this process be forthcoming about the intended uses of the measurement system.

It is also essential that a system designed ostensibly for formative purposes not be used for summative purposes—at least not without involving those who have contributed to the (earlier) formative process. Some organizations begin the design and implementation process by making explicit the intention that the measurement results will only be used formatively for a 3-year period of time, for example. That can generate the kind of buy-in that is required to develop meaningful measures and convince participants that the process is actually useful to them. Then, as the uses of the information are broadened, it may be more likely that managers will see the value of a system both formative and summative.

Changing the purposes of a performance measurement system clearly changes the incentives for participants and affects the likelihood that gaming (Propper & Wilson, 2003) will occur as data are collected and reported. We discuss gaming as a response to performance measurement in Chapter 10.

Identify the Resources Available for Designing, Implementing, and Maintaining the Performance Measurement System

Organizations planning performance measurement systems often face substantial resource constraints. One of the reasons for embracing performance measurement is to do a better job of managing the (scarcer) available resources. If a performance measurement system is mandated by external stakeholders (a central agency, an audit office, or a board of directors, as examples), there may be considerable pressure to plunge in without fully planning the design and implementation phases.

Under such conditions, the development work may be assigned to an ad hoc committee of managers, analysts, co-op or intern students, other temporary employees, or consultants. Identifying possible performance measures is usually iterative, time-consuming work, but is only a part of the process. The work of implementing the measures (identifying data that correspond to the performance constructs and collecting data for the measures), preparing reports, and maintaining and renewing the system is the key difference between a process that offers the *appearance* of having a performance measurement system in place (website, progress reports, testimonials by participants in the process) and a process that actually results in using performance data on a continuing basis to improve the programs in the organization. Although a "one shot" infusion of resources can be very useful as a way to get the process started, it is not sufficient to sustain the system. Measuring and reporting performance takes ongoing commitments of resources, including the time of persons in the organization.

Training for staff who will be involved in the design and implementation of the performance measures is important. On the face of it, a minimalist approach to measuring performance is straightforward. "Important" measures are selected, perhaps by an ad hoc committee, data are marshaled for those measures, and required reports are produced. But a commitment to designing and implementing a performance measurement system that is *sustainable* requires an understanding of the process of connecting performance measurement to managing with performance data (Kates, Marconi, & Mannle, 2001).

In some jurisdictions, the creation of legislative mandates for public performance reporting has resulted in organizational responses that meet the legislative requirements but do not build the capacity to sustain performance measurement. Performance measurement is intended to be a means rather than an end in itself. Unless the organization is committed to using the information to manage performance, it is unlikely that performance measurement will be productively integrated into the organization.

Key to developing a performance measurement system is having the data that measure the essential constructs linked to the organization's performance (see the discussion of logic modeling later in this chapter). In an ideal performance measurement system, both cost and results data are available and can be compared. An important driver behind the movement to develop programmed budgets in the 1960s was, in fact, the expectation that cost-effectiveness ratios could be constructed. However, the lack of both budgetary flexibility and information management capacities in most public sector organizations resulted in a significant barrier to being able to fully implement programmed planned budgeting systems.

Most public sector organizations now have accounting systems that permit managers to cost out programs. Information systems are more flexible than in the past, and the budgetary and expenditure data are more complete. Some organizations have also developed the capacity to cost out individual activities within each program.

Depending on the **program technologies** that are implicit in the transformation of resources into results, measuring outcomes can be a significant challenge. Typically, **low-probability program technologies** will offer more challenges than **medium** or **high-probability technologies**. For example, measuring outcomes for a therapy program intended to ameliorate depression in adolescents will be more difficult than measuring the outcomes of a river and stream-focused salmon enhancement program.

In situations where there are financial barriers to validly measuring outcomes, it is common for performance measures to focus on outputs. Outputs are commonly easier to measure, and the data are more readily available. As well, managers are usually more willing to have output data reported publicly because outputs are typically much easier to attribute to a specific program or even program activity. Although outputs are important as a way to report work done, they cannot be substituted for outcomes; the assumption that if outputs are produced, outcomes must have been produced is usually not defensible (see the discussion of measurement validity versus the validity of causes and effects in Chapter 4).

Take the Time to Understand
Organizational History Around Similar Initiatives

Performance measurement is not new. In Chapter 8 we learned that in the United States, local governments began measuring the performance of services in the first years of the 20th century (Williams, 2003). Since then, there have been several waves of reform that focused on measuring results. New public management emerged in part from efforts by Western democratic governments to eliminate fiscal deficits in the 1970s and 1980s.

In most public organizations, current efforts to develop performance measures come on top of other, previous attempts to improve the efficiency and effectiveness of their operations. Managers who have been a part of previous change efforts, particularly unsuccessful ones, have a history that will affect their willingness to support current efforts to measure performance. It is important to understand the organizational memory of change attempts in the past and to gain some understanding of *why* previous efforts to make changes have or have not succeeded. The organizational lore around these changes is as important as a dispassionate view, in that participants' *beliefs* will be the reality that the current change needs to address.

A significant issue for many public sector organizations is the retirement of many baby boomers who exercise their option to leave early, facilitating downsizing goals that governments have put into place. These employees will often have an understanding of the organization and its history. In organizations that have a history of successful change initiatives, losing the people who were involved can be a liability for designing and implementing a performance measurement system. Their participation in the past may have been important in successfully implementing change initiatives. On the other hand, if an organization has a history of questionable success in implementing change initiatives, organizational turnover may actually be a benefit.

Develop Logic Models for the Programs for Which Performance Measures Are Being Developed, and Identify the Key Constructs to Be Measured

In Chapter 2 we discussed logic models as a way to make explicit the intended cause and effect linkages in a program or even an organization. We discussed several different styles of logic models and pointed out that selecting a logic modeling approach depends in part on how explicit one wants to be about intended cause and effect linkages. A key requirement of logic modeling that explicates causes and effects is the presentation of *which* outputs are connected to *which* linking constructs, and *which* linking constructs are connected to *which* outcomes.

Key to constructing and validating logic models with stakeholders is identifying and stating clear objectives for programs (Kravchuk & Schack, 1996). Although this requirement might seem straightforward, it is one of the more challenging aspects of the logic modeling process. Often, program or organizational objectives are put together to satisfy the expectations of stakeholders, who may not agree among themselves about what a program is expected to accomplish. One way that these differences are sometimes resolved is to construct objectives that are general enough so as to appear to meet competing expectations. Although this solution is expedient from an organizational political standpoint, it complicates the process of measuring performance.

Criteria for sound program objectives were discussed in Chapter 1. Briefly, objectives need to state an expected change or improvement if the program works (for example, reducing the number of drug-related crimes), an expected magnitude of change (reduce the number of drug-related crimes by 20%), a target audience/population (reduce the number of drug related crimes by 20% in Harrisburg, Pennsylvania), and a time frame for achieving the intended result (reduce the number of drug-related crimes by 20% in Harrisburg, Pennsylvania in 2 years).

Although logic models do constrain us in the sense that they assume programs are open systems that are stable enough to be depicted as a static model, they are very useful as a means of identifying constructs that are candidates for performance measurement. Martin and Kettner (1996) have identified three major foci for performance measures: program efficiency (comparing inputs to outputs), program quality (whether outputs meet some specified quality standard), and program effectiveness (whether intended outcomes have been achieved). They suggest that a good performance measurement system needs to track all of these various program attributes, since each will be important to at least some program stakeholders.

The open systems metaphor also invites us to identify environmental factors that could affect the program, including those that impact our outcome constructs. Although most performance measurement systems do not measure factors that are external to the program or organization, it is worthwhile including such constructs as candidates for measurement. Measuring these environmental factors (or at least accounting for their influences qualitatively) allows us to begin addressing attribution questions.

Identify Any Constructs That Apply Beyond Single Programs

Organizational logic models can be seen as an extension of program logic models but because they typically focus on a higher-level view of programs or business lines, the constructs will be more general. The balanced scorecard is one type of organizational performance measurement system that includes a cause and effect model of key organizational-level constructs. Table 9A.1 in the appendix illustrates an organizational logic model for the British Columbia Ministry of Human Resources. The ministry is primarily concerned with providing income assistance and moving income assistance recipients into job training programs as a transition to employment. Table 9A.1 is complex, but if one wants to see how the operations of this entire organization fit together, an organizational logic model is a parsimonious way to show this visually and to identify constructs that might be candidates for constructing performance measures.

Performance measurement systems are sometimes expected to offer measures of performance that transcend single government departments and measure sectoral performance (for example, the benchmark report by the Oregon Progress Board, 2003, and the policy framework of the Government of Alberta, 2004).

Many social problems cannot easily be assigned to one administrative department. An example might be a program to reduce the number of single mothers on state social assistance. A social assistance department will usually have a mandate to provide funds to persons and families who are deemed to be in need. However, some single mothers may also have employment skills deficits, child care needs, needs for education, and/or needs for personal counseling to plan their transition from welfare to sustained employment. In nearly all jurisdictions, this range of programs will be provided, but are almost always provided by several separate departments.

Developing sectoral performance measures for this program would involve coordination of programming across departments and a sharing of responsibility and accountability for the overall program objective. If permitted to focus simply on the objectives of each government department during the design of the system, each contributor will have more of a tendency to select objectives that are conservative, that is, do not commit the department to be responsible for the overall sectoral outcome. In particular, if legislation has been passed which emphasizes departments being individually accountable, then sectoral objectives may well be overlooked.

A similar problem arises for nonprofit organizations. In Canada and the United States, many funding organizations (e.g., governments, private foundations, United Way) are opting for a performance-based approach to their relationship with the funded organizations that deliver programs and services. Increasingly, funders are demanding outcome-focused performance measurement as a condition for grants funding. Governments that have opted for contractual relationships with nonprofit service providers are developing performance contracting requirements that specify deliverables and often tie funding to the provision of evidence that these results have been achieved.

Nonprofit organizations are usually quite small and are focused on the amelioration of a community problem or issue that has attracted the commitment of organization members and volunteers. Being required to bid on contracts and account for the performance results of the money they have received is added onto existing administrative requirements and many of these organizations have limited capacity to do any of these tasks. Campbell (2002) has pointed out that in settings where the desired outcomes span several nonprofit providers, it is essential that there be some collaboration among funders and providers to agree on ways of directly addressing these outcomes. If providers compete and funders continue to address parts of a problem, the same sectoral disregard that was suggested for government departments will happen in the nonprofit sector.

One issue that can easily be overlooked as performance measures are being developed is the **levels of analysis problem** (McDavid, 2001). Suppose a government department develops a set of performance measures that is intended to indicate how the organization as a whole is doing. If the actual performance results suggest that the organization is meeting its overall objectives, it might be tempting to conclude that the programs that contribute to the objectives are also effective. That would be a mistake because success at one level does not warrant a conclusion that performance at other levels is also comparable. It is possible to have programs that are not meeting their objectives but, overall, the organization is meeting its objectives. Likewise, we cannot use program success alone to indicate organizational success, nor can we use individual employee performance measures to tell us whether programs or the organization are meeting their objectives.

Ideally, individual and group objectives should connect with program objectives which should in turn connect with organizational objectives. Performance needs to be measured at *all* these levels to be able to effectively manage organizational performance.

One additional issue with respect to organization-level and sectoral measures of performance is who should take responsibility for gathering the data and reporting interpretations of it. Since reporting responsibilities can be linked to expectations of accountability in some organizations, ownership of these measures becomes an important organizational-political issue. We discuss the political dimensions of performance measurement in Chapter 10.

Involve Prospective Users in Reviewing Logic Models and Constructs in the Proposed Performance Measurement

Developing logic models of programs and/or the organization as a whole is an iterative process. Although the end product is intended to represent the programmatic and causal reasoning that transforms resources into results, it is essential that logic models be reviewed and validated with organizational participants and other stakeholders. Buy-in at this stage of the development process will familiarize prospective users with the key constructs in the model(s) and set the agenda for developing performance measures. Program managers in particular will have an important stake in the system. Their participation in validating the logic models increases the likelihood that performance measurement results will be useful for program improvements.

Typically, logic models identify outputs and outcomes that are linked in intended causal relationships. Depending on the purposes of the performance measurement process, some constructs will be more important than others. For example, if a logic model for a job training and placement program operated by a community nonprofit organization has identified the number of persons who complete the training as an output and the number who are employed full-time 1 year after the program as an outcome, program managers would likely emphasize the output as a valid measure of program performance—in part because they have more control over that construct. But the funders might want to focus on the permanent employment results because that is really what the program is intended to do.

By specifying the intended causal linkages, it is possible to review the relative placement of constructs in the model and clarify which ones will be a priority for measurement. In our example, managers might be more interested in training program completions since they are necessary for any other intended results to occur. Depending on the clients, getting persons to actually complete the program can be a major challenge in itself. If the performance measurement system is intended to be summative as well, then measuring the permanent employment status of program participants would be important—although there would be a question of whether the program produced the observed employment results.

Measure the Constructs That Have Been Identified as Parts of the Performance Measurement System

We learned in Chapter 4 that the process of translating constructs into observables is **measurement**. For performance measurement, secondary data sources are the principal means of measuring constructs. Because these data sources already exist, their use is generally seen to be cost-effective. There are, however, several issues that must be kept in mind when using secondary data sources.

1. Can the existing data be adapted to fit constructs in the performance measurement system? In many performance measurement situations, the challenge is to adapt what exists, particularly data readily available via information systems, to what is needed to translate performance constructs into measures. Often, existing data have been collected for purposes that are not related to measuring and reporting on performance. Using these data raises validity questions: Do they really measure what the performance measurement designers say that they measure? Or, do they distort or bias the performance construct so that the data are not credible?

2. Do existing data sources sufficiently encompass the constructs that need to be measured? The issue here is whether our intended performance measures are matched by what we can get our hands on in terms of existing data sources.

A separate, but related, issue is whether existing data sources permit us to triangulate our measurements of key constructs. In different words, can we measure a given construct in two or more independent ways, ideally with different methodologies? Generally, triangulation increases confidence that the measures are valid.

3. Will using existing data sources change the way they are collected? Managers and other organizational members generally respond to incentives. If a performance measure becomes the focus of summative program or service assessments, and if the data for that measure are collected by organizational participants, it is conceivable that the data will be manipulated to indicate "improved" performance.

An example of this type of situation from policing was an experiment in Orange County, California, to link salary increases in the police department to reduced reporting rates for certain kinds of crimes (Staudohar, 1975). The agreement between the police union and management specified clear linkages between percentage reductions in four types of crimes and the magnitude of salary increases.

The experiment "succeeded." Crime rates in the four targeted crimes decreased just enough to maximize the wage increases. Correspondingly, crime rates *increased* for several related types of crimes. A concern in this case is whether the crime classification system may have been manipulated by participants in the experiment, given the incentive to "reduce" crimes in order to maximize salary increases.

If primary data sources are being used, several issues should also be kept in mind:

a) Are there *continuing* resources to enable collecting, coding, and reporting the data? If not, then situations can develop where the initial infusion of resources to get the system started may include funding to collect outcomes data (conduct a client survey, for example), but beyond this point, there will be gaps in the performance measurement system where these data are no longer collected.

b) Are there issues of sampling procedures, instrument design, and implementation that need to be reviewed or even done externally? In different words, are there methodological requirements that need to be established to ensure the credibility of the data?

c) Who will actually collect and report the data? If managers are involved, is there any concern that their involvement could be seen to be in conflict with incentives they face? Will managerial involvement undermine the credibility of these data, particularly if the performance results may be used summatively?

Often, when managers review the performance measures that are being proposed, they will be looking for the measures that represent their work and zero in on any issues that these raise. Indeed, if a draft of the proposed performance measures does not feature any that pertain to their programs, they may conclude that they are being deliberately excluded and are therefore vulnerable in future budgets. It is essential to have a rationale for each measure, and some overall rationale for featuring some measures but not others. Organization executives may need to be involved in settling any managerial disagreements.

In Chapter 4 we introduced measurement validity and reliability criteria to indicate the methodological requirements for sound measurement processes. Those criteria are rooted in the social sciences (Goodwin, 1997), and satisfying them is generally premised on having the resources to properly establish validity and reliability. In many performance measurement situations, there are no resources and limited time to determine whether each measure is defensible in methodological terms. Performance measurement is fundamentally about finding indicators that plausibly connect constructs with data. In terms of the kinds of validity discussed in Chapter 4, persons or teams that are developing and implementing performance measures usually pay attention to face validity (on the face of it, does the measure do an adequate job of representing the construct?), content validity (how well does the measure or measures represent the range of content implied by the construct?), and response process validity (have the participants in the measurement process taken it seriously?). Assessing internal structure, concurrent, predictive, convergent, and discriminant validity are often beyond the methodological resources in performance measurement settings. The reliability of performance measures is often assessed with a judgmental estimate of whether the measure and the data are accurate, that is, are collected and recorded so that there are no important errors in the ways the data represent the events or processes in question. In some jurisdictions, performance measures are audited for reliability (Texas State Auditor's Office, 2002).

An example of assessing the reliability and validity of measures of program results would be a social service agency that has included the number of client visits as a performance measure for the funder of their counseling program. Suppose that initially, the agency and the funder agree that the one measure is sufficient since payments to the agency are linked to the volume of work done and client visits are deemed to be a reasonably accurate measure for that purpose. To assess the validity and reliability of that measure, one would need to know how the data are recorded (by the social worker or by the receptionist, for example) and how the files are transferred to the agency database (manually or electronically as part of the intake process for each visit). Are there under or over counting biases in the way the data are recorded? Do telephone consultations count as client visits? What if the same client visits the agency repeatedly to a point where other prospective client appointments are less available? Should a second measure of performance be added that tracks the number of clients served (improving content validity)? Will that create a more balanced picture and create incentives to move clients through the treatment process? What if clients change their names—does that get taken into account in recording the number of clients served? Each performance measure or combination of measures for each construct will have these types of practical problems that must be addressed if the data in the performance measurement system are to be credible.

In jurisdictions where public performance reporting is mandated, a significant issue is reducing the number of performance measures for a ministry or department so that the performance report is readable and reasonably short. Typically, the number of performance measures in public reports is somewhere between 10 and 20, meaning that a lot of programs will not be represented in the public image of the department as reflected in the report.

A useful way to manage the matter of managers wanting their programs to be represented publicly is to commit to constructing internal performance reports. Internal reports are consistent with the balancing of formative and summative uses of performance measurement systems. It is our belief that unless a performance measurement system is used primarily for internal performance management, it is unlikely to be sustainable. Internal performance measures can more fully reflect each program and are generally seen to better represent the accomplishments of programs.

One additional measurement issue is whether measures and the data that correspond to the measures should be quantitative. In Chapter 5, we discussed the important contributions that qualitative evaluation methods can make to program evaluations. We included an example of how qualitative methods can be used to build a performance measurement and reporting system (Sigsgaard, 2002). There is a meaningful distinction between the information that is conveyed by words and numbers. Words can provide us with texture, feelings, and a more vivid understanding of situations. Words can qualify numbers, interpret numbers, and balance presentations that rely on numerical information. Most importantly, words can indicate qualities—how a program was experienced by particular clients as opposed to the number of clients served, for example.

In performance measurement systems, it is desirable to have both quantitative and qualitative measures/data. Stakeholders who take the time to read a mixed presentation can then learn more about performance. But in many situations, particularly where annual targets are set and external reporting is mandated, there is a strong bias toward numerical information, since targets are nearly always stated numerically. If the number of persons on social assistance is expected to be reduced by 10% in the 2004–2005 fiscal year, the most relevant data will be numerical. Whether the program meets its target or not, the percentage reduction in numbers of persons on social assistance provides no information about the process whereby that happened.

Performance measurement systems that focus primarily on providing information for formative uses should include deeper and richer measures than those used for public reporting. Qualitative information can provide managers with feedback that is very helpful in adjusting program processes to improve results. As well, qualitative information can reveal to managers the client experiences that accompany the process of measuring quantitative results.

Qualitative information, presented as cases or examples that illustrate a pattern that is reported in the quantitative data, can be a powerful way to convey the meaning of the numerical information. Although single cases can only illustrate, they communicate very effectively. For political decision makers, for example, case studies can be essential to convey the meaning of performance results—many political leaders are more comfortable with information that conveys results focused on individuals. Fundamentally, they relate to their constituents that way, so performance reports that do the same will be accessible to them.

Record, Analyze, Interpret and Report the Performance Data

Prospective users should be challenged to offer their interpretations of simulated patterns of performance information (Davies & Warman, 1998). One potential problem with any performance measurement system is the ambiguity of its measures. In the Oregon benchmarking project cited earlier, the affordability of housing was offered as an indicator of the well-being of the state (presumably, of the broad social and economic systems in the state). If housing prices are trending downwards, does that mean that things are getting worse or better?

From an economic perspective, declining housing prices could mean that: demand is decreasing in the face of a steady supply; demand is decreasing while supply is increasing; demand and supply are both increasing, but supply is increasing more quickly; or demand and supply are both decreasing, but demand is decreasing more quickly. Each of these scenarios suggests something different about the well-being of the economy. To complicate matters, each of these same scenarios would have different interpretations if we were to take a social rather than an economic perspective.

The point is that prospective users should be offered scenarios in which different trends and levels of measures are posed. If trends or levels have ambiguous interpretations—"it depends"—then it is quite likely that when the performance measurement system is developed, the same ambiguities will arise as reports are produced. Fundamentally, ambiguous measures invite conflicting interpretations of results and will tend to weaken the credibility of the system.

In addition to simulating different patterns of information for prospective users, it is important to ascertain what kinds of comparisons are envisioned with performance data. A common comparison is to look for trends over time, and make judgments based on interpretations of trends.

Another comparison is across similar administrative units. For example, a department that offers programs in different regions of a state may want to compare performance measures across the regions to see if there are "high" or "low" levels with respect to selected measures. In British Columbia, the Provincial Workers' Compensation Board uses its periodic surveys of clients to measure their satisfaction with the claims process and then compares satisfaction levels across regional offices of the Board.

A third type of comparison is with benchmarks, standards, or targets. In some program or service areas (hospital services, for example), it is common to use standards to assess waiting times for services. When physicians refer patients for testing or for medical procedures, waiting time can become a critical factor, especially where initial diagnoses indicate a progressive disease.

Existing benchmarks and standards, with the exception of highly visible projects like the Oregon benchmarking project, have tended to focus on measures of *process* or *outputs* rather than outcomes. Accreditation standards, for example, tend to focus on: the qualifications, numbers, and skills of staff members; the quality and conditions of treatment facilities; and perhaps, the procedures for treatment. Few benchmarks exist that specify the expected outcomes from treatments or services.

It is common for performance measures to be compared to annual targets. Performance reporting that is intended for public accountability purposes will typically include comparisons between performance targets and actual results. For example, a municipal government graffiti management program might have an objective of reducing the number of public buildings defaced by graffiti. If the 2004 target was a maximum of 5% of buildings with graffiti (measured by a year-end physical survey of all public buildings), the actual survey results could be compared to the target. If the survey revealed that 10% of public buildings had graffiti on them, the program manager (and other stakeholders) might decide to investigate the gap between the target and actual result. What explains the difference? Following this performance result up would entail asking why the observed result occurred—a question typically in the domain of program evaluation.

Setting targets can become a contentious process. If salaries of senior managers are linked to achieving targets, there will be pressure to make sure that targets are achievable. If reporting targets and achievements are part of an adversarial political culture, there will again be pressure to make targets conservative (Davies & Warman, 1998). Norman (2001) has suggested that performance measurement has a tendency to result in under-performance, for these reasons.

Buy-in is an incremental process. Managers will want to see what actually happens with the performance measures and the reports that are produced before they are willing to fully accept this change. Acceptance can also be eroded. If there is turnover in the organization's leadership, and the new executive shifts the balance from formative to summative uses of the performance results without involving managers in this shift, it is quite likely that resistance to the system will develop.

There is an issue of access to the performance data. In some organizations, the performance measurement function has been separated from line management entirely—managers do not have access to data, instead, they receive reports. Excluding managers and other organizational members from having access to performance data tends to reinforce a cultural norm that such information is power and a source of control. Related to access is the question of whether users can prepare their own reports, in addition to reports that are mandated. Are they given the opportunity to analyze the data included in existing reports to corroborate or disconfirm interpretations of the data?

Finally, how are reports prepared? Is there a regular cycle of reporting? Is there a process whereby reports are reviewed and critiqued internally before they are released to users? Often, legislative auditors have internal vetting processes wherein the authors of reports are expected to be able to defend the report before their peers before the report is released. This challenge function is valuable as a way of assessing the defensibility of the report and anticipating the reactions of stakeholders.

In many states in the United States and most provinces in Canada, public performance reports are mandatory. That requirement is usually stated in legislation and clearly puts a summative cast over the whole process. In British Columbia, for example, the *Budget Transparency and Accountability Act* (BTAA, 2000) specifies that ministries and all provincially funded agencies (including school districts, hospital districts, colleges and universities) will prepare annual performance reports to be tabled in the legislature in June of each year. In the United States, the 1993 *Government Performance and Results Act* now requires (as of 1999) that all federal departments must have a performance measurement and reporting system in place (Office of Management and Budget, 1993).

Although public performance reporting is clearly one way to meet accountability expectations, the real challenge is to develop and sustain a balance between external accountability uses of performance information and internal performance management uses. Given the nature of political cultures wherein adversarial political processes are the norm, there can be difficulties in acknowledging the shortcomings brought out in performance reports. Although building a capacity to learn from mistakes is clearly an intended purpose for many performance measurement systems, building and sustaining cultures that support learning organizations (Garvin, 1993) needs to take into account the political dimensions of organizational cultures. We address this issue more fully in Chapter 10.

Regularly Review Feedback From the Users and, if Needed, Make Changes to the Performance Measurement System

Users and organizational needs for performance data will change over time. Implementing a system with a fixed structure (logic models and measures) at one point in time will not ensure the durability or continued use of the system. There is a balance between the need to maintain continuity of performance measures, on the one hand, and the need to reflect changing organizational objectives, structures, and uses of the system, on the other hand (Kravchuk & Schack, 1996). In many performance measurement systems, there are measures that are replaced periodically and measures that are works in progress. The State of Oregon, for example, regularly updates and changes its performance measures to meet changing needs, even though the Oregon benchmarks reports are intended to convey long-term trends in key performance indicators (Oregon Progress Board, 2003).

Continuity in the measures increases the capacity of measures to be compared over time. Measures displayed as a time series can, for example, show trends in environmental factors, as well as changes in intended outputs and outcomes. By comparing environmental trends with outcome trends, it may be possible to adjust results to take into account influences of plausible rival hypotheses on particular outcome measures. Although this process depends on the length of the time series and is often judgmental, it does permit analysts to use some of the same tools that would be used by program evaluators. In Chapter 3, recall that in the York crime prevention program evaluation the unemployment rate in the community was an external variable that was included in the evaluation to assist the evaluators in determining whether the program was the cause of observed changes in the reported burglary rate.

But continuity can also make a system less relevant over time. Suppose, for example, that a performance measurement system was designed to pull data from several different databases and the original computer programming task to make this work was expensive. There may well be a desire not to go back and repeat this work, simply because of the resources involved. Likewise, if a performance measurement system is based on a logic model that becomes outdated, then the measures will no longer fully reflect what the program(s) or the organization is trying to accomplish. But going back to redo the logic model (which can be a time-consuming, iterative process) may not be feasible, given the resources available.

The price of such a decision might be a gradual reduction in the uses of the system, which may not be readily detected. If the system, for example, had been intended as a formative and summative tool, its gradual erosion might well be viewed with equanimity by managers who would rather not be judged by the performance measurement results.

With all the activity to design and implement performance measurement systems, there has been surprisingly little effort to date to evaluate the effectiveness of these systems. We discuss what is known about the ways in which performance measurement is used in Chapter 10, but it is appropriate here to suggest some practical steps to generate feedback that can be used to modify and better sustain performance measurement systems:

• Develop channels for user feedback. This is intended to create a process that will allow the users to provide feedback and revise, review, and update the performance measures. Furthermore, this is intended to help identify when corrections are required and how to address errors and misinterpretations.

● Create an expert review panel. Performance measurement is conducted on an ongoing basis, and this expert panel review should provide feedback and address issues and problems over a long-term time frame.

● There needs to be an independent assessment of buy-in and use by managers and staff.

● An enduring concern with performance measurement systems is whether the information is credible. We discuss the incentives for gaming, in Chapter 10, but here it is important to mention the growing movement to audit performance measurement systems and the reports that are produced from them. In Canada, for example, the Canadian Comprehensive Auditing Foundation (CCAF) has published a set of principles intended to guide public performance reporting. Collectively, these principles are intended to ensure that a performance report is useful, credible, and based on accurate information. Externally auditing the performance reporting process is viewed by some as an essential part of ensuring the longer-term credibility of the system. Davies and Warman (1998) point to the importance of auditing in the context of the performance reports of the (British) National Meteorological Office:

> An independent audit, then, is not a luxury, it is a *necessity*. The credibility of the whole system of agencies is put at risk if the data from one is found to be unverified and open to dispute. Where performance-related bonuses are linked with outcomes, it is unreasonable to expect staff concerned to be responsible for the measurement and reporting of results in an objective manner when the very same results will determine their own pay. (p. 47)

● Within 5 years of implementing a performance measurement system, there should be an evaluation of its intended and unintended contributions to the organization.

● SUMMARY

The 12 criteria for designing and implementing performance measurement systems discussed in this chapter impose some demanding requirements on the process. It is quite likely that in any given situation, one or more of these criteria will be hard to address. Does that mean that unless performance measurement systems are designed and implemented with these 12 criteria in view, that the system will fail? No, but it is reasonable to assert that each criterion is important and does enhance the likelihood of success.

Among the criteria, six are essential. Each contributes something necessary for successful design and implementation:

- Sustained leadership—without it, the process will drift and eventually halt
- Communications—essential to developing a common understanding of the process and increasing the likelihood of buy-in
- Clear expectations for the system—be open and honest about the purposes behind this process so that stakeholders and problems can be dealt with directly
- Resources sufficient to free up the time and expertise needed—by taking resources from other programs to measure and report performance, the process is viewed as a competitor and is often given short shrift
- Logic models that identify the key program and organizational constructs—the process of logic modeling disciplines the selection of constructs and performance measures
- A measurement process that succeeds in producing valid measures in which stakeholders have confidence—too few performance measurement systems pay attention to measurement criteria that ultimately determine the perceived usefulness of the system

These six criteria can be thought of as individually necessary but together they are not guaranteed to predict a successful design and implementation process. Performance measurement is a craft. In that respect it is similar to program evaluation. There is considerable room for creativity and professional judgment as organizations address the challenges of measuring results.

DISCUSSION QUESTIONS ●

1. Assume that you are a consultant to the head of a government agency (1,000 employees) that delivers social service programs to families. The families are poor and most of them have one parent (the mother) who is either working for relatively low wages or is on social assistance. In your role, you are expected to give advice to the department head that will guide the organization into the process of developing and implementing a performance measurement system. What advice would you give about getting the

Appendix 9A.1 Logic Model for Ministry of Human Resources, British Columbia, Canada

		Implementation		Intended Outcomes		
Inputs	Components	Implementation Objectives	Outputs	Linking Constructs	Shorter Term	Longer Term
Ministry expenditures No. of FTEs Strategic partnerships: • HRDC	Employment Programs	To provide individualized employment plans and services for able-bodied and persons with disabilities	No. of clients referred to Job Placement Program No. referred to Training for Jobs pilot program No. receiving employment supports and specialized job training/placement No. of persons with disabilities utilizing support, training and placement services	No./% of Job Placement clients who are off income assistance after 6 months No./% of Training for Jobs clients off income assistance after 6 months No./% of persons with disabilities with employment income	Improved job skills Overcoming barriers to employment Increased self-reliance	Long-term, sustainable employment for individuals A continued dynamic labor market Alleviation of shortage of skilled labor Broader range of workforce participation by persons with disabilities Breaking the cycle of intergenerational dependency
• Minister's Council on Employment for Persons with Disabilities	Temporary Assistance	To assist those unable to work because of a short-term medical condition, because they are a single parent caring for a young child or they are caring for a disabled family member	No. of clients completing self-directed work search No. meeting employment plan goals No. completing employment plans No. of clients diverted to employment	No. of job placements	Shorter length of time that a person is out of the workforce Successful transition back to employment	
• Ministries that provide services to those with disabilities	Continuous Assistance	To assist individuals who are not expected to gain financial independence through employment	No. of clients referred to Community Assistance Program No. of multiple-barriered clients assisted to find volunteer opportunities	Volunteer community participation by clients Client employment	Involvement in the community	Self-reliance Independence

Implementation					Intended Outcomes	
Inputs	Components	Implementation Objectives	Outputs	Linking Constructs	Shorter Term	Longer Term
Partnerships with nonprofit sector and private sector	Supplementary Assistance	To assist eligible clients to provide supports for people in need	No. of clients receiving emergency accommodation; No. provided with short-term assistance when forced from their homes by disasters; No. provided with health services; No. provided with transit passes	Reduced need for other government support	Increased self-sufficiency	More opportunity to contribute to society
	Executive and Support Services	To provide staff with a clear understanding of their roles and responsibilities and provide training opportunities	No. of positions with core competencies defined; No. of health and safety programs provided; No. of employees with EPDPs[a] in place	Ministry more responsive to employees	Productive employees; Innovative employees	Ministry will be able to more efficiently achieve its goals
	Employment and Assistance Appeal Tribunal	To provide individuals with a streamlined, fair, and responsive appeal system and provide streamlined service delivery	No. of clients accessing appeal system; Percentage of program and service contracts that are performance-based	Reduced appeal times; Results-based service delivery	Ministry is accountable, transparent, and responsive	Clients and citizens will feel confident that the ministry operates fairly and openly

a. Employee Performance Development Plans.

process started? What things should he or she do to increase the likelihood of success in implementing performance measures? How should the department head work with managers and staff to get them onside with this process? Try to be realistic in your advice—assume that there will not be new resources to develop and implement the performance measurement system.

2. Performance measurement systems are usually intended to improve the efficiency and effectiveness of programs. But very few organizations have taken the time to assess whether their performance measurement systems are actually making a difference. Suppose that the same organization that was depicted in question 1 had implemented its performance measurement system. The department head now wants to find out (assume it is 3 years later) whether the system has actually improved the efficiency and effectiveness of the agency's programs. Suppose that you were giving this person advice about how to design a project to assess whether the performance measurement system had "delivered." Think of this as an opportunity to apply your program evaluation skills to finding out whether this performance measurement system was successfully implemented. What would be possible criteria for the success of the system? How would you set up a research design that would allow you to see whether the system had the intended incremental effects?

● REFERENCES

Campbell, D. (2002). Outcomes assessment and the paradox of nonprofit accountability. *Nonprofit Management & Leadership, 12*(3), 243–259.

Davies, M., & Warman, A. (1998). Auditing performance indicators: The meteorological office case study. *Journal of Cost Management,* January/February, 43–48.

de Lancer Julnes, P., & Holzer, M. (2001). Promoting the utilization of performance measures in public organizations: An empirical study of factors affecting adoption and implementation. *Public Administration Review, 61*(6), 693–708.

de Waal, A. A. (2003). Behavioral factors important for the successful implementation and use of performance management systems. *Management Decision, 41*(8), 688–697.

Garvin, D. A. (1993). Building a learning organization. *Harvard Business Review, 71*(4), 78–90.

Goodwin, L. D. (1997). Changing conceptions of measurement validity. *Journal of Nursing Education, 36*(3), 102–107.

Government of Alberta. (2004). *Government of Alberta strategic business plan: Budget 2004 on route on course.* Edmonton, AB: Alberta Finance.

Kates, J., Marconi, K., & Mannle, Jr., T. E. (2001). Developing a performance management system for a federal public health program: The Ryan White CARE Act Titles I and II. *Evaluation and Program Planning, 24*(2), 145–155.

Kotter, J. P. (1995). Leading change: Why transformation efforts fail. *Harvard Business Review, 73*(2), 59–67.

Kravchuk, R. S., & Schack, R. W. (1996). Designing effective performance-measurement systems under the Government Performance and Results Act of 1993. *Public Administration Review, 56*(4), 348–358.

Legislative Assembly of British Columbia. (2000). *Budget Transparency and Accountability Act* (SBC 2000, Chapter 23). Victoria, BC: Queen's Printer.

Martin, L. L., & Kettner, P. M. (1996). *Measuring the performance of human service programs.* Thousand Oaks, CA: Sage.

Mayne, J. (2001). Addressing attribution through contribution analysis: Using performance measures sensibly. *Canadian Journal of Program Evaluation, 16*(1), 1–24.

McDavid, J. C. (2001). Program evaluation in British Columbia in a time of transition: 1995–2000. *Canadian Journal of Program Evaluation, Special Issue,* 3–28.

Newcomer, K. E. (1997). Using performance measurement to improve public and nonprofit programs. In K. E. Newcomer (Ed.), *New directions for evaluation* (No. 75, p. 102). San Francisco: Jossey-Bass.

Norman, R. (2001). Letting and making managers manage: The effect of control systems on management action in New Zealand's central government. *International Public Management Journal, 4*(1), 65–89.

Norman, R., & Gregory, R. (2003). Paradoxes and pendulum swings: Performance management in New Zealand's public sector. *Australian Journal of Public Administration, 62*(4), 35–49.

Office of Management and Budget. (1993). *Government Performance Results Act of 1993.* Retrieved May 2005, from http://www.whitehouse.gov/omb/mgmt-gpra/gplaw2m.html#h10

Oregon Progress Board. (2003). *Is Oregon making progress? The 2003 benchmark performance report.* Salem: State of Oregon.

Propper, C., & Wilson, D. (2003). The use and usefulness of performance measures in the public sector. *Oxford Review of Economic Policy, 19*(2), 250–267.

Senge, P. M. (1990). *The fifth discipline: The art and practice of the learning organization.* New York: Doubleday/Currency.

Sigsgaard, P. (2002). MCS approach: Monitoring without indicators. *Evaluation Journal of Australasia, 2*(1), 8–15.

Staudohar, P. D. (1975). An experiment in increasing productivity of police service employees. *Public Administration Review, 35*(5), 518.

Texas State Auditor's Office. (2002). *An audit report on fiscal year 2001 performance measures at 14 entities.* Austin, TX: Texas State Auditor's Office.

Thor, C. G. (2000). The evolution of performance measurement in government. *Journal of Cost Management,* May/June, 18–26.

Wildavsky, A. B. (1979). *Speaking truth to power: The art and craft of policy analysis.* Boston: Little Brown.

Williams, D. W. (2003). Measuring government in the early twentieth century. *Public Administration Review, 63*(6), 643–659.

CHAPTER 10

USING AND SUSTAINING PERFORMANCE MEASUREMENT SYSTEMS

INTRODUCTION ●

In Chapter 9, we discussed the design and implementation of performance measurement systems and outlined 12 steps for guiding the process. Implementation implies that performance measurement data are being collected regularly, are being analyzed, and are being reported. If the system is intended primarily for formative uses, reporting may be informal and open-ended; that is, analyses and reports are prepared as needed. If the reports are prepared to meet external accountability requirements, they have a summative intent. In Texas, for example, annual performance reports are tabled in the legislature with the intention that they be available for use by legislative committees, individual legislators, and the public.

Generating performance results and communicating them is necessary but not sufficient to ensuring that performance information is actually used. This chapter will examine ways that performance information is intended to be used and is actually used, as well as factors important to the sustainability of a performance measurement system.

Among intended uses, improving **accountability** as well as efficiency and effectiveness are central (Government Performance Results Act, 1993; Halachmi, 2002). These are broad terms and will need to be unpacked to understand what advocates of performance measurement are saying about ways that this information can be used. There has not been a great deal of systematic research on how users themselves see the potential for performance measurement, so we will summarize key findings from an ongoing research project in British Columbia, Canada to show how legislators themselves see the ways that they intend to use performance reports.

Actual uses of performance measures have been studied for federal, state and local governments, but the studies that have been completed leave gaps in our understanding. Leviton (2003) points out that many existing studies of the ways that evaluations are used employ methodologies that would not be adequate for an evaluation study—most rely entirely on self-reports and their attendant biases. Studies of the actual uses of performance information tend to be limited in similar ways (Leviton, 2003).

In this chapter we examine five problems that can impede the success and sustainability of efforts to measure and report performance. If these problems are not adequately addressed, there is an increased risk that performance measurement will not live up to its promise.

● COMPARING IDEAL AND REAL
ORGANIZATIONS: THE ROLE OF
ORGANIZATIONAL POLITICS IN DESIGNING AND
IMPLEMENTING PERFORMANCE MEASUREMENT SYSTEMS

Designing and implementing performance measurement systems can be seen as a part of a broader strategy to adopt performance management. The performance management cycle introduced in Chapter 1 begins with strategic planning (setting clear goals and objectives) and moves on to designing policies and programs that flow from the strategic objectives, implementing policies and programs, evaluating them (using program evaluation and performance measurement), reporting the results (internally and/or externally), and finally, connecting the evaluation results to the next round of strategic planning. Performance measurement contributes to performance management by providing information on the results of programs and policies. It feeds the reporting and accountability phase of performance management that in turn is linked to the formulation and modification of strategic goals and objectives for the organization.

But, the performance model introduced in Chapter 1 is normative. It outlines what *ought* to happen in organizations that have implemented performance management. Real world situations are usually quite different and are often more complex. When we design and implement performance measurement systems, we must go beyond ideals, and consider the "psychological, cultural, and political implications of organizational change" (de Lancer Julnes, 1999, p. 49). De Lancer Julnes and Holzer (2001) have distinguished a rational/technical framework and a political/cultural framework as key to understanding the successful adoption, implementation, and use of performance measures. Keeping this in mind can help illuminate the differences between how performance measures are *intended* to be used and the manner and extent to which they are *actually* used.

The Rational/Technocratic Framework

Rational factors focus on the technical issues that drive the design and implementation of performance measurement systems: *resources; knowledge and information; a goal orientation in the organization* (having clear goals and objectives); and *external requirements that provide a mandate or enable the process* (legislation might be an example). Much of the literature that advocates the adoption of performance measurement systems, and performance management systems in which performance measurement is embedded, tends to emphasize these rational factors—the orientation of this literature is that successful performance measurement involves addressing and resolving these challenges and successful implementation will follow. Akin to the "organizations as machines" metaphor elaborated by Morgan (1997), organizations are seen as complex systems guided by rational and technical considerations which can be addressed by the right combination of mandate, resources, and know-how. People are assumed to behave in ways that support a goal-oriented, planning-focused, information-driven culture that thrives by setting goals; designing and implementing programs that will achieve those goals; measuring how well the programs work; and making whatever changes are needed to improve the efficiency and effectiveness of programs and policies. Contemporary models of learning organizations (see Morgan, 1997; Senge, 1990) share many characteristics with rational performance management-focused organizations.

The Political/Cultural Framework

The rational cycle that is intended to illustrate an ideal commitment to performance management becomes much more complex when we take into account the political factors that affect each stage in the process. De Lancer Julnes and Holzer (2001) identify political/cultural factors that will affect the adoption, implementation, and use of performance measurement: *internal interest groups; external interest groups and unions; risk taking;* and *management and nonmanagement attitudes.* Similarly, Morgan (1997) offers a political metaphor for understanding organizations. Unlike the rational/technical view, a political lens will "see" the coalition politics, the bargaining, the conflicts, and the compromises that characterize most if not all organizational decisions and initiatives. Figure 10.1 takes the performance management cycle introduced in Chapter 1 and adds to it the political factors that affect each stage of the performance management process. The boxes in Figure 10.1 summarize the ways that organizational politics and the behaviors of individuals and groups can influence the rational actions at each stage of the performance management cycle.

The political framework of performance management moves beyond what rationally *ought* to be, to what *is.* Successful performance management is more than following prescriptions—it involves real people, with motivations, goals, and a capacity to respond to incentives. Understanding this dimension of performance management is essential for its long-term sustainability. The design, implementation, and use of performance measurement systems are all affected by both rational/technical *and* political/cultural factors. As we explore intended and actual uses of performance measurement systems, we will see

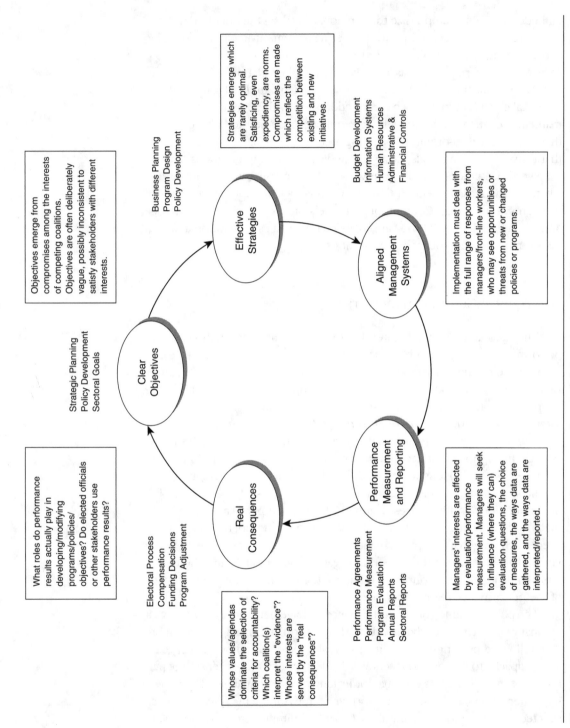

Objectives emerge from compromises among the interests of competing coalitions. Objectives are often deliberately vague, possibly inconsistent to satisfy stakeholders with different interests.

Business Planning
Program Design
Policy Development

Strategies emerge which are rarely optimal. Satisficing, even expediency, are norms. Compromises are made which reflect the competition between existing and new initiatives.

Budget Development
Information Systems
Human Resources
Administrative &
Financial Controls

Implementation must deal with the full range of responses from managers/front-line workers, who may see opportunities or threats from new or changed policies or programs.

Strategic Planning
Policy Development
Sectoral Goals

Effective Strategies

Aligned Management Systems

Clear Objectives

Performance Measurement and Reporting

Real Consequences

What roles do performance results actually play in developing/modifying programs/policies/ objectives? Do elected officials or other stakeholders use performance results?

Electoral Process
Compensation
Funding Decisions
Program Adjustment

Performance Agreements
Performance Measurement
Program Evaluation
Annual Reports
Sectoral Reports

Whose values/agendas dominate the selection of criteria for accountability? Which coalition(s) interpret the "evidence"? Whose interests are served by the "real consequences"?

Managers' interests are affected by evaluation/performance measurement. Managers will seek to influence (where they can) evaluation questions, the choice of measures, the ways data are gathered, and the ways data are interpreted/reported.

Figure 10.1   Public Sector Performance Management: Impacts of Organizational Politics

that many analysts have not included the political/ cultural factors in their models. This has tended to produce a view of performance measurement that relies on unrealistic expectations for organizations that take on the challenge of implementing such a system. Problems that emerge may be blamed on insufficient implementation efforts when, in fact, the problems stem from a simplistic view of how organizations really operate.

INTENDED USES OF PERFORMANCE ● MEASUREMENT SYSTEMS

As new public management has moved to center stage, performance measurement systems have been implemented as a key part of this movement. Intended uses of performance information reflect broad expectations for what program evaluations can do. As we review several key summaries of intended uses, we will see that performance measurement is expected to contribute as well to a wide range of organizational and governmental objectives.

The Government Performance and Results Act (United States Congress, Public Law 103-62, 1993; hereafter GPRA) is the legislative mandate to implement performance measurement and reporting in the United States federal government. Section 2 of the act summarizes the stated purposes of the act:

b) Purposes.–The purposes of this Act are to–

  (1) improve the confidence of the American people in the capability of the Federal Government, by systematically holding Federal agencies accountable for achieving program results;

  (2) initiate program performance reform with a series of pilot projects in setting program goals, measuring program performance against those goals, and reporting publicly on their progress;

  (3) improve Federal program effectiveness and public accountability by promoting a new focus on results, service quality, and customer satisfaction;

  (4) help Federal managers improve service delivery, by requiring that they plan for meeting program objectives and by providing them with information about program results and service quality;

  (5) improve congressional decision making by providing more objective information on achieving statutory objectives, and on the relative effectiveness and efficiency of Federal programs and spending; and

  (6) improve internal management of the Federal Government.

The clear message is that the U.S. federal government is committing itself to a process of moving toward results-based management. Setting objectives, measuring performance, and comparing actual results to intended results are keys to being more accountable. The act also suggests that a results-based performance measurement system will facilitate comparisons among programs to assess their relative efficiency and effectiveness. Recall that in Chapter 1 we introduced efficiency and effectiveness as two of the principal questions that drive program evaluations. The GPRA envisions creating a means to use performance measurement to answer the same questions.

Hatry (1999, p. 158) suggests 10 uses of performance information, which again include the themes of improving accountability, efficiency, and effectiveness:

1. Respond to elected officials' and the public's demands for accountability

2. Help formulate and justify budget requests

3. Help make operational resource allocations and in raising funds

4. Trigger in-depth examinations of why performance problems (or successes) exist and what corrections are needed

5. Help motivate personnel to continue to make program improvements

6. Formulate and monitor the performance of contractors and grantees

7. Provide data for . . . in depth program evaluations

8. Support strategic and other long-term planning efforts (by providing baseline information and later tracking progress)

9. Communicate better with the public to build public trust

10. Above all, help provide better services more effectively

Hatry underscores the value of performance measurement for budgetary and resource allocations. Connecting program results to resources is a major emphasis in the Office of Management and Budget's (OMB) Program Assessment Rating Tool (PART), which is intended to assess and rate federal programs (U.S. General Accounting Office, 2004). The PART assessment process can be seen as building on the GPRA's focus on results. In fact, the General Accounting Office (GAO) has suggested that the requirements for the PART process are intended to closely couple congressional budgetary allocations for programs with intended performance results, and in doing so, replace the more loosely linked resource-results pairings under the GPRA (U.S. General Accounting Office, 2004).

Ammons (2001) describes four types of measures in performance measurement systems for local governments: workload or output measures, efficiency measures, effectiveness or outcome measures, and productivity measures. He suggests that measuring performance is a part of striving for excellence—managers can use these measures to improve accountability, link budgets to results (improve efficiency and cost-effectiveness), support both formative and summative decisions within organizations, and monitor and hold contractors accountable.

Both Ammons (2001) and Hatry (1999), then, suggest that performance information can be used in resource allocation decisions. Mandated public reporting of performance results can be seen as a summative use of performance results. We will later discuss the incentives for organizational managers who are expected to be involved in designing and implementing performance measures that are used formatively and summatively.

Elected Officials as Intended
Users of Performance Information

Elected officials are intended to be principal users of performance information. The widespread adoption of performance measurement systems at all three levels of government in the United States and Canada suggests an expectation that elected officials will use the performance information that is available to them, to be better informed about program results, and in some cases, be able to compare results to expenditures.

We know surprisingly little about the ways that legislators actually use performance information. The broad goals of performance measurement (enhancing accountability, efficiency, and effectiveness) are generally treated as self-evident reasons for designing and implementing performance measurement systems. In advocating for performance measurement systems, including legislating their existence and purposes, it has been assumed that once the information is available, it will be used by principal stakeholders.

Findings From a Study of Legislators'
Expected Uses of Performance Reports

Recently, members of the legislature in the province of British Columbia (BC) were surveyed as part of a 3-year project to assess the ways that performance reports are actually used by legislators. The survey was intended to be a baseline measure, and focused on their expected uses of performance reports. Prior to the 2003 survey, the BC Government had passed legislation, the Budget Transparency and Accountability Act (Legislative Assembly of British Columbia, 2000; hereafter BTAA), mandating performance reporting for all government agencies, but legislators had not yet seen their first performance report when the survey was done.

The survey was distributed to all 79 members of the legislative assembly and 36 of them completed the survey, for a response rate of 46% (McDavid & Huse, 2003). Among the sections in the survey, one asked legislators the extent to which they expected to use the annual performance reports for 15 different activities. For each survey item, legislators could rate his or her expected level of use on a scale from 1 to 5 (1 was "will not use at all," 3 was "will use some," and 5 was "will use to a great extent."

Table 10.1 summarizes the results. In the table, expected uses of performance reports have been grouped to reflect key themes in the literature. All 15 expected uses were rated higher than "some use." Legislators expect that they will use performance reports for a wide range of activities. Although there are no sizeable differences among the five groups of expected uses, *efficiency/ effectiveness* has the highest overall weighted group mean (3.86). The second highest group of expected uses was *communication/information* (3.79), then *accountability* uses (3.61), and then *budgetary decision* uses (3.41).

The optimism of BC legislators is also reflected in their responses to a question asking them to identify the biggest single downside risk of performance reports (Table 10.2). The largest category of responses was that there were "no downside risks."

The findings from this study suggest that legislators expect to use performance reports in many different ways. Some of the expected uses (*program/policy decisions* and *budgetary decisions*) imply that the legislators believe performance information can be substituted for program evaluations. Although analysts have generally recognized the complementary and mutually supportive natures of program evaluation and performance measurement (Hatry, 1999; Mayne, 2001), there has been a tendency for advocates to substitute performance measurement for program evaluation (Perrin, 1998). If performance measurement systems are going to be sustainable, they need to be constructed with realistic expectations about the ways they can actually be used.

Promoting Usage by Auditing Performance Reports

In the United States, some state governments have committed themselves to auditing the quality of performance information. In Texas, for example, the state auditor's office plays an important role in assuring stakeholders that the performance information included in quarterly and annual reports is accurate (Aristigueta, 1999). We will look more closely at Texas as an exemplar for performance measurement later in this chapter.

Table 10.1   British Columbia Legislators' Expected Uses of Public Performance Reports

Expected Uses of Reports	Mean	n	Group Mean
Efficiency/Effectiveness			
Provide deputy ministers and CEOs with information for better managing their organizations.	3.52	35	
Identify performance problems in ministries and Crowns.	3.92	36	3.86
Improve the quality of government services.	3.89	35	
Assess the efficiency and effectiveness of ministries and Crowns.	4.11	34	
Communication/Information			
Improve communications between elected and administrative officials.	3.52	36	
Improve my understanding of ministries and Crowns.	3.92	35	3.79
Show constituents what ministries and Crowns are doing.	3.94	34	
Program/Policy Decisions			
Develop or revise policies and programs.	3.60	33	3.70
Make decisions about the continuation of policies and programs.	3.79	33	
Accountability			
Hold the Government accountable to the citizens of British Columbia.	3.74	36	
Hold ministers and Crown Boards of Directors accountable.	3.67	36	3.61
Hold deputy ministers and Crown CEOs accountable.	3.43	36	
Budgetary Decisions			
Allocate budgetary resources among ministries.	3.38	35	
Reduce the costs of programs and services.	3.32	32	3.41
Support Executive Council decisions.	3.53	31	

In Canada, provincial auditors have played a key role in urging legislative assemblies across the country to implement performance measurement systems. Half of the 10 Canadian provinces have passed legislation that mandates performance measurement and public reporting (Manitoba Office of the Auditor General, 2000). The Canadian Comprehensive Auditing

Table 10.2 British Columbia Legislators' Survey: Expected Single Biggest Benefit and Single Biggest Downside of Having Annual Performance Reports from Ministries and Crown Corporations

Legislator Uses of Annual Service Plan Reports	Members of the Legislative Assembly
Single Biggest Benefit	n = 29
For accountability/improved accountability	41.4%
Can assess/evaluate actual versus intended results	17.2%
Inform legislators and/or the public	13.8%
Encourage a focus on results	10.3%
Improve the budgeting process	10.3%
Other	6.9%
Total	99.9%[1]
Single Biggest Downside Risk	n = 28
No downside/risks	25.0%
Missing the targets that were set	21.4%
Going through the motions only	17.9%
The time and costs to produce reports	7.1%
Reduces organizational flexibility	7.1%
Too much organizational openness	7.1%
Biased reporting of results	3.6%
Other	10.7%
Total	99.9%[2]

1. Does not total 100% due to rounding.

2. Does not total 100% due to rounding.

Foundation (CCAF), a nonprofit advisory body that plays a major role in advising the provincial auditors on trends and best practices, has made the promotion of performance measurement and accountability one of its principal activities (CCAF-FCVI, 2004). Most recently, the CCAF has facilitated the development of nine reporting principles, or guidelines, designed to "help governments advance the quality of their formal reporting on performance, in keeping with the results-oriented and values-based approaches they are taking" (CCAF-FCVI, 2002, p. i). A further goal of this effort, once governments become more comfortable and experienced in using the principles, will be to have the annual performance reports audited, to provide independent assurance that the reports are fair, accurate, and written so that any reader can see how the agency performed. Promoting an audit function for governments that mandate performance reporting is intended to increase the credibility of the information in the reports and increase the likelihood that stakeholders will come to rely on the reports and use them in decision making.

ACTUAL USES OF PERFORMANCE MEASUREMENT ●

We summarize several studies that have examined the ways that performance measurement is actually used in public and nonprofit organizations. In the United States, research has been done for all three levels of government (federal, state, and local). Most of the research is based on surveys of intended users of performance measures or performance reports, although there have been several studies based on case studies of particular jurisdictions. In Canada, there is relatively little research, and the studies that have been done are limited in scope.

American Studies of Actual Uses of Performance Measures

American jurisdictions that have been studied generally report mixed levels and types of uses for performance information (Aristigueta, 1999; Streib & Poister, 1999). In this section, we will review several studies of the U.S. federal, state, and local governments.

U.S. Federal Government Uses of Performance Measures

How program decision makers use performance information gathered is shown in the following example of an assessment of a federally funded job training partnership agency in Chicago (Heinrich & Lynn, 1999). The agency contracts out most of its training and job placement activities and requires all contractors to provide performance information that can be used to monitor and to determine renewal of contracts. Using a pooled time-series cross-sectional sample of contractors, Heinrich and Lynn tested a multivariate model that predicted whether contracts have been renewed or not. Two of their findings are worth reporting: first, lower contractor costs per job placement, an efficiency measure, predicted successful contract renewals; second, the most significant predictor of contract renewal was job placement performance—the *poorer* the contractor's record, the more likely they were to have their contract renewed (Heinrich & Lynn, 1999). They speculate that agency politics influences the linkage between placement performance and contract renewal—a suggestion that accords with the general point made in Figure 10.1 that political and cultural variables will affect all stages of the "rational" performance management cycle.

Uses of Performance Information in U.S. States

U.S. states have been surveyed several times by researchers interested in measuring how performance information is actually used. Moynihan and Wallace (2001) surveyed all 50 U.S. states as part of the Maxwell School's Government Performance Project. The survey focused on the extent to which "[p]erformance information is frequently used by executive branch officials in decision-making" (Moynihan & Wallace, 2001, p. 8). Respondents were officials responsible for strategic planning and performance measurement.

Forty-nine of the 50 states responded to the survey. There are three key findings: states where leaders become publicly involved in strategic planning and goal setting tend to report higher levels of performance information uses; states where the performance information is more accessible to the public tend to use it more; and states where the reported credibility of performance information is higher tend to report higher levels of usage (Moynihan & Wallace, 2001).

Minnesota passed a law in 1993 that mandated performance measurement and reporting by all major state departments and agencies. The first performance reports were completed in 1994. The law provided for an assessment of the uses of the performance reports by the legislative auditor. In a 1995 study (State of Minnesota Office of the Legislative Auditor, 1995), surveys and interviews were conducted with the senior public servants in all reporting agencies as well as the internal agency committees that were involved in the development of the performance reports. Legislators were not interviewed. The legislative auditor concluded that although there was some usage of some of the reports, the results were disappointing. Few of the reports were given much time with respective legislative committees. The study recommended significant changes in the process of preparing the reports including a much shorter report with more graphics. The legislature voted in 1997 to repeal the act that mandated annual public performance reporting.

The Minnesota Department of Transportation (MN/DOT) was one of the state agencies that was required to develop performance measures and report results annually. MN/DOT decided to integrate its performance measurement and reporting requirements into a comprehensive performance management system (Feit, 2003). This process was started in 1995 but eventually stalled as the organization struggled to measure all its important activities. At one point over 700 performance measures had been established in efforts to create some kind of overall scorecard for the department. When a new administration came to power in 1999, the MN/DOT initiative was re-started, but this time about 125 measures were selected. The major innovation was using a dashboard model to display and color-code the measures. In all, 45 dashboard gauges were created to display all the measures. Each gauge had three zones: green, amber, and red (denoting, respectively, "normal," "needed attention," or "needed immediate attention"). The dashboard metaphor has succeeded in communicating performance results in ways that are easily understandable and increasing the likelihood that they will be used.

Feit (2003) suggests that one of the key lessons learned by the MN/DOT experience is the importance of consistent, long-term leadership in this process. Interestingly, even after 8 years of implementation efforts, the department is just now linking funding requests to projected service levels. An example would be requesting funding sufficient to deliver on a commitment to have bare lanes within 10.9 hours state-wide, after a snowstorm. To support this budgeting process, the department will have to use **activity-based accounting** so that each service is separately costed out (Feit, 2003). The final step that will be implemented is to link manager rewards to objective performance information, and realizing that step will mean a fully implemented performance management system has been achieved for MN/DOT.

Performance Measurement Use in Texas

The State of Texas has been measuring performance of its agencies since 1991 (Aristigueta, 1999). Having more than 20 years experience with state-wide performance measurement, Texas provides us with a good example of the longer-term effects of implementing and using results-focused performance measures.

Texas has integrated performance measurement into a comprehensive performance management system that has four stages: planning, budgeting, implementation, and evaluation (Aristigueta, 1999). Beginning in 1998–1999, public performance reporting has been mandatory, including both quarterly and annual reports for state agencies.

The state auditor has a key role in ensuring the credibility of the performance measures that are an essential part of the performance management process. In addition to publishing reports that are intended to assist state agencies in performance measurement development, collection, use, and reporting, the state auditor annually assesses the reliability and accuracy of performance measures. In one case study summary of the Texas performance measurement system, the state auditor determined that 80% of the performance measurement data was accurate (Tucker, 2002).

Overall, external assessments of the Texas system have been very positive. The Government Accounting Standards Board (GASB) State of Texas Report (Tucker, 2002) summarizes Texas' achievements this way:

> The State of Texas is an excellent example of a state government that has effectively institutionalized the use of performance measures. Performance measures have been used for more than twenty years, and the most notable improvements have been the legislative requirements for including performance measures in the printed budget beginning in 1991 and the actual use of performance measures in the budget allocation process. The most recent initiative has survived changes in the executive administration and appears to be thoroughly institutionalized in the state government. (p. 2)

A similar assessment of the Texas system is offered in Aristigueta (1999):

> Performance data is used extensively in various Texas decision-making and accountability systems. Agencies use the information to shape their strategic plans and request funds for legislative appropriations. They also, to varying degrees and in different ways, use the information for strategic and operational management as well as program and employee evaluations. (p. 123)

Both of these assessments are based on interviews with stakeholders, many of whom had a stake in how well the system was seen to be implemented and used. A different perspective on the Texas performance measurement system is offered in the 2001 Audit Report by the Texas State Auditor's Office. Even though the GASB case study suggested that over 80% of the measures are accurate, the state auditor reaches a different conclusion, having audited the actual data collection and reporting process in 14 state agencies:

> Serious deficiencies exist in the methods 12 of the 14 audited entities use to collect, calculate, and report key performance measures. The deficiencies are a combination of inadequate controls over the performance measurement process, entities' not following their definition of the measures, and entities' not being able to support their results. . . . These

deficiencies compromise the usefulness of the State's performance measures system as a decision-making tool. Decision makers cannot rely on reported results for 53 percent of the audited measures, all of which were key measures, that the 14 entities reported for fiscal year 2001. This is the highest rate of unreliability since the State Auditor's office began certifying performance measures in 1994. (Texas State Auditor's Office, 2002, p. i)

The lesson from this case is that assessments of the efficacy of performance measurement systems that rely on interviews with key informants offer perspectives based on one line of evidence. As Leviton (2003) has pointed out, relying on interviews alone to assess evaluations is not adequate if our goal is defensible findings.

Local Government Uses of Performance Measures

Local governments have also been surveyed to find out whether and how performance information is being used. Streib and Poister (1999) did a survey of senior local government officials in U.S. communities with a greater than 25,000 population to find out whether and how performance measurement systems had been developed. Although they achieved a 57% response rate overall (695 survey responses out of 1,218 mailed) they learned that only 38% of those responding indicated that their jurisdiction used performance measures. Their study focused on local governments that indicated that they had a centralized city-wide performance measurement system (159 cities). Streib and Poister, using Bouckaert's (1993) typology, assessed the perceived validity, legitimacy, and functionality of performance measurement systems. The survey items that measured functionality are most directly tied to actual uses of performance measures. For each of the 12 potential uses, respondents were asked to indicate whether impact was substantial, moderate, slight, or none. A majority of managers indicated that there was a substantial or moderate impact for performance information being used to:

- Improve the accountability of individual managers
- Improve the level of employee focus on organizational goals
- Improve service quality
- Improve the quality of decisions
- Change the focus of programs
- Improve community relations
- Change budget allocations to departments and programs

On the other hand, the majority of respondents indicated no impact or slight impacts for: reductions in the costs of city operations; improvements in the relationship between administrators and elected officials; improving the objectivity of personnel performance appraisals; and improving the level of employee motivation.

Summary of Actual Uses of Performance Reports

The evidence for actual uses of performance reports suggests that although there is a broad movement to implement performance measurement at federal, state, and local levels, actual implementation is quite mixed. Many jurisdictions have publicly committed to measuring performance, and a superficial reading of websites or even survey results might suggest widespread implementation. But, more detailed studies of implementation and actual uses suggest that there is a gap between what is intended and what actually occurs.

An additional comment is in order in this section. The sample of actual use studies included here is intended to illustrate the range of research that has been done to date. The discerning reader will note that most studies rely on interviews (usually survey-based) and self-reports of uses. Most respondents are stakeholders and often are managers who have some direct responsibility for designing and implementing their performance measurement systems. As we see in Chapter 11, managers who participate in program evaluations or performance measurement processes face various incentives, and act accordingly. Asking a program manager to be the principal client for a summative program evaluation, for example, might well force that person to act in ways that protect his or her program—compromising the neutrality of his or her views on whether the program should be reduced in scale. Asking persons who have played a key role in developing a performance measurement system whether and how it is used creates a similar problem—their responses in a survey would tend to reflect their own investment in making the system work.

● PROBLEMS AND ISSUES IN IMPLEMENTING AND SUSTAINING PERFORMANCE MEASUREMENT SYSTEMS

It is important to step back from the process of designing and implementing performance measurement systems to consider some of the broader challenges that affect the likelihood that this kind of initiative is here to stay. Table 10.3 summarizes challenges that can make a difference as to whether and how well performance measurement succeeds in public and nonprofit

Table 10.3	Challenges in Implementing and Sustaining Performance Measurement Systems

1. Lack of fit between organizational needs and performance measurement solutions.

2. The complexity of accountability relationships in public and nonprofit organizations.

3. The attribution problem—replacing program evaluation with performance measurement.

4. The levels of analysis problem—assuming that organizational performance tells us about program performance.

5. Gaming—anticipating and responding to the unintended consequences of performance measurement.

organizations. Each of the five issues in Table 10.3 is elaborated in the discussion below.

Lack of Fit Between Organizational Needs and Performance Measurement Solutions

Kravchuk and Shack (1996) have pointed out that committing an organization to performance measurement has implications for the ways that information is gathered, filtered, and used by managers. Their concern is that the structured nature of performance measurement processes will tend to filter out information sources that might otherwise be used to make decisions. Another way to look at this issue is to acknowledge that complex public sector organizations may be faced with a paradox. On the one hand, performance measurement is intended to facilitate the results-based management that frees organizations from "red tape" and other rigidities. But the nature of performance measurement systems, particularly where they try to capture the complexity of programming in larger public sector organizations, is that they tend to be rigid and can become a means of centralizing decision making (Kravchuk & Schack, 1996).

Most importantly, designing and implementing performance measurement systems have implications for the ways that managers function in public sector organizations:

Decision-makers should, therefore, not view performance measurement as an excuse to relax their vigil; measures are not a substitute for effective ongoing program management and periodic formal evaluations. An excessive reliance on measures can reinforce certain pathologies for which modern management has been severely criticized, such as a mechanistic approach to decision-making, obsessively managing by the numbers, over simplification, distancing themselves from the work, and "group-think." (Kravchuk & Schack, 1996, p. 356)

Nonprofit organizations are also increasingly being asked to account for the funds they consume, by developing and implementing performance measures. These measures tend to be focused on outcomes, and for many organizations, measuring these outcomes is challenging and forces nonprofit organizations to reach for results that may well be beyond their control to deliver. Managers who want to use performance measures would prefer to focus on results they can manage; forcing a focus on outcomes may result in divorcing the performance measurement system from managerial information needs.

The Complexity of Accountability Relationships

What *is* accountability? Are there kinds of accountability that compete or even contradict each other? Which ones do performance measurement enhance? In this section we review several different models of accountability that can be applied to public and nonprofit organizations: the *results-based accountability model,* the *hierarchical model,* and the *multi-organizational model.* Accountability models are normative, that is, they tell us what *should* or *ought* to happen in organizations that aspire to efficient and effective administrative standards and practices. But the politics and culture of organizations and whole political systems have significant and potentially confounding impacts on what ought to happen. Marrying *ought* to *is* will give us a different picture of the likelihood of success for implementing and sustaining performance measurement systems.

Accountability for Results

Performance measurement is intended to focus organizational managers and other stakeholders on results—ideally, outcomes. If organizations are managing for results, then accountability is about being accountable for results. When an agency sets objectives, develops performance measures for those objectives, and sets yearly targets, we have an ideal scenario for measuring results, comparing results to targets and, if necessary, accounting for the differences. Presumably, program and organizational managers have agreed on the targets, so they should be prepared to live with the actual results and take responsibility for the good, the bad, and the ugly.

Accountability in Systems Where Third Parties Deliver the Programs

Contracting for services is now a key part of the way governments meet their mandates. Increasingly, social services are being delivered by agencies that are under contract with government departments. Contractors are now expecting provider agencies to account for their results; performance contracting is becoming an important way to extend results-based management principles to third-party delivery of services.

Agencies under contract are often nonprofit organizations in the business of delivering services because their members have a normative commitment to contributing to the welfare of society. For them, performance contracting represents a major shift in values from a funder–provider relationship based on trust, to one where providers are expected to compete for funds and are also expected to deliver specified results. Performance contracting requires them to develop the capacity to measure and report results, and this is a challenge for many smaller nonprofit organizations (see, for example, Hall, Phillips, Meillat, & Pickering, 2003).

What complicates these accountability relationships is the fact that many provider agencies receive funding from multiple sources, some of it from contracts, some of it from grants. From a provider perspective, accountability requirements for different funders may not coincide and may require parallel regimes to measure and report performance. The duration of contracts (typically a year or more) and the focus on delivering services can mean that agencies do not have the resources to build the capacity to track and document their performance. Further, because agencies typically see themselves as accountable to multiple stakeholders, their view of accountability may be quite different from a typical funder's perspective. Funders that want performance measures focused on outcomes may encounter opposition from service providers who argue that they cannot be held responsible for outcomes that are affected by factors beyond their control.

Sawhill and Williamson (2001) have suggested that successful outcomes-focused performance measurement in community-based agencies requires a view of service delivery that involves both interrelated provider agencies and their funders. Few, if any, social service problems conform to existing organizational boundaries or mandates. A collaborative approach among service providers and funders would facilitate funding and services that meet a fuller range of needs for a client population. But a collaborative approach requires stepping back from the current single-agency–focused competitive bidding model, which tends to compartmentalize both service delivery and funding.

Input and Process-Focused
Accountability: The Hierarchical Model

Accountability for results is part of a normative, rational view of the whole performance management cycle—a view that does not fully reflect the political and cultural dimensions of organizations, local governments, provinces, states, and national governments. The reality is that accountability relationships are more complex and even contradictory for public and nonprofit organizations. In Canadian provinces and the federal government, for example, there continues to be a tradition of ministerial responsibility that implies that the elected minister who heads up each public organization is responsible for all that happens in his or her organization.

Ministerial responsibility is coupled with ministerial accountability. Because ministers are elected officials and are expected to be accountable, the managers in the hierarchy below them are in turn accountable to the minister. This upward-focused hierarchical accountability is succinctly summarized as follows:

> Accountability is the obligation to account for responsibilities conferred. In the public sector, this means that each manager, in both ministries and Crown corporations, is accountable to a superior for managing the responsibilities and resources provided. At its highest level, it means that government is similarly accountable to the Legislative Assembly for its performance in managing the responsibilities and resources entrusted to it. (Auditor General of British Columbia & Deputy Ministers' Council, 1996, p. 23)

Upward-focused accountability has a major impact on the culture of organizations. In parliamentary systems and Western democracies wherein the opposition party or parties are a key part of the political culture, ministers of the Crown (in Canada, for example) or cabinet secretaries (in the United States) generally want to avoid being seen to make mistakes. Given the importance of being seen to be running ministries or departments that are not doing politically embarrassing things, an important norm in the cultures of public organizations is to rely on rules and procedures and expect that these will be followed. "Following the rules" is often the way we characterize "bureaucratic" behavior in organizations. Charges of being rule-bound and inflexible are often coupled with charges of being inefficient and ineffective. Indeed that was one of the substantial criticisms of public bureaucracies leveled by Osborne and Gaebler (1992).

Another way to look at this situation is to recognize that we have the potential for an accountability paradox for public sector organizations. In a culture wherein "avoid making mistakes" is a key norm, most managers will respond to the incentives associated with this norm. Entrepreneurial behavior or initiatives, because they usually involve trying new things that tend to be more risky, will be less likely to be rewarded than behavior that demonstrates a capacity to deal with issues and problems in ways that do not create public embarrassment for the political leaders in the organization.

But new public management norms emphasize managing for results, and along the way, taking the risks needed to innovate and improve efficiency and effectiveness. Accountability for results has quite a different focus for managers than accountability for process. Doing one can be seen as trading off with the other, particularly in organizations where resources are scarce.

The paradox is that our normative organizational culture (what public organizations *ought* to be doing in Western democracies to respond to the challenges of new public management) urges accountability for results whereas the contemporary political culture is still situated in a view of accountability that is centuries old (Hatt, 2004). Accountability for results may well be an acceptable model as long as mistakes are not visible. But as soon as a mistake becomes visible the accountability that is expected is focused on process. Organizations that have moved to a results-based management can be caught not having sufficient resources focused on process (Good, 2003).

If performance information is being used formatively, that is, within the organization for program improvements, then feedback on what is being done well and what is not being done well can be a useful tool for managers. The risks of being seen to have made a mistake are generally not as high, so managers will tend to rely on this kind of information—very often they are substantially involved in its creation in the first place. But summative uses of such information where performance results are made public create the potential for criticism and in political environments where opposition parties, interest groups, and the media are looking for evidence of government mismanagement, performance data can become a liability. Under such conditions, organizational participants have an incentive *not* to implement or use performance measures, or if they do, to make these measures tell the best possible story.

Accountability Where Organizations or Governments Share Program Responsibilities

Neither the results-based model nor the hierarchical model is fully adequate to address accountability issues raised by a growing number of multi-organizational arrangements to design and deliver programs. Horizontal management arrangements wherein departments coordinate their efforts to achieve common objectives are becoming more common (Bakvis & Juillet, 2004; Pal & Teplova, 2003). Governments are entering into partnership arrangements across levels of government and with the private sector to produce and deliver goods and services. Partnerships are intended in part to extend financial resources beyond what any one partner would be able to bring to the table. Tuohy (2003) has called these arrangements loosely coupled networks. Networks have a strong aspect of horizontality that is intended to create the flexibility to respond to problems and issues that might otherwise fall between the cracks of existing bureaucracies.

An example of such a network has been developed in Vancouver, British Columbia to tackle a complex array of social, economic, and legal problems in Vancouver's downtown eastside neighborhood (Bakvis & Juillet, 2004). In the past two years, the BC Provincial Government, the City of Vancouver, and the Federal Government of Canada, together with a number of nonprofit organizations have created the Vancouver Agreement. Although this "organization" has succeeded in constructing a means of tackling the complex of problems in this community, existing accountability relationships that rely on organizational hierarchies are a constraint for the stakeholders. Because funds are mixed together to implement programs and because the processes that transforms these fund into results are dynamic, individual funders are not able to track the success of their contributions. Accountability relationships are multiple, tangled, and obscured (Tuohy, 2003). There is growing recognition that multi-organizational arrangements for tackling complex social problems are efficacious, but also a genuine concern that these networks are not amenable to the same kinds of scrutiny we expect from other public sector organizations.

Attributing Outcomes to Programs

The typical scenario for outcome-focused performance measures is that changes in the measurement values over time or across districts or client groups are *assumed* to tell us something about what the program accomplished. We do not usually build into performance measurement systems the kinds of comparisons that yield counterfactual "program vs. no-program" scenarios—the paradigmatic case for program evaluations that was suggested in Table 8.1.

Attribution becomes more of a problem the further away we get from program outputs. In other words, attributing the behavior of longer-term outcome variables to a program is far more challenging than attributing the variations in outputs or linking constructs to the program.

If program managers are expected to be involved in developing, implementing, and using performance measures (particularly measures that focus on outcomes), the *intended uses* become important. There will inherently be more incentives for managers to take "ownership" of measures that are used formatively.

Summative uses of outcome results create situations where managers, knowing that they often do not control the environmental factors that can affect their program outcomes, perceive a dilemma. If they are expected to develop and use performance measures, they can be called to account if they fail to do so. But if they do develop and use performance measures, they could be held responsible for poor performance data that reflect factors that they cannot control. In most public organizations, new public management initiatives notwithstanding, procedural rules and restrictions still mean that even if managers could see ways of changing programs or program environments to increase the probability of success, they have limited freedom to do so.

The consequence of this dilemma may be that managers develop performance measures so that "on paper" the system exists, but find ways of delaying full implementation until opportunities arise that can be used to redirect efforts away from performance measurement in the organization. Recognition of this situation is a key part of planning for performance measurement. Resolving it depends on altering the incentives for those who are expected to develop, implement, and use performance measures.

There is a related problem for organizations that deliver human service programs. In Chapter 2, we introduce the idea that programs vary in the robustness of their intended cause and effect linkages. Programs that are based on engineering knowledge, such as a highway maintenance program, typically incorporate technologies that have a high probability of success. If the program is implemented as planned, it is highly likely that the intended outcomes will occur.

At the other end of the continuum, there are many programs that operate with **low-probability technologies**. In general, these programs focus on human service issues, and involve efforts to ameliorate or change human conditions, knowledge, attitudes, or behaviors. Even if these programs are fully implemented, we often observe mixes of program successes and failures. Our knowledge of what works tends to be far less certain than is true of **high-probability technologies**. There are usually important environmental variables that affect both program activities and outcome variables. In many situations it is not possible to mitigate the influences of these variables—low-probability technologies tend to be much more vulnerable to external influences.

For program managers who are involved in delivering programs that incorporate low-probability technologies, their best efforts may not succeed, simply because we do not know how to navigate the factors that can impact on program outcomes. For these kinds of programs, being accountable for outcomes is daunting. Aside from the challenges of measuring intended outcomes, program managers will point out that outcomes are not really under their control. An example might be a social welfare agency that has a program focused on single mothers. The objective of the program is to support these women in their efforts to obtain permanent jobs. The logic model for the program emphasizes training and child care support during the training. But the single mothers may have more needs than this program can meet. Some may need to deal with substance abuse problems or deal with former partners who continue to harass the family. Some may need to deal with psychological issues from their own family of origin.

Single programs typically will not be able to address this range of issues, so to hold any one of them accountable for the program objective is not appropriate. Even if a comprehensive program *could* be designed and implemented, the state of the art of our knowledge of how to achieve the objective is likely to mean that there will be a lot of partial successes and failures in program outcomes. Holding managers to account for their *efforts* is appropriate, but holding them to account for the outcome may easily result in gaming behaviors.

The Levels of Analysis Problem

In the discussions on performance measurement, we have generally referred to situations where *program* performance is being measured. The logic modeling approaches discussed in Chapter 2, the research design considerations included in Chapter 3, and the measurement issues in Chapter 4 are focused on evaluating programs. However, performance measurement can occur at different levels within and between public and nonprofit organizations.

We can measure the performance of individuals, the performance of programs or administrative units, the performance of organizations, the performance of sectors (the health and social services sector of a government, for example), and the performance of whole governments. In some performance measurement systems where public reporting has been mandated, there is a problem with the conceptual framework for performance measurement. Where government departments are expected to report on their performance there appears to be an assumption that if we measure the performance of whole departments, we can assume that program performance within those organizations is known. If a department is meeting its performance targets, we might conclude that its programs are also performing well. Although this assumption simplifies the task of measuring performance, it is *not* the case that one level of analysis is equivalent to the other. Knowing how ministries are performing is not equivalent to knowing how their programs are performing, or vice versa.

This *level of analysis* problem can be extended. Performance measurement can also focus on how well individuals are doing in an organization or a program. Again, it might seem reasonable to assume that if an organization is performing well, then the people are also performing well. But as managers know, there is more to effective organizations than the programs and the people who deliver them. It is fallacious to assume that knowledge of performance at one level implies knowledge of performance at other levels. To fully measure performance, it will be necessary to measure it at *each level* in the organization (individual, program, organization as a whole) in order to obtain a credible picture of how well the organization is doing.

The Problem of Gaming Performance Measures

Performance measurement is intended to get organizations and managers to focus on the results of what they do. To do so, objectives for programs must be clearly articulated, and valid measures that connect with the constructs in those objectives need to be developed. Because programs are a part of the political process in our society, clearly stated objectives that can assist managers in knowing what they are expected to accomplish, can also be viewed as a liability. Many programs have different stakeholder groups who may not even share the same expectations for the program, so there are advantages in articulating objectives that encompass a range of possible interpretations. That way, programs cannot be criticized for excluding an important issue or stakeholder group.

Performance measures are intended to link constructs to observables and provide evidence that can be used to compare and even judge performance. The measures themselves become a powerful incentive for managers, and affect the way objectives are actually translated into organizational actions.

Examples of Performance Measures
That Have Created Unintended Behaviors

There is a rich history of examples that suggest that once measures are established and used to assess performance, they can produce unintended organizational behaviors (Frank, 1958; Kerr, 1975; Ridgway, 1956). Some of the stories of performance measures gone awry are humorous, even if they are apocryphal. Courty and Marschke (2003) cite a Soviet example of an industry that produced chandeliers—the performance measure focused on weight and the result was the heaviest chandeliers in the world. Perrin (1998) cites a similar example from Poland: the furniture industry was given a target of so many tonnes of furniture—the result was the heaviest furniture in the world.

In a more extended example, Otley (2003) recounts a story based on his early experience as a British mining engineer. His first project was to develop a computer model of a coal mine. Using an existing model of how a single coal face operated, he quickly extended this model to a whole mine. Validating the model involved comparing the model's predicted mine outputs with data from the actual mine. The model predicted average output quite well, but could not predict the variability in output. Since the model was intended in part to assist in the design of an underground transportation system, peak loads needed to be accurately estimated.

Otley assumed that he had made some kind of programming error—he spent several weeks searching for such an error, to no avail. He decided to look at the actual data themselves to see if anything emerged. The weekly data had patterns. The mining output data showed that for a typical Monday through Thursday, actual tonnes of coal produced conformed pretty closely to a budgeted target for each day. But on Friday, the actual tonnes could be anything from much more to much less than the daily average. It turned out that the mine managers knew that for every day of the week but Friday, they could report an output to headquarters that was close to the budgeted output because the actual tonnes were only totaled up on Fridays. To reconcile their reported figures with the weekly total (being on budget with actual production was their performance measure), they used the Friday figure creatively.

The mine managers had created an additional way of assuring that they met the weekly production targets. At the bottom of the mine shaft there was a bunker that was supposed to be used to store coal that could not be transported to the surface during a given day. The bunker was supposed to be emptied on Friday, so that it could be used to buffer the next week's daily production—the hoist that brought the coal to the surface was a bottleneck and the bunker was a way to work with this problem. But Otley discovered that the bunker was often full on Monday mornings—the managers reckoned that having a full bunker to start the week meant that they had a leg up on that week's quota and since the penalty for under-producing was greater than any consequence for overproducing, they responded to the incentives.

Mine managers had developed ways to game the performance measure for which they were accountable. For Otley's modeling process, the output data were not sufficiently accurate to be useful.

A further example: In Canadian hospitals, there is currently a shortage of hospital beds (Conference Board of Canada, 2004), particularly for urban areas of the country. A key performance measure now widely used is the length of stay for each service or diagnostic grouping. The focus is on minimizing the length of stay and comparing length of stay for diagnostic groupings with peer hospitals.

This practice has an important effect on patients, particularly those who are recovering from surgery. Working to that performance measure has significantly altered the way patients are expected to cope with postoperative recovery. In addition, this measure does *not* focus on patient outcomes—they are generally more difficult to measure and are therefore not generally used.

In his discussion of the development of performance measurement systems in Great Britain, Shaw (2000) cites an example of a measure put in place to specify maximum waiting list times for medical operations. In response to this performance measure, health providers may begin placing patients on a "consultant review" list, where they are seen by medical consultants to determine whether and when they need to be placed on the official waiting list. In effect, this process is creating a hidden waiting list for operations, where waiting times are not being monitored. In turn, this situation raises a concern that actual public service needs are being under-reported and hence unrecognized by both citizens and politicians (Shaw, 2000).

In Britain in the early 1990s, the Thatcher government mandated that UK universities would be ranked periodically on their research quality. Funding allocations for the subsequent time period would depend on relative ranks on several measures, a principal one being the number of publications (Pollitt, 2000). Universities quickly adapted to this regime and, within a short period of time, new journals had sprung up, publications had increased, and some departments had started the practice of recruiting "high producers" just before the next assessment, to augment their own scores on the performance measures.

Understanding the Incentives for Gaming

When performance measures are used to shape organizational and individual behavior, we can expect people to respond to these measures and adapt their behaviors to them. As measures are introduced, persons affected by them will assess whether they need to pay attention to them; some performance measurement systems function purely at a symbolic level, with no real consequences to stakeholders if performance varies up or down over time.

If there are consequences attached to performance results, gaming is likely to occur. People will figure out how to adjust their behaviors to reflect desirable levels of the measures that affect their work and their own well-being. Most measures can capture only parts of an intended objective, so working to performance measures may leave gaps between what is sought and what is achieved—performance is "on target" but it misses important behaviors and creates others that can even undermine the intended outcomes.

Gaming is dynamic. In their discussion of the Job Training and Partnership Act in the United States, Courty and Marschke (2003) show that over its 20-year history there have been several important changes in the key performance measures to offset the behaviors of contracting organizations. They suggest that designing and implementing a performance measurement system is an evolutionary process—there is no point where it's safe to rest on one's laurels.

A premise of the new public management movement has been to decentralize management so that managers have the freedom to do their jobs efficiently and effectively. Performance measurement is intended to support this by focusing on results and encouraging managers to innovate and improve the ways they convert resources into results. But gaming increases centralization (Courty & Marschke, 2003). That suggests that over time, organizations that have adopted performance measurement systems will experience pressures to increase the level of control over the ways that the measures are applied and interpreted—in effect, increasing centralization.

How to mitigate these problems? Ridgway (1956) suggests that relying on one measure is the least preferable alternative. Gaming would most likely occur where one measure is used. Using multiple measures, each with its own validity and reliability profile, offers triangulation and reduces the likelihood that any one measure will become the focus of gaming behaviors. Ridgway also goes on to suggest that multiple weighted measures (combined into some kind of index) are ideal for measuring performance. However, finding weights that accord with the organizational expectations, especially in the public sector, is usually very challenging.

The bottom line is that a commitment to measuring performance and using this information will affect the behaviors of those affected by the ways the data "perform." Behaviors that are consonant with intended outcomes are desirable. But it is always possible that unintended and perhaps undesirable behaviors will also result. The higher the stakes, the more likely that gaming will occur.

SUMMARY ●

Most models of implementing performance measurement are premised on the process being primarily rational and technical. However, implementing and sustaining performance measurement systems must take into account the political and cultural factors in all organizations. Improving accountability, efficiency, and effectiveness are consistently cited as expected uses of performance measurement systems. In fact these expectations have been built into some legislative mandates for performance measurement.

Elected officials are expected to be users of performance information in systems that produce performance reports. In one empirical study, elected legislators had high expectations about the ways that they will use performance reports. They see performance reports helping them to: improve accountability; communicate with stakeholders; improve efficiency and effectiveness; make program and policy decisions; and make budget decisions.

Actual uses, as reported in surveys, suggest that performance measurement systems have not yet realized their potential in many jurisdictions. Implementation takes time—in one study of the Minnesota Department of Transportation, it has taken 8 years for the performance measurement system to be able to provide reliable and accurate information on performance to customers and stakeholders so that they can discuss choices and priorities.

There are two accountability models embedded in our democratic culture: hierarchical accountability and accountability for results. Neither model covers all situations in which we demand that elected and appointed officials be held accountable. Holding managers accountable for results does not relieve them of the burden of being accountable for processes and inputs.

Increasing numbers of public sector and nonprofit organizations are expected to function as partners, either with other public organizations or with private sector organizations. There is currently no accountability model that adequately covers these situations.

Designing, implementing, and sustaining performance measurement systems is complex and involves organizational change. Performance measures can be used formatively or summatively and the uses can and do change over time. Because measuring performance can have both program and even personal consequences for managers and executives, gaming is a fact of organizational life. Gaming affects the validity of performance information and the behaviors of organizational participants. Gaming is dynamic, that is, it evolves over time. Responses to gaming will reduce it, but probably never eliminate it.

● DISCUSSION QUESTIONS

1. What are the key differences between the technical/rational view of implementing performance management in organizations and the political/cultural view?

2. Some commentators have suggested that failures of performance measurement systems to live up to their promises are due to poor or inadequate implementation. This view suggests that if organizations properly implement performance measurement, paying attention to what is really needed to get it right, performance measurement will be successful. Another view is that performance measurement itself is a flawed idea and that no amount of attention to implementation will solve its problems. What are your views on this issue?

3. Will auditing performance reports increase their usefulness? Why?

4. If you were making recommendations to the governor of the State of Texas about ways of improving their performance measurement system, what recommendations would you make?

5. The record on uses of performance reports suggests that they are used by decision makers to some extent, but there is room for improvement. What three things would you suggest to make performance reports more useful for elected decision makers?

6. What is the *levels of analysis* problem in performance measurement systems?

7. What does it mean for organizational managers to "game" performance measures? What are some ways of reducing the occurrence of this problem?

● REFERENCES

Ammons, D. N. (2001). *Municipal benchmarks: Assessing local performance and establishing community standards (2nd ed.).* Thousand Oaks, CA: Sage.

Aristigueta, M. P. (1999). Managing for results in state government. Westport, CT: Quorum Books.

Auditor General of British Columbia & Deputy Ministers' Council. (1996). *Enhancing accountability for performance: A framework and an implementation plan— Second joint report.* Victoria: Queen's Printer for British Columbia.

Bakvis, H., & Juillet, L. (2004). *The horizontal challenge: Line departments, central agencies and leadership.* Ottawa, ON: Canada School of Public Service.

Bouckaert, G. (1993). Measurement and meaningful management. *Public Productivity & Management Review, 17*(1), 31–43.

CCAF-FCVI. (2002). *Reporting principles: Taking public performance reporting to a new level.* Ottawa, ON: Canadian Comprehensive Auditing Foundation.

CCAF-FCVI. (2004). *Report to members 2003.* Ottawa, ON: Canadian Comprehensive Auditing Foundation.

Conference Board of Canada. (2004). *Understanding health care cost drivers and escalators.* Ottawa, ON: The Conference Board of Canada.

Courty, P., & Marschke, G. (2003). Dynamics of performance-measurement systems. *Oxford Review of Economic Policy, 19*(2), 268–284.

de Lancer Julnes, P. (1999). Lessons learned about performance measurement. *International Review of Public Administration, 4*(2), 45–55.

de Lancer Julnes, P., & Holzer, M. (2001). Promoting the utilization of performance measures in public organizations: An empirical study of factors affecting adoption and implementation. *Public Administration Review, 61*(6), 693–708.

Feit, D. (2003). Measuring performance in the public sector. *Cost Management, 17*(2), 39–45.

Frank, A. G. (1958). Goal ambiguity and conflicting standards: An approach to the study of organization. *Human Organization, 17*(1), 8–13.

Good, D. A. (2003). *The politics of public management: The HRDC audit of grants and contributions.* Toronto, ON: University of Toronto Press.

Halachmi, A. (2002). Performance measurement, accountability, and improved performance. *Public Performance and Management Review, 25*(4), 370–374.

Hall, M., Phillips, S., Meillat, C., & Pickering, D. (2003). *Assessing performance: Evaluation practices and perspectives in Canada's voluntary sector.* Toronto, ON: Canadian Centre for Philanthropy.

Hatry, H. P. (1999). *Performance measurement: Getting results.* Washington, DC: Urban Institute Press.

Hatt, L. (2004). *Examining the role of legislative committees in using performance information: A comparative study.* University of Victoria, Unpublished report.

Heinrich, C. J., & Lynn, L. E. (1999). *Governance matters: The influence of program structure and management on Job Training Partnership Act (JTPA) performance.* Chicago: The Irving B. Harris Graduate School of Public Policy Studies.

Kerr, S. (1975). On the folly of rewarding A, while hoping for B. *Academy of Management Journal, 18*(4), 769–783.

Kravchuk, R. S., & Schack, R. W. (1996). Designing effective performance-measurement systems under the Government Performance and Results Act of 1993. *Public Administration Review, 56*(4), 348–358.

Legislative Assembly of British Columbia. (2000). *Budget Transparency and Accountability Act* (SBC 2000, Chapter 23). Victoria, BC: Queen's Printer.

Leviton, L. C. (2003). Evaluation use: Advances, challenges and applications. *American Journal of Evaluation, 24*(4), 525–535.

Manitoba Office of the Auditor General. (2000). *Inter-jurisdictional comparison on trends and leading practices in business planning and performance measurement.* Winnipeg, Manitoba: Queen's Printer.

Mayne, J. (2001). Addressing attribution through contribution analysis: Using performance measures sensibly. *Canadian Journal of Program Evaluation, 16*(1), 1–24.

McDavid, J. C., & Huse, I. (2003). *Public performance reporting: Legislator perceptions of performance reporting principles and assurance of credibility.* Unpublished report, University of Victoria, Canada.

Morgan, G. (1997). *Images of organization* (2nd ed.). Thousand Oaks, CA: Sage.

Moynihan, D. P., & Wallace, P. (2001). *When does performance information contribute to performance information use? Putting the factors in place.* Working paper, Maxwell School of Syracuse University, Campbell Public Affairs Institute.

Osborne, D., & Gaebler, T. (1992). *Reinventing government: How the entrepreneurial spirit is transforming the public sector.* Reading, MA: Addison-Wesley.

Otley, D. (2003). Management control and performance management: Whence and whither? *British Accounting Review, 35*(4), 309–326.

Pal, L. A., & Teplova, T. (2003). *Rubik's cube? Aligning organizational culture, performance measurement, and horizontal management.* Ottawa, ON: Carleton University, School of Public Policy and Administration.

Perrin, B. (1998). Effective use and misuse of performance measurement. *American Journal of Evaluation, 19*(3), 367–379.

Pollitt, C. (2000). How do we know how good public services are? In B. G. Peters & D. J. Savoie (Eds.), *Governance in the Twenty-first Century: Revitalizing the public service* (pp. 119–152). Montreal: Canadian Centre for Management Development, McGill-Queen's University Press.

Ridgway, V. F. (1956). Dysfunctional consequences of performance measurements. *Administrative Science Quarterly, 1*(2), 240–247.

Sawhill, J. C., & Williamson, D. (2001). Mission impossible?: Measuring success in nonprofit organizations. *Nonprofit Management and Leadership, 11*(3), 371–386.

Senge, P. M. (1990). *The fifth discipline: The art and practice of the learning organization.* New York: Doubleday/Currency.

Shaw, I. (2000). *Evaluating public programmes: Contexts and issues.* Aldershot; Burlington, VT: Ashgate.

State of Minnesota Office of the Legislative Auditor. (1995). *Development and use of the 1994 agency performance reports.* Saint Paul, MN: Program Evaluation Division.

Streib, G. D., & Poister, T. H. (1999). Assessing the validity, legitimacy, and functionality of performance measurement systems in municipal governments. *American Review of Public Administration, 29*(2), 107–123.

Texas State Auditor's Office. (2002). *An audit report on fiscal year 2001 performance measures at 14 entities.* Austin: Texas State Auditor's Office.

Tucker, L. (2002). *Use and the effects of using performance measures for budgeting, management, and reporting.* Case study: State of Texas. Retrieved July 21, 2004, from http://www.seagov.org/sea_gasb_project/state_tx.pdf

Tuohy, C. H. (2003). Agency, contract, and governance: Shifting shapes in accountability in the health care arena. *Journal of Health Politics, Policy and Law, 28*(2,3), 195–215.

United States Congress (1993). *Government Performance and Results Act of 1993.* Public Law 103-62 (August 3). 107 STAT. 285.

U.S. General Accounting Office. (2004, January). *Performance budgeting: Observations on the use of OMB's Program Assessment Rating Tool for the fiscal year 2004 budget.* Retrieved August 11, 2004, from http://www.gao.gov/new.items/d04174.pdf

CHAPTER 11

PROGRAM EVALUATION AND PROGRAM MANAGEMENT

Joining Theory and Practice

INTRODUCTION ●

Program evaluation is intended to be a flexible and situation-specific means of answering questions, testing hypotheses, or describing program processes. Evaluations can focus on a broad range of issues, spanning program resources to program outcomes, and generally are intended to yield information that reduces the level of uncertainty about the issues that prompted the evaluation.

As we learned in Chapter 1, program evaluations can be **formative**; that is, aimed at producing findings, conclusions and recommendations that are intended to improve the program. Formative program evaluations are typically done with a view to offering program and organizational managers information that they can use to improve the existing program. Generally, questions about the continuation of support for the program itself are not part of formative evaluation agendas.

Program evaluations can also be **summative**, that is, intended to render judgments on the program. Summative evaluations can focus on issues that are similar to those included in formative evaluations (program effectiveness, for example) but the *intention* is to produce information that can be used to make decisions about the program's future, such as whether to reallocate resources elsewhere, or whether to terminate the program.

The purposes of an evaluation clearly affect relationships between evaluators and managers. Formative evaluations, because they are intended to provide program-improvement information, typically do not raise questions that could threaten the program's future. Generally, managers are more likely to view formative evaluations as "friendly" evaluations, and are more likely to be willing to cooperate with the evaluators. They have an incentive to do so because the evaluation is intended to assist them without raising questions that could result in major changes, including reductions to or even the elimination of a program.

Summative evaluations, because they typically raise critical questions about the program's future, are generally viewed quite differently. Program managers face different incentives in providing information or even cooperating with the evaluators. Their programs are at stake, and although the program evaluation process can be distinguished from personnel evaluation, many managers find it difficult to separate their well-being from the well-being of their programs.

From an evaluator's standpoint, the experience of conducting a formative evaluation can be quite different from conducting a summative evaluation. Typically, program evaluators depend on program managers to provide key information, and to arrange access to people, data sources, and other sources of evaluation information. Securing and sustaining cooperation is affected by the purposes of the evaluation—managerial reluctance might well be expected where the stakes include the future of the program itself.

● CAN MANAGEMENT AND EVALUATION BE JOINED?

How does program evaluation, as a part of the performance management cycle, relate to program management? Under what conditions, if any, can we expect managers to be involved, or even take the lead, in evaluating their own programs? Are program evaluation and program management compatible roles in public and nonprofit organizations?

Wildavsky (1979), in his seminal book *Speaking Truth to Power,* introduces his discussion of management and evaluation this way:

> Why don't organizations evaluate their own activities? Why don't they seem to manifest rudimentary self-awareness? How long can people work in organizations without discovering their objectives or determining how well they are carried out? I started out thinking that it was bad for organizations not to evaluate, and I ended up wondering why they ever do it. Evaluation and organization, it turns out, are somewhat contradictory. (p. 212)

Wildavsky's principal concern is in reconciling what he sees as the requirements of evaluation with the requirements of organizations, and managing organizations. He wants to know who will evaluate and who will administer. He also wants to know how power will be divided between them (Wildavsky, 1979).

When he contrasts management and evaluation, Wildavsky has in mind summative evaluations. His image of evaluations reflected the (then) dominant belief that program evaluation can lead to finding better programs, and even finding different objectives for future programs. The Federal Government of Canada, for example, defined program evaluation this way:

> Program evaluation in federal departments and agencies should involve the systematic gathering of verifiable information on a program and demonstrable evidence on its results and cost-effectiveness. Its purpose should be to periodically produce credible, timely, useful and objective findings on programs appropriate for resource allocation, program improvement and accountability. (Office of the Comptroller General of Canada, 1981, p. 3)

If evaluations are to be used to re-allocate resources as well as improve programs, organizations must have the capacity to respond to evaluations that have both formative and summative faces. This suggests an image of organizations that are amenable to re-thinking existing commitments—managers would need to balance attachment to their programs with attachment to the evaluation process and results. The *rational/technical* view of organizations (de Lancer Julnes & Holzer, 2001) that we discuss in Chapter 10 suggests that within such organizations, decision making is to be based on evidence, managers and workers are to behave in ways that do not undermine a results-focused culture, and summative evaluations are to be welcomed as a part of regular management processes.

Wildavsky's view of organizations as settings where speaking truth to power is a challenge is similar to the *political/cultural* image of organizations offered by de Lancer Julnes and Holzer (2001). Wildavsky views the respective roles of evaluators and managers as being painted in contrasting colors. Evaluators are described as people who question assumptions, people who are skeptical, people who are detached, people who view organizations/programs as instruments, and people who ultimately focus on the social needs that the program serves, rather than on organizational needs.

By contrast, organizational/program managers might be characterized as people who are committed to their programs, people who are advocates for what they do and what their programs do, and people who do not want to see their resources diminished.

How do organizations resolve the question of who has the power and authority to make decisions, and control the flows of information? One might imagine, for example, that if evaluators were given their way (and rational/technical norms dominated the organizational culture), programs might be implemented as experiments or quasi-experiments, with clearly stated objectives, opportunities for comparisons to no-program groups, baseline measurements, and sufficient control over the implementation process to ensure the construct validity of the program as an independent variable (or variables). This view of trying out new programs was the essence of Donald Campbell's experimenting society (Watson, 1986).

By contrast, managers would implement programs to meet organizational and client needs, and objectives would be stated in ways that facilitate flexible interpretations of what was important, depending on the audience. Managers would want program objectives to be able to withstand the scrutiny of stakeholders with different values and expectations. Experimentation creates political problems: What do you tell prospective clients who want the program, but cannot get access to it because they are members of a "control group"? What do you tell the persons to whom you report (and the elected officials to whom they report) when client groups question either the lack of flexibility in the service (to maintain construct validity) or its lack of availability (to increase internal validity)?

One solution to the organizational location problem is to make program evaluation an external function. Thus, evaluators would be a part of an agency that was not under the administrative control of the organization's managers. In British Columbia, for example, the Secretary of Treasury Board at one point outlined a plan for creating a centralized evaluation capacity in the government (Wolff, 1979). Treasury Board analysts, housed in that central agency, would have conducted evaluations of line department programs with a view to preparing reports for Treasury Board managers.

The plan was never implemented, in part because line departments strongly objected to the creation of a central evaluation unit that would not be accountable to line department executives. In fact, at that point, some departments were developing in-house evaluation units, which were intended to perform functions that executives argued would be duplicated by any centralized evaluation unit.

In 1985, a survey of British Columbia Government analysts and managers who had at least some responsibility for conducting program evaluations in their respective ministries queried their views on who should be responsible for program evaluations (Koyl, 1985). Table 11.1 lists the 107 responses to that question. There was very little support for central agencies alone in that role (3 responses out of 107), and limited support for any central agency involvement (16 responses out of 107). The largest single category of responses was for program management to be responsible for program evaluation (57 responses out of 107, or 53%).

● PROSPECTS FOR BUILDING CULTURES THAT SUPPORT EVALUATION

In the last 25 years there has been a broad movement in the field of evaluation toward finding ways of knitting evaluation and management together. Instead of seeing evaluation as an activity that challenges management, we are encouraged to believe that evaluators can work with managers to define and execute evaluations that combine the best of what both parties bring to that relationship. Utilization-focused evaluation, for example, is premised on producing evaluations that managers and other stakeholders will use—and ensuring that means developing a working relationship between evaluators and managers. Patton (1997) characterizes the role of the evaluator this way:

Table 11.1   Who Should Be Responsible for Program Evaluation in the British Columbia Government?

	n	*Percentage*
Ministry evaluation branches	15	14.1
Program management	57	53.3
Ministry executive	2	1.9
Evaluation branch plus program management	10	9.3
Central agencies	3	2.8
Evaluation branch plus central agencies	1	1.0
Program management plus central agencies	4	3.7
Evaluation branch plus program management plus central agencies	7	6.5
Ministry executive plus central agencies	1	1.0
Program management plus others	4	3.7
Other	3	2.8
Total	107	100.0

The evaluator facilitates judgment and decision-making by intended users rather than acting as a distant, independent judge. Since no evaluation can be value-free, utilization-focused evaluation answers the question of whose values will frame the evaluation by working with clearly identified, primary intended users who have responsibility to apply evaluation findings and implement recommendations. In essence, I shall argue, evaluation use is too important to be left to evaluators. (p. 21)

Love (1993) has elaborated a position that asserts that internal evaluation shops are not only the dominant trend in public and nonprofit organizations, but can conduct both formative and summative evaluations. He outlines six stages in the development of internal evaluation capacity, beginning with ad hoc program evaluations and ending with cost-benefit analyses:

1. Ad hoc evaluations focused on single programs

2. Process-focused regular evaluations

3. Goal setting, measurement of program outcomes, monitoring, adjustment

4. Evaluations of program effectiveness, improving organizational performance

5. Evaluations of technical efficiency and cost-effectiveness

6. Cost-benefit analyses

These six stages can be seen as a gradual transformation of the intentions of evaluations from formative to summative purposes. Love does acknowledge the importance of an internal working environment where organizational members are encouraged to participate, where *trust* of evaluators and their commitment to the organization is part of the culture.

In suggesting possible ways that evaluation could be joined to organization and management, Wildavsky (1979) also emphasized the importance of trust:

> Like the model community of scholars, the self-evaluating organization would be open, truthful and explicit. It would state its conclusions in public, show how they were determined, and give others the opportunity of refuting them. Costs and benefits of alternative programs for various groups in society would be indicated as precisely as available knowledge would permit. Everything would be above board. Are there ways of securing the required information? Can the necessary knowledge be created? Will the truth make men free? Whatever the answers to these questions might be, each depends on trust among social groups and within organizations. Acceptance of evaluation requires a community of shared values. (p. 234)

Is it possible to construct organizations in which the trust emphasized by Wildavsky and Love is a dominant feature of the culture? In such organizations, program evaluators would be encouraged to ply their trade earnestly, speak the truth as they see it, and could count on organizational participants, including program managers, to be open to the implications of evaluation results.

● LEARNING ORGANIZATIONS AS SELF-EVALUATING ORGANIZATIONS

Morgan (1997), in *Images of Organizations*, elaborates an organizational metaphor that suggests that the organization can be seen as a brain. Within that broad metaphor, he elaborates a metaphor for "**learning organizations**."

Using the work of Senge (1990), Morgan suggests that learning organizations must develop capacities to:

- Scan and anticipate change in the wider environment to detect significant variations
- Develop an ability to question, challenge, and change operating norms and assumptions
- Allow an appropriate strategic direction and pattern of organization to emerge

Key to establishing a learning organization is what Morgan (1997) calls double-loop learning; that is, learning how to learn. Organizations must get outside their established structures and procedures and instead focus on processes to create information, which in turn can be used to constantly assess the status quo and make changes. Double-loop learning not only involves a capacity to correct behaviors so that a given objective is attained, but also involves a capacity to assess whether objectives are still appropriate.

Garvin (1993) has suggested five "building blocks" for creating learning organizations. Table 11.2 summarizes these "building blocks" and their associated activities. By reviewing these steps, one can see a key role for evaluations. In fact, evaluation suffuses such organizations with emphasis on evidence-based decision making, a questioning attitude with respect to the status quo, experimentation (and evaluations of the results of such experiments), and treating knowledge as a resource.

Empowerment Evaluation and the Learning Organization

Fetterman (Fetterman, 2001; Fetterman, Kaftarian, & Wandersman, 1996) argues that one way to contribute to the development of a learning organization is through the process of *empowerment evaluation*. **Empowerment evaluation** is defined as the use of evaluation concepts, techniques, and findings to help program managers and staff evaluate their own programs and thus improve practice and foster self-determination in organizations. In empowerment evaluations, program staff conduct the evaluation while the external program evaluator acts as a facilitator and coach, keeping the exercise credible and on-track by providing advice, reality-checks, and quality controls. The intent of empowerment evaluation is to make evaluation part of the normal planning and management of programs and to help managers and staff feel more in charge of their own destinies. "Too often," argue Fetterman et al. (1996),

Table 11.2   Building Blocks for Creating Learning Organizations

1. Systematic problem solving
   - Tackling problems using a sequence of hypothesis-generating hypothesis-testing actions
   - Insisting on data rather than assumptions
   - Attention to details: "how do we know that is true?"

2. Experimentation
   - Small, controlled modifications and tests of existing programs
   - Searching for and testing new knowledge (pilot and demonstration projects)
   - Managers must have both the incentives and the skills to experiment

3. Learning from past experience
   - Systematically recording, displaying, and reviewing the evidence from past performance
   - Both this information and the skills to use and interpret it need to be widely distributed in the organization

4. Learning from others
   - "Steal ideas shamelessly"
   - Find out who is "the best," learn why they are, and adapt their practices to your organization

5. Transferring knowledge
   - Knowledge must be spread quickly and efficiently throughout the organization
   - Knowledge is treated as a resource

external evaluation is an exercise in dependency rather than an empowering experience: in these instances the process ends when the evaluator departs, leaving participants without the knowledge or expertise to continue for themselves. In contrast, an evaluation conducted by program participants is designed to be ongoing and internalized in the system, creating the opportunity for capacity building (p. 9).

Fetterman views evaluation as fundamentally a formative exercise. He argues that the assessment of a program's worth is not an endpoint in itself but part of an ongoing process of program improvement. Fetterman (2001) acknowledges, however, that

the value or strength of empowerment evaluation is directly linked to the purpose of the evaluation.... Empowerment evaluation makes a significant contribution to internal accountability, but has serious limitations in the area of external accountability.... An external audit or assessment would be more appropriate if the purpose of the evaluation was external accountability. (p. 145)

In addition, Fetterman stresses that empowerment evaluation can only be successful in the right kind of organizational environment, one which is guided by a commitment to truth and honesty. Program managers and staff must have the latitude to experiment, in an environment conducive to sharing successes and failures and an atmosphere that is self-critical, trusting, and supportive. "Without any of these elements in place," he argues, "the exercise may be self-serving and of limited utility" (Fetterman, 2001, p. 6).

What are the prospects for creating learning organizations in contemporary public and nonprofit settings? Given the current organizational and political cultures of jurisdictions in Canada and the United States, it is possible that some elements of learning organizations can be implemented in some organizations. But our dominant approach to structuring the way we provide programs and services relies heavily on hierarchical organizations. As we saw in Chapter 10, even though new public management exhorts us to pay attention to results, and not get bogged down in bureaucratic red tape, our political culture continues to emphasize the importance of not making mistakes, and where a mistake has been detected, being able to demonstrate that the process that led to that decision is defensible. Organizational managers are now expected to be accountable for both process and results. We believe that where these two norms compete, there is a tendency for process accountability to win out. The attendant effect of encouraging risk-avoidance strategies and behaviors generally discourages experimentation and learning from mistakes, which is a key component of learning organizations.

CAN PROGRAM MANAGERS ● EVALUATE THEIR OWN PROGRAMS?

Increasingly, program managers are expected to play a major role in evaluating their own programs. In many situations, particularly for managers in nonprofit organizations, resources to conduct evaluations are very scarce. But expectations that programs will be evaluated (and that information will be provided that can be used by funders to make decisions about the program's future) are growing. Designing and implementing performance measurement systems generally presumes a key role for managers. In Chapter 10, we discuss the ways that relying on performance measures summatively can produce unintended consequences—managers will respond to the incentives that are implied by performance measures, and shape their behavior and organizational activities to reflect what the measures say is valued.

Clearly, expecting managers to evaluate their own programs, given the incentives alluded to above, can result in biased program evaluations.

Indeed, a culture can be built up around the evaluation function such that evaluators are expected to be *advocates* for programs. Departments and agencies under such conditions would use their evaluation capacity to defend their programs, structuring evaluations and presenting results so that programs are seen to be above criticism.

Evaluations produced by organizations under such conditions will tend to be viewed outside the organization with skepticism. Funders, or analysts who are employed by the funders, will work hard to expose weaknesses in the methodologies used and cast doubt on the information in evaluation reports. In effect, adversarial relationships can develop, which serve to "expose" weaknesses in evaluations, but are generally not conducive to building self-evaluating or learning organizations.

The reality is that expecting program managers to evaluate their own programs, particularly where evaluation results are likely to be used in funding decisions, is likely to produce evaluations that reflect the incentives in such situations. They are not necessarily credible even to the managers themselves. Program evaluation, as an organizational function, becomes distorted and contributes to a view that all evaluations are biased.

Parenthetically, Nathan (2000), who has worked with several top American policy research centers, points out that internal evaluations are not the only ones that may reflect incentives that bias evaluation results. "Even when outside organizations conduct evaluations," he argues, "the politics of policy research can be hard going. To stay in business, a research organization (public or private) has to generate a steady flow of income. This requires a delicate balance in order to have a critical mass of support for the work one wants to do and at the same time maintain a high level of scientific integrity" (p. 203). Nevertheless, such incentives are likely to be more prevalent and stronger with internal evaluations.

Should managers *participate* in evaluations of their own programs? Generally, scholars and practitioners who have addressed this question have tended to favor managerial involvement. Love (1993) envisions evaluators working closely with program managers to produce evaluations on issues that are of direct relevance to managers. Patton (1997) stresses that among the fundamental premises of utilization-focused evaluation, the first is the commitment to working with the intended users to ensure that the evaluation actually gets used. Among the range of users, program managers are key stakeholders. Moreover, Fetterman (2001) argues that the best data are secured through close interaction with and observation of program managers and staff, because they are typically the most knowledgeable about their program and its strengths and weaknesses.

An important dissenting voice in the chorus that advocates evaluator-manager contact and even collaboration is Scriven's (1997) view that program evaluators should keep their distance from the organizations and people with whom they work. Getting too close to program managers amounts to compromising the ***objectivity*** of the evaluation process, and undermines the key contribution that evaluators can make: speaking the truth and offering an unbiased view of a program.

For Scriven, objectivity is an important feature of what evaluators can offer to their clients. To claim that an evaluation is objective is a strong statement to stakeholders that they can rely on the evaluation results and recommendations. Objectivity has been a criterion for high quality evaluations historically (Office of the Comptroller General of Canada, 1981), and continues to have a "scientific" appeal to practitioners and clients.

HOW SHOULD EVALUATORS RELATE TO MANAGERS? ● STRIVING FOR OBJECTIVITY IN PROGRAM EVALUATIONS

One issue that has been debated in discussions of "good" evaluation practice is whether program evaluations can and should strive to be objective. For Scriven (1997), objectivity is defined as "with basis and without bias" (p. 480), and an important part of being able to claim that an evaluation is objective is to maintain distance between the evaluator and what is being evaluated. There is a crucial difference, for Scriven, between being an evaluator and being an evaluation consultant. The former relies on **validity** as one's stock-in-trade, and objectivity is a central part of being able to claim that one's work is valid. The latter work with their clients and stakeholders but, according to Scriven, in the end they cannot offer analysis, conclusions, or recommendations that are not tainted by interactions and all the biases that they entail.

In addition to Scriven's view that objectivity is a key part of evaluation practice, other professions have asserted and continue to assert that professional practice is or at least ought to be objective. In the 2003 edition of the Government Auditing Standards (GAO, 2003), government auditors are enjoined to perform their work this way:

> Professional judgment requires auditors to exercise professional skepticism, which is an attitude that includes a questioning mind and a critical assessment of evidence. Auditors use the knowledge, skills, and experience called for by their profession to diligently perform, in good faith and with integrity, the gathering of evidence and the *objective evaluation of the sufficiency, competency, and the relevancy of evidence.* (p. 51; italics added)

Should evaluators claim that their work is also objective? Objectivity has a certain cachet, and as a practitioner, it would be appealing to be able to assert to prospective clients that one's work will be objective. Indeed, in situations where evaluators are competing with auditors for clients, claiming objectivity could be an important factor in convincing clients to purchase evaluator services.

● CAN PROGRAM EVALUATORS BE OBJECTIVE?

If giving managers a stake in evaluations compromises evaluator and evaluation objectivity, then it is important to unpack what is entailed by claims that evaluations or audits are objective. Is Scriven's definition of objectivity defensible? Is objectivity a meaningful criterion for high-quality program evaluations? Could we defend a claim to a prospective client that our work would be objective?

Scriven (1997) suggests a metaphor to understand the work of an evaluator: when we do program evaluations, we can think of ourselves as expert witnesses. We are, in effect, called to "testify" about a program, we offer our expert opinions, and the "court" (our client) can decide what to do with our contributions.

Scriven takes the courtroom metaphor further when he asserts that in much the same way that witnesses are sworn to tell "the truth, the whole truth, and nothing but the truth," (Scriven, 1997, p. 496) evaluators can rely on a common-sense notion of the truth as they do their work. If such an oath "works" in courts (Scriven believes it does), then despite the philosophical questions that can be raised by a claim that something is true, we can and should continue to rely on a common-sense notion of what is true and what is not.

Scriven's main point is that program evaluators should be prepared to offer objective evaluations and that to do so, it is essential that we recognize the difference between conducting ourselves in ways that promote our objectivity and ways that do not. Even those who assert that there cannot be any truths in our work are, according to Scriven, uttering a self-contradictory assertion: they wish to claim the truth of a statement that there are no truths.

Although Scriven's argument has a common-sense appeal, it encounters problems when we examine it more closely. There are essentially two main weaknesses in the approach he takes.

First, Scriven's metaphor of evaluators as expert witnesses fails in important respects. In courts of law, expert witnesses are routinely challenged by their counterparts and by opposing lawyers. Unlike Scriven's evaluators, who do their work, offer their report, and then absent themselves to avoid possible compromises of their objectivity, expert witnesses in courts undergo a high level of scrutiny. Even where expert witnesses have offered their version of the truth, it is often not clear whether that is *their* view or the views of a party to a legal dispute. Expert witnesses can sometimes be "purchased."

Second, witnesses speaking in court can be severely penalized if it is discovered that they have lied under oath. For program evaluators, it is far less likely that sanctions will be brought to bear even if it could be demonstrated that an evaluator did not speak the truth. Undoubtedly, an evaluator's place in the profession can be affected when the word gets around that he or she has been "bought" by a client, but the reality is that in the practice of program evaluation, clients can and do shop for evaluators who are likely to "do the job right." "Doing the job right" can mean that evaluators are paid to *not* speak "the truth, the whole truth, and nothing but the truth."

Looking for a Defensible Definition of Objectivity

Given the problems with Scriven's uses of metaphors and the attendant questions that that raises about his view of objectivity, are there other definitions of objectivity that are more tractable in terms of assisting our practice of program evaluation? The Federal Government of Canada's Office of the Comptroller General (OCG) was among the government jurisdictions that advocated the importance of objectivity in evaluations. In one statement, objectivity is defined this way:

> Objectivity is of paramount importance in evaluative work. Evaluations are often challenged by someone: a program manager, a client, senior management, a central agency or a minister. *Objectivity means that the evidence and conclusions can be verified and confirmed by people other than the original authors.* Simply stated, the conclusions must follow from the evidence. Evaluation information and data should be collected, analyzed and presented so that if others conducted the same evaluation and used the same basic assumptions, they would reach similar conclusions. (Treasury Board of Canada Secretariat, 1990, p. 28; italics added)

This definition of objectivity emphasizes the **reliability** of evaluation findings and conclusions. This implies, at least in principle, that the work of one evaluator or one evaluation team could be repeated, with the same results, by a second evaluation of the same program.

A Natural Science Definition of Objectivity

The OCG criterion of repeatability is similar to the way scientists understand objectivity. Findings and conclusions, in order to be accepted by the discipline, *must* be replicable.

There is, however, a critical difference between program evaluation practice and scientific practice. In the sciences, the methodologies and procedures that are used to conduct research and report the results are *intended* to facilitate replication. Methods are scrutinized by one's peers and if the way the work has been conducted and reported passes this test, it is then "turned over" to the community of researchers, where it is subjected to independent efforts to replicate the results. If a particular set of findings cannot be replicated by independent researchers, the community of research peers eventually discards the results as an artifact of the setting or the scientist's biases. The initial reports of cold fusion reactions (Fleischmann & Pons, 1989) prompted additional attempts to replicate the reported findings, to no avail. Fleischman and Pons' research methods proved to be faulty and cold fusion did not pass the test of replicability.

For scientists, objectivity has two important elements, both of which are necessary. Methods and procedures need to be constructed and applied so that the work done, as well as the findings, are open to scrutiny by one's peers. Although the *process* of doing a science-based research project does not by itself make the research objective, it is a necessary condition for objectivity. **Scrutability** of methods facilitates repeating the research. If findings can be replicated independently, the community of scholars engaged in similar work confers objectivity on the research.

This definition of objectivity does not imply that objectivity confers "truth" on scientific findings. Indeed, the idea that objectivity is about scrutability of methods and repeatability of findings is consistent with Kuhn's (1962) notion of paradigms. Kuhn suggested that communities of scientists who share a "world view" are able to conduct and interpret research results. Within a paradigm, "normal science" is about solving puzzles that are implied by the theoretical structure that undergirds the paradigm. "Truth" is agreement, based on research evidence, among those who share a paradigm.

In program evaluation practice, much of what we call methodology is tailored to a particular setting. Increasingly, we are taking advantage of mixed qualitative/quantitative methods (Hearn, Lawler, & Dowswell, 2003) when we design and conduct evaluations, and our own judgment as professionals plays a key role in how evaluations are designed and data are gathered, interpreted, and reported. Owen and Rogers (1999) make this point when they state:

No evaluation is totally objective: it is subject to a series of linked decisions [made by the evaluator]. Evaluation can be thought of as a point of view rather than a statement of absolute truth about a program. Findings must be considered . . . within the context of the decisions made by the evaluator in undertaking the translation of issues into data collection tools and the subsequent data analysis and interpretation. (p. 306)

Although the Office of the Comptroller General (Treasury Board of Canada Secretariat, 1990) criterion of *repeatability* in principle might be desirable, it is rarely applicable to program evaluation practice.

Implications for Evaluation Practice

Where does this leave us? Scriven's (1997) criteria for objectivity—with basis and without bias—are not defensible in as much as they depend on the "objectivity" of a single evaluator.

Not even in the natural sciences, where the subject matter and methods are far more conducive to Scriven's definition, do researchers rely on one scientist's assertions about "truth" and "objectivity." Instead, the **scientific method** demands that results be stated so that they *can* be corroborated or disconfirmed, and it is via that process that "objectivity" is conferred on one's work.

The practice of program evaluation is at odds with claims that we evaluators can be objective in the work we do. We very rarely have opportunities to repeat an evaluation to check the reliability of the findings. And we use methods in ways that demand substantial inputs of our own professional judgment. Evaluation is not a science. Instead it is a craft that mixes together methods with artistry to produce products that are tailored to contexts, and almost always have unique characteristics.

CRITERIA FOR BEST PRACTICES ●
IN PROGRAM EVALUATION: ASSURING
STAKEHOLDERS THAT EVALUATIONS ARE HIGH QUALITY

Many professional associations that represent the interests and views of program evaluators have developed codes of ethics or best practice guidelines. A review of several of these guideline documents indicates that there are no specific mentions of objectivity among the criteria suggested for good evaluations (AERA, 2000; American Evaluation Association, 1995; Australasian Evaluation Society, 2002; Organization for Economic Cooperation and Development, 1998).

Earlier mentions of objectivity as a desirable feature of program evaluations (Canadian Evaluation Society, 1989; Office of the Comptroller General of Canada, 1990) and the more recent position taken by Scriven (1997) are not generally reflected in professional standards and guidelines. There are, however, government organizations that in their guidelines for assessing evaluation reports do discuss the issue of objectivity (see, for example, Office of Management and Budget, 2004; Treasury Board of Canada Secretariat, 2004).

It would seem that although some program evaluators and perhaps some clients/stakeholders are prepared to make objectivity a criterion for sound practice, the evaluation profession as a whole is not. Instead, professional evaluation organizations tend to mention the accuracy and credibility of evaluation information (American Evaluation Association, 1995; Organization for Economic Cooperation and Development, 1998), the honesty and integrity of evaluators and the evaluation process (AERA, 2000; American Evaluation Association, 1995; Australasian Evaluation Society, 2002), the completeness and fairness of evaluation assessments (American Evaluation Association, 1995; Australasian Evaluation Society, 2002), and the validity and reliability of evaluation information (American Evaluation Association, 1995).

In addition, professional guidelines emphasize the importance of declaring and avoiding conflicts of interest (Australasian Evaluation Society, 2002; Canadian Evaluation Society, n.d.) and the importance of impartiality in reporting findings and conclusions (American Evaluation Association, 1995). As well, guidelines tend to emphasize the importance of competence in conducting evaluations (American Evaluation Association, 1995; Canadian Evaluation Society, n.d.) and the importance of upgrading evaluation skills (American Evaluation Association, 1995; Canadian Evaluation Society, n.d.).

Collectively, these guidelines cover many of the characteristics of evaluators and evaluations that we might associate with objectivity: accuracy, credibility, validity, reliability, completeness, fairness, honesty, and integrity. But, as we have seen, objectivity is more than just having good evaluators or even good evaluations—it is a process that involves corroboration of one's findings by one's peers. Our profession is so diverse and includes so many different epistemological and methodological stances that asserting objectivity would not be supported by many evaluators.

But, the evaluation profession does not exist alone in the current world of professionals who claim expertise in evaluating programs. The movement toward connecting evaluation to accountability expectations in public sector and nonprofit organizations has created situations where evaluation professionals, with their diverse backgrounds and standards, are compared to accounting professionals, who generally have a much more uniform view of their professional standards. Because the public sector auditing community in particular has predicated objectivity of their practice, it is arguable that they have a marketing advantage with prospective clients (see Everett, Green, & Neu, in press; Radcliffe, 1998). Further, with some key governmental departments asserting that in assessing the quality of evaluations one of the key criteria should be objectivity of the findings, that criterion confers an advantage on practitioners who claim that their process and products are objective.

What should evaluators tell prospective clients who, having heard that the auditing profession makes claims about their work being objective, expect the same from a program evaluation? If we tell clients that we cannot produce an objective evaluation, there may be a risk of their going elsewhere for assistance. On the other hand, claims that we can be objective are not supported by the norms of the profession overall.

Perhaps the best way to respond is to offer criteria that cover much of the same ground as is covered if one conducts evaluations with a view to their being "objective." Criteria like accuracy, credibility, honesty, completeness, fairness, impartiality, avoiding conflicts of interest, competence in conducting evaluations, and a commitment to staying current in skills are all relevant. They would be among the desiderata that scientists and others who *can* make defensible claims about objectivity would include in their own practice. The criteria mentioned are, also, among the principal ones included by auditors and accountants in their own standards (GAO, 2003).

To sum up, current guidelines and standards that have been developed by professional evaluation associations do not claim that program evaluations should be objective. Correspondingly, as practicing professionals, we should not be making such claims in our work. That does not mean that we are without standards and, indeed, we *should* be striving to be honest, accurate, fair, impartial, competent, highly skilled, and credible in the work we do. If we are these things, we can justifiably claim that our work meets the same professional standards as work done by scholars and practitioners who might claim to be objective.

ETHICS AND EVALUATION PRACTICE •

The evaluation guidelines, standards, and principles that have been developed for the evaluation profession all speak, in different ways, to ethical practice. Although evaluation practice is not guided by a set of professional norms that are enforceable (Rossi, Lipsey & Freeman, 2004), ethical guidelines are an important reference point for evaluators. Increasingly, organizations that involve people (clients or employees, for example) in research are expected to take into account the rights of their participants as study objectives are framed, measures and data collection are designed and implemented, results are interpreted, and findings are disseminated. In universities, human research ethics committees routinely scrutinize research plans to ensure that they do not violate the rights of participants. In both the United States and Canada, there are national policies or regulations that are intended to protect the rights of persons who are participants in research (U.S. Department of Health and Human Services, 2001; Medical Research Council of Canada, 2003).

Newman and Brown (1996) have undertaken an extensive study of evaluation practice to establish ethical principles that are important for evaluators in the roles they play. Underlying their work are ethical principles, which they trace to Kitchener's (1984, 1985) discussions of ethical norms. Table 11.3 summarizes ethical principles that are taken in part from Newman and Brown (1996) and from the Tri-Council Policy on the Ethical Conduct for Research Involving Humans (Medical Research Council of Canada, 2003), and shows how each one corresponds to a major section in the American Evaluation Association's "Guiding Principles for Evaluators" (1995).

The ethical principles summarized in Table 11.3 are not absolute. Each one needs to be weighted in the context of a particular evaluation project, and balanced with other ethical considerations. For example, the "Keeping promises" principle suggests that contracts, once made, are to be honored, and normally that is the case. But suppose that an evaluator makes an agreement with the executive director of a nonprofit agency to conduct an evaluation of a major program that is delivered by the agency. The contract specifies that the evaluator will deliver three interim progress reports to the executive director, in addition to a final report. As the evaluator begins her work, she learns from several agency managers that the executive director has been re-directing money from the project budget for office furniture, equipment, and his own travel expenses—none of these being connected with the program that is being evaluated. In her first interim report, she brings these concerns to the attention of the executive director, who denies any wrong doings, and makes it clear that the interim reports are not to be shared with anyone else. The evaluator discusses this situation with her colleagues in the firm in which she is employed and decides to inform the chair of the board of directors for the agency. She has broken her contract but has

Table 11.3   Relationships Between the American Evaluation Association Principles and Ethical Principles for Evaluators

American Evaluation Association Guiding Principles	Ethical Principles for Evaluators
Systematic inquiry	Maximizing benefits
Evaluators conduct systematic, data-based inquiries about the subject of evaluation	Minimizing harms
	Balancing harms and benefits
Competence	
Evaluators provide competent performance to stakeholders	Minimizing harms
Integrity/honesty	Being honest
Evaluators ensure the honesty and integrity of the entire evaluation process	Keeping promises
	No conflicts of interest
Respect for people	
Evaluators respect the security, dignity, and self-worth of the respondents, program participants, clients, and other stakeholders with whom they interact	Free and informed consent
	Privacy and confidentiality
	Respect for vulnerable persons
Responsibilities for the general and public welfare	Procedural justice—ethical reviews of projects are fair, independent and transparent
Evaluators articulate and take into account the diversity of interests and values that may be related to the general and public welfare	Distributive justice—persons are not discriminated against, and there is respect for vulnerable persons

called upon a broader, and in this case, higher, principle that speaks to the honesty and integrity of the evaluation process.

In the appendix to this chapter, we have included a case that provides you with an opportunity to make a choice for an evaluator who works in a government department. The evaluator is in a difficult situation and has to decide what decision she should make, balancing ethical principles and her own well-being as the manager of an evaluation branch in that department. There is no right answer to this case. Instead, it gives you an opportunity to see how challenging ethical choice-making can be, and it gives you an opportunity to make a choice and build a rationale for your choice.

● SUMMARY

The relationship between managers and evaluators is affected by the incentives that each party faces in their relationship. If evaluators have been commissioned to conduct a summative evaluation, it is likely that program managers will defend their programs. Expecting managers under these conditions to participate as neutral parties in an evaluation ignores the potential for conflicts of commitments that will affect the accuracy and completeness of information that managers provide about their own programs.

Formative evaluations, where it is generally possible to project a "win-win" scenario for managers and evaluators, offer incentives to managers to be forthcoming so that they benefit from an assessment based on an accurate and complete understanding of their programs. Historically, a vast majority of evaluations have been formative. Although advocates for program evaluation and performance measurement imply that evaluations can be used for resource allocation/re-allocation decisions, it is rare to have an evaluation that does that. There has been a gap between the promise and the performance of evaluation functions in governments, in that regard.

Promoting quality standards for evaluations has and continues to be an important indicator of the professionalization of evaluation practice. Although objectivity has been a feature of "good" evaluations in the past, professional associations have all opted not to include objectivity among the criteria that define high quality evaluations.

Currently, and into the future, evaluators and accountants will be connected with efforts by government and nonprofit organizations to be more accountable. In some situations, evaluation professionals and accounting professionals will compete for work with clients. Because the accounting profession continues to assert that their work is objective, evaluators will have to address the issue of how they characterize their own practice, so that clients can be assured that the work of evaluators meets standards of rigor, defensibility, and ethical practice.

● DISCUSSION QUESTIONS

1. Why are summative evaluations more challenging to do than formative evaluations?

2. How should program managers be involved in evaluations of their own programs?

3. What is a learning organization, and how is the culture of a learning organization supportive of evaluation?

4. What would it take for an evaluator to claim that her/his evaluation was objective? Given those requirements, is it possible for any evaluator to say that his/her evaluation is objective? Under what circumstances, if any?

5. Suppose that you are a practicing evaluator and you are discussing a possible contract to do an evaluation for an agency. The agency director is very interested in your proposal, but in the discussions, says that he wants an objective evaluation. If you are willing to tell him that your evaluation will be objective, you have the contract. How would you respond to this situation?

6. Other professions like medicine, law, accounting, and social work have guidelines for professional practice that can be enforced against individual practitioners, if need be. Evaluation has guidelines, but they are not enforceable. What would be the advantages and disadvantages of the evaluation profession having enforceable practice guidelines?

APPENDIX 1
A Case Study in Ethical Decision Making

Fiona's Choice: An Ethical Dilemma for a Program Evaluator

Fiona Barnes did not feel well as the deputy commissioner's office door closed behind her. She walked back to her office wondering why bad news seems to come on Friday afternoons. Sitting at her desk, she went over the events of the last several days, and the decision that lay ahead of her. This was clearly the most difficult situation she had encountered since her promotion to the position of Director of Evaluation in the Department of Human Services.

Fiona's predicament had begun the day before, when the new commissioner, Fran Atkin, had called a meeting with Fiona and the deputy commissioner. The governor was in a difficult position: in his recent election campaign, he had made potentially conflicting campaign promises. He had promised to reduce taxes and had also promised to maintain existing health and social programs, while balancing the state budget.

The week before, a loud and lengthy meeting of the commissioners in the state government had resulted in a course of action intended to resolve the issue of conflicting election promises. Fran Atkin had been persuaded by the governor that she should meet with the senior staff in her department, and after the meeting, a major evaluation of the department's programs would be announced. The evaluation would provide the governor with some post-election breathing space. But the evaluation results were predetermined—they would be used to justify program cuts. In sum, "a compassionate" but substantial reduction in the department's social programs would be made to ensure the department's contribution to a balanced budget.

As the new commissioner, Fran Atkin relied on her deputy commissioner, Elinor Ames. Elinor had been one of several deputies to continue on under the new administration and had been heavily committed to developing and implementing key programs in the department, under the previous administration. Her success in doing that had been a principal reason why she had been promoted to deputy commissioner.

On Wednesday, the day before the meeting with Fiona, Fran Atkin had met with Elinor Ames to explain the decision reached by the Governor, downplaying the contentiousness of the discussion. Fran had acknowledged some discomfort with her position, but she believed her department now had a mandate. Proceeding with it was in the public's interest.

Elinor was upset with the governor's decision. She had fought hard over the years to build the programs in question. Now she was being told to dismantle her legacy—programs she believed in that made up a considerable part of her budget and person-year allocations.

In her meeting with Fiona on Friday afternoon, Elinor had filled Fiona in on the political rationale for the decision to cut human service programs. She also made clear what Fiona had suspected when they had met with the commissioner earlier that week—the outcomes of the evaluation were predetermined: they would show that key programs where substantial resources were tied up were not effective and would be used to justify cuts to the department's programs.

Fiona was upset with the commissioner's intended use of her branch. Elinor, watching Fiona's reactions closely, had expressed some regret over the situation. After some hesitation, she suggested that she and Fiona could work on the evaluation together, "to ensure that it meets our needs and is done according to our standards." After pausing once more, Elinor added, "Of course, Fiona, if you do not feel that the branch has the capabilities needed to undertake this project, we can contract it out. I know some good people in this area."

Fiona was shown to the door and asked to think about it over the weekend.

Fiona Barnes took pride in her growing reputation as a competent and serious director of a good evaluation shop. Her people did good work that was viewed as being honest, and they prided themselves on being able to handle any work that came their way. Elinor Ames had appointed Fiona to the job, and now this.

Your Task

Analyze this case and offer a resolution to Fiona's dilemma. Should Fiona undertake the evaluation project? Should she agree to have the work contracted out? Why?

In responding to this case, consider the issues on two levels: look at the issues taking into account Fiona's personal situation and the "benefits and costs" of the options available to her; and look at the issues from an organizational standpoint, again weighting the "benefits and the costs." Ultimately, you will have to decide how to weight the benefits and costs from both Fiona's and the department's standpoints.

REFERENCES ●

AERA. (2000). *The ethical standards of the American Educational Research Association.* Retrieved August 17, 2004, from http://www.aera.net/uploadedFiles/About_AERA/Ethical_Standards/EthicalStandards.pdf

American Evaluation Association. (1995). Guiding principles for evaluators. *New Directions for Program Evaluation* (66), 19–26.

Australasian Evaluation Society. (2002). *Guidelines for the ethical conduct of evaluations.* Lyneham, Australia: Australasian Evaluation Society.

Canadian Evaluation Society. (1989). *The value in evaluation: A statement for managers.* Ottawa, ON: CES.

Canadian Evaluation Society. (n.d.). *CES Guidelines for Ethical Conduct.* Retrieved August 17, 2004, from http://www.unites.uqam.ca/ces/ethicse.html

de Lancer Julnes, P., & Holzer, M. (2001). Promoting the utilization of performance measures in public organizations: An empirical study of factors affecting adoption and implementation. *Public Administration Review, 61*(6), 693–708.

Everett, J., Green, D., & Neu, D. (2005). Independence, objectivity and the Canadian CA profession. *Critical Perspectives on Accounting, 16*(4), 415–440.

Fetterman, D. M. (2001). *Foundations of empowerment evaluation.* Thousand Oaks, CA: Sage.

Fetterman, D. M., Kaftarian, S. J., & Wandersman, A. (1996). *Empowerment evaluation: Knowledge and tools for self-assessment & accountability.* Thousand Oaks, CA: Sage.

Fleischmann, M., & Pons, S. (1989). Electrochemically induced nuclear fusion of deuterium. *Journal of Electroanalytical Chemistry, 261*(2A), 301–308.

GAO. (2003). *Government auditing standards: 2003 revision.* GAO-03–673G. Washington, DC: Government Printing Office.

Garvin, D. A. (1993). Building a learning organization. *Harvard Business Review, 71*(4), 78–90.

Hearn, J., Lawler, J., & Dowswell, G. (2003). Qualitative evaluations: Combined methods and key challenges. *Evaluation, 9*(1), 30–54.

Kitchener, K. S. (1984). Intuition, critical evaluation and ethical principles: The foundation for ethical decisions in counseling psychology. *Counseling Psychologist, 12*(3), 43–56.

Kitchener, K. S. (1985). Ethical principles and ethical decisions in student affairs. In H. J. Canon & R. D. Brown (Eds.), *Applied ethics in student affairs* (*New Directions for Student Services, 30,* 17–30). San Francisco: Jossey-Bass.

Koyl, M. (1985). *A survey and analysis of program evaluation in the Province of B.C. and recommendations for an appropriate policy and guidelines.* Unpublished Master's report, University of Victoria, British Columbia, Canada.

Kuhn, T. S. (1962). *The structure of scientific revolutions.* Chicago: University of Chicago Press.

Love, A. J. (1993). Internal evaluation: An essential tool for human service organizations. *Canadian Journal of Program Evaluation, 8*(2), 1–15.

Medical Research Council of Canada. (2003). *Tri-council policy statement: Ethical conduct for research involving humans.* Ottawa, ON: Public Works and Government Services Canada.

Morgan, G. (1997). *Images of organization* (2nd ed.). Thousand Oaks, CA: Sage.

Nathan, R. P. (2000). *Social science in government: The role of policy researchers* (Updated ed.). Albany, NY: Rockefeller Institute Press.

Newman, D. L., & Brown, R. D. (1996). *Applied ethics for program evaluation.* Thousand Oaks, CA: Sage.

Office of Management and Budget. (2004). *What constitutes strong evidence of a program's effectiveness?* Retrieved July, 2004, from http://www.whitehouse.gov/omb/part/2004_program_eval.pdf

Office of the Comptroller General of Canada. (1981). *Guide on the program evaluation function.* Ottawa, ON: Treasury Board of Canada Secretariat.

Office of the Comptroller General of Canada. (1990). *Program evaluation methods: Measurement and attribution of program results.* Ottawa, ON: OCG.

Organization for Economic Cooperation and Development. (1998). *Best practice guidelines for evaluation.* Paris: OECD.

Owen, J. M., & Rogers, P. J. (1999). *Program evaluation: Forms and approaches* (International ed.). London, Thousand Oaks, CA: Sage.

Patton, M. Q. (1997). *Utilization-focused evaluation: The new century text* (3rd ed.). Thousand Oaks, CA: Sage.

Radcliffe, V. S. (1998). Efficiency audit: An assembly of rationalities and programmes. *Accounting, Organizations and Society, 23*(4), 377–410.

Rossi, P. H., Lipsey, M. W., & Freeman, H. E. (2004). *Evaluation: A systematic approach* (7th ed.). Thousand Oaks, CA: Sage.

Scriven, M. (1997). Truth and objectivity in evaluation. In E. Chelimsky & W. R. Shadish (Eds.), *Evaluation for the 21st century: A handbook* (pp. xiii, 542). Thousand Oaks, CA: Sage.

Senge, P. M. (1990). *The fifth discipline: The art and practice of the learning organization.* New York: Doubleday/Currency.

Treasury Board of Canada Secretariat. (1990). *Program evaluation methods: Measurement and attribution of program results* (3rd ed.). Ottawa, ON: Deputy Comptroller General Branch, Government Review and Quality Services.

Treasury Board of Canada Secretariat. (2004). *Guide for the review of evaluation reports.* Retrieved August 12, 2004, from http://www.tbs-sct.gc.ca/eval/tools_outils/4001752_e.asp

U.S. Department of Health and Human Services. (2001). *Code of Federal Regulations: Part 46—Protection of Human Subjects.* Retrieved August 12, 2004, from http://www.hhs.gov/ohrp/humansubjects/guidance/45cfr46.htm

Watson, K. F. (1986). Programs, experiments, and other evaluations: An interview with Donald Campbell. *Canadian Journal of Program Evaluation, 1*(1), 83–86.

Wildavsky, A. B. (1979). *Speaking truth to power: The art and craft of policy analysis.* Boston: Little Brown.

Wolff, E. (1979). *Proposed approach to program evaluation in the Government of British Columbia.* Victoria, BC: Treasury Board.

CHAPTER **12**

THE NATURE
AND PRACTICE OF
PROFESSIONAL JUDGMENT
IN PROGRAM EVALUATION

Good judgment is based on experience. Unfortunately, experience is based on poor judgment.

—Anonymous

INTRODUCTION ●

Throughout this book, we have referred to the importance of professional judgment in the practice of evaluation. Our view is that evaluators rely on their professional judgment in all evaluation settings. Although most textbooks in the field, as well as most academic programs that prepare evaluators for careers as practitioners, do not make the acquisition or practice of sound professional judgment an explicit part of evaluator training, this does not change the fact that professional judgment is an integral part of our practice.

To ignore or minimize the importance of professional judgment suggests a scenario that has been described by Schön (1987):

> In the varied topography of professional practice, there is the high, hard ground overlooking a swamp. On the high ground, manageable problems lend themselves to solutions through the application of research-based theory and technique. In the swampy lowland, messy, confusing problems defy technical solutions. . . . The practitioner must choose. Shall he remain on the high ground where he can solve relatively unimportant problems according to prevailing standards of rigor, or shall he descend to the swamp of important problems and non-rigorous inquiry? (p. 3)

THE NATURE OF THE EVALUATION ENTERPRISE ●

Evaluation can be viewed as a structured process that creates and synthesizes information that is intended to reduce the level of uncertainty for stakeholders about a given program or policy. It is intended to answer questions or test hypotheses, the results of which are then incorporated into the information bases used by those who have a stake in the program or policy.

What Is Good Evaluation Practice? Methodological Considerations

Views of evaluation research and practice, and in particular about what they ought to be, vary widely. At one end of the spectrum, advocates of a highly structured approach to evaluations tend to emphasize the use of research designs that ensure sufficient internal and statistical conclusions validity that the key causal relationships between program and outcomes can be isolated and tested. According to this view, experimental designs are the benchmark of sound evaluations, and departures from this ideal are associated with problems that either require specifically designed (and usually complex) methodologies to resolve, or are simply not resolvable—at least to a point where plausible threats to internal validity are controlled.

In the United States, major federal departments have recently implemented systems that encourage that evaluation projects, whenever possible, be based on experimental designs. For example, the revised guidance for the Program Assessment Rating Tool (PART) used by the Office of Management and Budget (OMB) to assess the effectiveness of programs states the following under the heading "What Constitutes Strong Evidence of a Program's Effectiveness?" (Office of Management and Budget, 2004):

> The revised PART guidance this year underscores the need for agencies to think about the most appropriate type of evaluation to demonstrate the effectiveness of their programs. As such, the guidance points to the randomized controlled trial (RCT) as an example of the best type of evaluation to demonstrate actual program impact. Yet, RCTs are not suitable for every program and generally can be employed only under very specific circumstances. (p. 1)

As well, in the United States the recently enacted *No Child Left Behind Act* (U.S. Department of Education, 2002) has, as a central principle, the idea that a key criterion for the availability of federal funds for school projects should be that a reform

> has been found, through scientifically based research to significantly improve the academic achievement of students participating in such program as compared to students in schools who have not participated in such program; or, has been found to have strong evidence that such program will significantly improve the academic achievement of participating children. (Section 1606a, 11A-B)

In Canada, a major federal department (Human Resources and Skills Development Canada) that funds evaluations of social service programs has specified in guidelines that randomized experiments are ideal for evaluations, but at minimum, evaluation designs must be based on quasi-experimental research designs that include comparison groups which permit before-and-after assessments of program effects between the program and the control groups (Human Resources Development Canada, 1998).

Problems With Experimentation
as a Criterion for Good Methodologies

Historically, advocates for experimental approaches have argued in part that the superiority of their position rests in the belief that sound, internally valid, research designs obviate the need for the evaluator to "fill in the blanks" with information that is not gleaned directly from the (usually) quantitative comparisons built into the designs. The results of a good experimental design are said to be defensible, and therefore, more valid as a basis for supporting decisions about a program or policy.

Although experimental evaluations continue to be done (Ford, Gyarmati, Foley, et al., 2003; Gustafson, 2003), during the 1960s and 1970s significant amounts of money were spent funding experimental evaluations of social programs in the United States and Canada. Examples include evaluations of the New Jersey Negative Income Tax Experiment (Pechman & Timpane, 1975) and in Canada, an evaluation of the Manitoba Minimum Income Experiment (Hum & Simpson, 1993). These and other large-scale evaluations exemplified a commitment to social experimentation as a way to create usable knowledge. Donald Campbell's ideal of the experimenting society (Watson, 1986), wherein new social programs would be rigorously evaluated as pilot programs before being fully implemented, served as part of a broad rationale for using experiments and quasi-experiments to evaluate social programs.

Large-scale social experiments tended to be complex and sometimes controversial evaluations. The Kansas City Preventive Patrol Experiment (Kelling, 1974) was an example of a major evaluation that relied on an experimental design which was intended to resolve a key policy question—in that case, whether the level of routine preventive patrol (assigned randomly to samples of police patrol beats in Kansas City, Missouri) made any differences to the actual and perceived levels of crime and safety in the selected patrol districts of Kansas City. Because the patrol levels were kept secret to conceal them from the citizens (and presumably potential law breakers), the experimental results encountered a substantial external validity problem—even if the findings supported the hypothesis that the level of routine preventive patrol had no significant impacts on perceived levels of safety and crime, or on actual levels of crime, how could any other police department announce that it was going to reduce preventive patrol without jeopardizing citizen confidence? Even in the experiment itself, there was evidence that the police officers who responded to calls for service in the reduced patrol beats did so with more visibility (lights and sirens) than normal—suggesting that they wanted to establish their visibility in the low-patrol beats (Kelling, 1974). In different words, there was a construct validity threat that was not controlled—compensatory rivalry whereby the patrol officers in the control beats acted to "beef up" the perceived level of patrol in the experimental beats.

The Importance of Determining Causality: The Core of the Evaluation Enterprise

Advocates for experimental research designs point out that since most evaluations include, as a core issue, whether the program was effective, experimental designs are the best way to answer these causal questions.

Scriven (in progress, unpublished) connects this to the resurgent U.S. government emphasis on experimental designs in evaluations. Scriven points out that it is, however, possible to generate valid causal knowledge many other ways and argues for a pluralism of methods in the evaluation of programs. A key point he makes is that human beings are "hard wired" to look for causal relationships in the world around them. In an evolutionary sense, we have a built-in capacity to observe causal connections. In Scriven's words:

> Our experience of the world and our part in it, is not only well understood by us but pervasively, essentially, perpetually a causal experience. A thousand times a day we observe causation, directly and validly, accurately and sometimes precisely. We see people riding bicycles, driving trucks, carrying boxes up stairs, turning the pages of books, picking goods off shelves, calling names, and so on. So, the basic kind of causal data, vast quantities of highly reliable and checkable causal data, comes from observation, not from elaborate experiments. Experiments, especially RCTs, are a marvelously ingenious extension of our observational skills, enabling us to infer to causal conclusions where observation alone cannot take us. But it is surely to reverse reality to suppose that they are the primary or only source of reliable causal claims: they are, rather, the realm of flight for such claims, where the causal claims of our everyday lives are the ground traffic of them. (pp. 6–7)

Alternative Perspectives on the Evaluation Enterprise

At the other end of the spectrum are approaches that eschew positivistic or postpositivistic approaches to evaluation for methods that are rooted in anthropology, or postmodern subfields of sociology or other disciplines. Advocates of these qualitative approaches have pointed out that the positivist view of evaluation is itself based on a set of values and assumptions about how to observe and measure the patterns of human interactions.

Quantitative methods cannot claim to control subjectivity and bias or eliminate the need for evaluators to use professional judgment. Smith (1994) argues that quantitative methods necessarily involve judgment calls:

> . . . decisions about what to examine, which questions to explore, which indicators to choose, which participants and stakeholders to tap, how to respond to unanticipated problems in the field, which contrary data to report, and what to do with marginally significant statistical results are judgment calls. As such they are value-laden and hence subjective. . . . Overall the degree of bias that one can control through random assignment or blinded assessment is a minute speck in the cosmos of bias. (pp. 38–39)

Moreover, advocates of qualitative approaches argue that quantitative methods miss the meaning of much of human behavior. Understanding intentions is critical to getting at what the "data" really mean, and the only way to do that is to embrace methods that treat individuals as unique meaning-generators who need to be understood on their own terms (Schwandt, 2000). Positions between these two ends of the evaluation spectrum have tended to be more pragmatic in that they recognize the value of being able to use structured designs, where they are feasible, but also recognize the value of employing a range of complementary approaches in a given situation to create information that is credible and, hence, likely to be used (Patton, 1997).

Reconciling Evaluation Theory With the Diversity of Practice

The practice of program evaluation is even more diverse than the range of normative approaches and methods that populate the textbook and coursework landscape. Experimental evaluations continue to be done and are still viewed by many practitioners as exemplars. Substantial investments in time and resources are typically required, and these limit the number and scope of evaluations that are able to randomly assign units of analysis (usually people) to program and control conditions.

Conducting experimental evaluations entails creating a structure that may produce statistical conclusions and internal validity, but fail to inform decisions about implementing the program or policy in nonexperimental settings (the Kansas City Preventive Patrol Experiment is an example of that). Typically, experiments are time limited, so participants adjust their behaviors to their expectations of how long the experiment will last, as well as to what their incentives are as it is occurring. Cronbach (1980), and more recently, Shadish, Cook, and Campbell (2002) have argued that an emphasis on internal validity alone can stunt the external validity of the evaluation, and limits its utility to decision makers.

The existence of controversies over the construction, execution, and interpretation of many of the large-scale social experiments that were conducted in the 1970s to evaluate programs and policies would suggest that very few methodologies are unassailable–even experimental research designs (Basilevsky & Hum, 1984). The *craft* of evaluation research, even research that is based on randomly assigned treatment and control conditions, is such that its practitioners do not concur about what exemplary practice is, even in a given situation.

The practice of evaluation is such that it is rare to have the resources and the control over the program setting needed to conduct even a quasi-experimental evaluation. Instead, typical evaluation settings are limited by significant resource constraints and the expectation that the evaluation process will somehow fit into the existing administrative process that has implemented a policy or a program. The burgeoning interest in performance measurement as an evaluation approach tends to be associated with an assumption that existing managerial and information technological resources will be sufficient to implement performance measurement systems, and produce information for formative *and* summative purposes.

Working in the Swamp: The Real World of Evaluation Practice

Typical program evaluation methodologies rely on multiple data sources to "strengthen" research designs that are case study designs (diagrammed in Chapter 3 as XO designs where X is the program and O is the observations on the outcome variable that is expected to be affected by the program). The program has been implemented at some time in the past and now the evaluator is expected to assess program effectiveness. There is no pretest and no control group—there are insufficient resources to construct these comparisons. Although multiple data sources permit triangulation of findings, they do not change the fact that the basic research design is the same—it is simply repeated for each data source (which is a strength), but is still subject to all the weaknesses of that design. In sum, typical program evaluations are conducted after the program is implemented, in settings where the evaluation team has to rely on evidence about the program group alone.

In such situations, some evaluators would advocate not using the evaluation results to make any causal inferences about the program. In other words, it is argued that such evaluations ought not to be used to try to address the question: "Did the program make a difference, and if so, what difference(s) did it make?"

But, many evaluations are commissioned with the need to know whether the program worked and why. Even formative evaluations often include questions about the effectiveness of the program (Cronbach, 1980; Weiss, 1998).

In situations where a client wants to know if and how the program was effective, and there is clearly insufficient time, money, and control to construct an evaluation design that meets criteria that are textbook-appropriate for answering those questions, evaluators have a choice. They can advise their client that wanting to know whether the program or policy worked and why is not appropriate, or they can proceed, knowing that their work will not be defensible by the standards that a substantial number of textbooks advocate.

Usually the work proceeds. When it does, does that make it poor practice? Should much of what we actually do as evaluators be purged from the collective experience of the profession because it does not meet particular normative (and not necessarily practical or workable) standards?

We would argue that it should not. We maintain that the way to answer causal questions without research designs that can convincingly rule out rival hypotheses is to acknowledge that in addressing issues like program effectiveness (which we take to be the central question in most evaluations), we cannot offer definitive findings. Instead, our findings, conclusions, and our recommendations, supported by the evidence at hand and by our ***professional judgment***, will reduce the uncertainty associated with the question.

We would also argue that in *all* evaluations, regardless of how sophisticated they are, evaluators use one form or another of professional judgments. The difference between the most sophisticated experimentally designed evaluation and the evaluation based on a case study design is the amount and the kinds of professional judgments that are entailed—not that the former is appropriate for assessing program effectiveness, and the latter is not.

Where our research designs are weak, we introduce our own experience and our own assessments which in turn are conditioned by our values, beliefs, and expectations. These become part of the basis on which we interpret the evidence at hand and are also a part of the conclusions and the recommendations. This professional judgment component in every evaluation means that we should be aware of what it is and learn how to cultivate sound professional judgment.

Common Ground Between Program Evaluators and Program Managers

The view that all evaluations incorporate professional judgments to a greater or lesser extent means that evaluators have a lot in common with program managers. Managers often conduct assessments of the consequences of their decisions—informal evaluations, if you will. These are not usually based on any systematic gathering of information but instead rely on a manager's own observations, values, beliefs, expectations, and experiences—their professional judgment. That these assessments are done "on the fly" and are often based on information that is gathered using research designs that do not warrant causal conclusions does not vitiate their being the basis for good management practice. Good managers become skilled at being able to recognize patterns in the complexity of their environments. Inferences from observed or sensed patterns (Mark, Henry, & Julnes, 2000) to causal linkages are informed by their experience and judgment, are tested by observing and often participating in the consequences of a decision, and in turn add to the fund of knowledge and experience that contributes to their sound professional judgment.

Situating Professional Judgment in Program Evaluation Practice

Scriven (1994) emphasizes the centrality of judgment in the synthesis of evaluation findings/lines of evidence to render an overall assessment of a program. For Scriven, the process of building toward and then rendering a holistic evaluation judgment is a central task for evaluators. Judgments can be improved by constructing rules or decision processes that make explicit how evidence will be assessed and weighted. Scriven suggests that generally, judgments supported by decision criteria are superior to those that are intuitive (Scriven, 1994).

Although evaluations typically use several different lines of evidence to assess a program's success, methodologies such as cost-utility analysis exist for weighting and amalgamating findings that combine multiple measures of program effectiveness (Levin & McEwan, 2001). However, they are data-intensive and aside from the health sector they are not widely used. The more typical situation is described by House and Howe (1999). They point out that the findings from various lines of evidence in an evaluation may well contain conflicting information about the worth of a program. In this situation, evaluators use their professional judgment to produce an overall conclusion. The process of rendering such a judgment engages the evaluator's own knowledge, values, beliefs and expectations. House and Howe (1999) describe this process:

> The evaluator is able to take relevant multiple criteria and interests and combine them into all-things-considered judgments in which everything is consolidated and related. . . . Like a referee in a ball game, the evaluator must follow certain sets of rules, procedures, and considerations—not just anything goes. Although judgment is involved, it is judgment exercised within the constraints of the setting and accepted practice. Two different evaluators might make different determinations, as might two referees, but acceptable interpretations are limited. In the sense that there is room for the evaluator to employ judgment, the deliberative process is individual. In the sense that the situation is constrained, the judgment is professional. (p. 29)

There are also many situations where evaluators must make judgments in the absence of clear methodological constraints or rules to follow. House and Howe (1999) go on to point out that "for evaluators, personal responsibility is a cost of doing business, just as it is for physicians, who must make dozens of clinical judgments each day and hope for the best. The rules and procedures of no profession are explicit enough to prevent this" (p. 30).

Program evaluators are currently debating whether and how our knowledge and our practice can be codified so that evaluation is viewed as a coherent body of knowledge and skills, and practitioners are seen to be competent (King, Stevahn, Ghere, & Minnema, 2001). This debate has focused in part on what is needed to be an effective evaluator—core competencies that provide a framework for assessing the adequacy of evaluation training as well as the adequacy of evaluator practice.

In a study of 31 evaluation professionals in the United States, practitioners were asked to rate the importance of 49 evaluator competencies (King et al., 2001) and then try to come to a consensus about the ratings, given feedback on how their peers had rated each item. The 49 items were grouped into four broad clusters of competencies: systematic inquiry (most items were about methodological knowledge and skills); competent evaluation practice (most items focused on organizational and project management skills); general skills for evaluation practice (most items were on communication, teamwork, and negotiation skills); and evaluation professionalism (most items focused on self-development and training, ethics and standards, and involvement in the evaluation profession).

Among the 49 competencies, one was "making judgments," and referred to an overall evaluative judgment, as opposed to recommendations, at the end of an evaluation (King et al., 2001, p. 233). It was rated the *second lowest* on average among all the competencies. This finding suggests that judgment is not perceived to be that important (although the item average was still 74.68 out of 100 possible points). King et al. (2001) suggested that "[s]ome evaluators agreed with Michael Scriven that to evaluate is to judge; others did not" (p. 245). The "reflects on practice" item, however, was given an average rating of 93.23—*a ranking of 17* among the 49 items. Schön (1987) makes reflection on one's practice the key element in being able to develop sound professional judgment. For both of these items, there was substantial disagreement among the practitioners about their ratings, with individual ratings ranging from 100 (highest possible score) to 20. The discrepancy between the low overall score for "making judgments" and the higher score for "reflects on practice" may be related to the difference between *making* a judgment, as an action, and *reflecting* on practice, as a personal quality.

We believe that judgment is woven throughout evaluation competencies. Instead of it being just one or perhaps two among a list of competencies, we see professional judgment being a part of the whole process of working with clients, framing evaluation questions, designing and conducting evaluation research, analyzing and interpreting the information, and communicating the findings, conclusions, and recommendations to stakeholders. If you go back to the outline of a program evaluation process offered in Chapter 1, or the outline of the design and implementation of a performance measurement system offered in Chapter 9, we believe that professional judgment is a part of *all* the steps in both processes. Further, we see different kinds of professional judgment being more or less important at different stages in evaluation processes. We will come back to the relationships between evaluation competencies and professional judgment later in this chapter.

● PROFESSIONALISM AND EVALUATION PRACTICE

The idea that evaluation is a profession or aspires to be a profession is an important part of contemporary discussions of the scope and direction of the enterprise (Altschuld, 1999). Modarresi, Newman, and Abolafia (2001) quote Leonard Bickman, who was president of the American Evaluation Association in 1997, in asserting that "we need to move ahead with professionalizing evaluation or else we will just drift into oblivion" (p. 1). Bickman and others in the evaluation field are aware that other related professions continue to carve out territory, sometimes at the expense of evaluators.

What does it mean to be a professional? What distinguishes a profession from other occupations? Eraut (1994) suggests that professions are characterized by: a core body of knowledge that is shared through the training and education of those in the profession; some kind of government-sanctioned license to practice; a code of ethics and standards of practice; and self-regulation (and sanctions for wrongdoings) through some kind of professional association to which members of the practice community must belong.

Professional Knowledge as Applied Theory

The core body of knowledge that is shared among members of a profession can be characterized as knowledge that is codified, publicly available (taught and learned by aspiring members of the profession) and is supported by validated theory (Eraut, 1994). One view of professional practice is that competent members of a profession apply this validated theoretical knowledge in their work. Competent practitioners are persons who have the credentials of the profession (including evidence that they have requisite knowledge) and have established a reputation for being able to translate theoretical knowledge into sound practice.

Professional Knowledge as Practical Know-How

An alternative view of professional knowledge is that it is the application of practical know-how to particular situations. The competent practitioner uses his or her experiential and intuitive knowledge to assess a situation and offer a diagnosis (in the health field) or a decision in other professions (Eraut, 1994). Although theoretical knowledge is a part of what competent practitioners rely on in their work, practice is much more than applying theoretical knowledge. It includes a substantial component that is learned through practice itself. Although some of this knowledge can be codified and shared (Schön, 1987; Tripp, 1993), part of it is tacit, that is, known to individual practitioners, but not shareable in the same ways that we share the knowledge in textbooks, lectures, or other publicly accessible learning and teaching modalities.

Polanyi (1958) described **tacit knowledge** as the capacity we have as human beings to integrate "facts" (data and perceptions) into patterns. He defined tacit knowledge in terms of the process of discovering theory: "This act of integration, which we can identify both in the visual perception of objects and in the discovery of scientific theories, is the tacit power we have been looking for. I shall call it tacit knowing" (Polanyi & Grene, 1969, p. 140).

For Polanyi, tacit knowledge cannot be communicated directly. It has to be learned through one's own experiences—it is by definition personal knowledge. Knowing how to ride a bicycle, for example, is in part tacit. We can describe to others how the physics and the mechanics of getting onto a bicycle and riding it works, but the experience of getting onto the bicycle, pedaling, and getting it to stay up is quite different from being told how to do so.

One implication of acknowledging that what we know is in part personal is that we cannot teach everything that is needed to learn a skill. The learner can be guided with textbooks, good examples, and even demonstrations, but that knowledge (Polanyi calls it impersonal knowledge) must be combined with the learner's own capacity to tacitly know—to experience the realization (or a series of them) that she understands how to use the skill.

Clearly, from this point of view, practice is an essential part of learning. One's own experience is essential for fully integrating impersonal knowledge into working knowledge. But because the skill that has been learned is in part tacit, when the learner tries to communicate it, she will discover that, at some point, the best advice is to suggest that the new learner try it and "learn by doing." This is a key part of craftsmanship.

Learning and practicing a craft depends on developing and honing sound professional judgment. Knowledge, skills, and experience are a part of craftsmanship as are the values and beliefs of the profession and the individual practitioner. Some of this can be taught and learned but a part will depend on how well the practitioner learns from her own experience.

Balancing Theoretical and Practical Knowledge in Professional Practice

The difference between the technical/rational and the practical know-how views of professional knowledge (Fish & Coles, 1998) has been characterized as the difference between *knowing that* (publicly accessible, propositional knowledge and skills) and *knowing how* (practical, intuitive, experientially grounded knowledge that involves wisdom, or what Aristotle called *praxis*) (Eraut, 1994).

These two views of professional knowledge highlight different views of what professional practice is and indeed ought to be. The first view can be illustrated with an example. In the field of medicine the technical/rational view of professional knowledge and professional practice continues to support efforts to construct and use expert systems—software systems that can offer a diagnosis based on a logic model that links combinations of symptoms in a probabilistic tree to possible diagnoses (Fish & Coles, 1998). By inputting the symptoms that are either observed or reported by the patient, the expert system (embodying the public knowledge that is presumably available to competent practitioners) can treat the diagnosis as a problem to solve. Clinical decision making employs algorithms that produce a probabilistic assessment of the likelihood that symptom, drug, and other technical information will support one or another alternative diagnoses.

The second view of professional knowledge as practical know-how embraces a view of professional practice as craftsmanship and artistry. Although it acknowledges the importance of experience in becoming a competent practitioner, it also complicates our efforts to understand the nature of professional practice. If practitioners know things that they cannot share and their knowledge is an essential part of sound practice, how do professions find ways of ensuring that their members are competent? Later in this chapter we will summarize ways that the evaluation profession can provide opportunities for novice practitioners to acquire the practical knowledge that they need to become competent.

UNDERSTANDING PROFESSIONAL JUDGMENT ●

What are the different kinds of professional judgment? How does professional judgment impact the range of decisions that evaluators make? Can we construct a model of how professional judgment affects evaluation decisions?

Fish and Coles (1998) have constructed a typology of four kinds of professional judgment in the health care field. We believe that these can be generalized to the evaluation field. Each builds on the previous one, and the extent and kinds of judgment differ across the four kinds. At one end of the continuum, practitioners apply *technical judgments* that are about specific issues involving routine tasks. Typical questions include: What do I do now? How do I apply my existing knowledge and skills to do this routine task? In an evaluation, an example of this kind of judgment would be how to select a random sample from a population of case files in a social service agency.

The next level is *procedural judgment* which focuses on procedural questions and involves the practitioner comparing the skills/tools that he or she has available to accomplish a task. Practitioners ask questions like: What are my choices to do this task? From among the tools/knowledge/skills available to me, which combination works best for this task? An example from an evaluation would be deciding how to contact clients in a social service agency—whether to use a survey (and if so, mailing, telephone, interview format, or some combination) or use focus groups (and if so, how many, where, how many participants in each, how to gather them).

The third level of professional judgment is *reflective*. It again assumes that the task or the problem is a given, but now the practitioner is asking: How do I tackle this task? Given what I know, what are the ways that I could proceed? Are the tools that are easily within reach adequate, or instead, should I be trying some new combination or perhaps developing some new ways of dealing with this task or problem? A defining characteristic of this third level of professional judgment is that the practitioner is reflecting on his or her practice and seeking ways to enhance his or her practical knowledge and skills.

An example from a needs assessment for child sexual abuse prevention programs in an urban school district serves to illustrate reflective judgment on the part of the evaluator in deciding on the research methodology. Classes from an elementary school are invited to attend a play acted by school children of the same ages as the audience. The play is called "No More Secrets" and is about an adult–child relationship that involves touching and other activities. At one point in the play, the "adult" tells the "child" that their touching games will be their secret. The play is introduced by a professional counselor and after the play, children are invited to write questions on cards that the counselor will answer. The children are told that if they have questions about their own relationships with adults, these questions will be answered confidentially by the counselor. The evaluator, having obtained written permissions from the parents, contacts the counselor, who, without revealing the identities of any of the children, indicates to the evaluator the number of potentially abusive situations among the students who attended the play. Knowing the proportion of the school district students that attended the play, the evaluator is able to roughly estimate the incidence of potentially abusive situations in the school-age population.

The fourth level of professional judgment is *deliberative*—it explicitly involves a practitioner's own values. Here the practitioner is asking: What ought I to be doing in this situation? No longer are the ends or the tasks fixed, but instead the professional is taking a broad view that includes the possibility that the task or problem may or may not be an appropriate one to pursue. Professionals at this level are asking questions about the nature of their practice and connecting what they do as professionals with their broader values and moral standards. The case study in Appendix 1 of Chapter 11 is an example of a situation that would involve deliberative judgment.

Modeling the Professional Judgment Process

Since professional judgment is an integral part of the evaluation process, it will influence a wide range of decisions that evaluators make in their practice. The four types of professional judgment that Fish and Coles (1998) describe suggest decisions of increasing complexity from discrete technical decisions to global decisions that can affect an evaluator's present and future roles as an evaluation practitioner. Figure 12.1 displays a model of the way that professional judgment is involved in evaluator decision making. The model focuses on a single decision—a typical evaluation would involve many such decisions of varying complexity. In the model, evaluator values, beliefs, and expectations, together with both shareable and practical (tacit) knowledge combine to create a fund of experience that is tapped for professional judgments. In turn, professional judgments influence the decision at hand.

We will present the model and then discuss it, elaborating on the meanings of the key constructs in the model.

Evaluator decisions have consequences. They may be small—choosing a particular alpha (α) level for tests of statistical significance will impact on which findings are noteworthy, given a criterion that significant findings are worth reporting; or they may be large—deciding not to conduct an evaluation in a situation where the findings are being specified in advance by a key

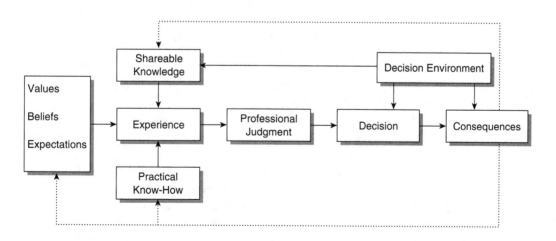

Figure 12.1   The Professional Judgment Process

stakeholder could affect the career of an evaluation practitioner. These consequences feed back to our knowledge (both our shareable and our practical know-how), values, beliefs and expectations. Evaluators have an opportunity to learn from each decision and one of our challenges as professionals is to increase the likelihood that we take advantage of such learning opportunities.

The model can be unpacked by discussing key constructs in it. Some constructs have been elaborated in this chapter already (shareable knowledge, practical know-how, and professional judgment), but it is worthwhile to define each one explicitly in one table. Table 12.1 summarizes the constructs in Figure 12.1 and offers a short definition of each. Several of the constructs will then be discussed further to help us understand what roles they play in the process of forming and applying professional judgment.

The Decision Environment

The particular situation or problem at hand influences how a program evaluator's professional judgment will be exercised. Each opportunity for professional judgment will have unique characteristics that will demand that it be approached in particular ways. For example, a methodological issue will require a different kind of professional judgment from one that centers on an ethical issue. Even two cases involving a similar question of methodological choice will have facts about each of them that will influence the professional judgment process. The extent to which the relevant facts about a particular situation are fully known or understood by the evaluator will affect the professional judgment process.

The decision environment provides constraints and incentives and costs and benefits, both real and perceived, that affect a professional's judgment. Some examples include the expectations of the client, the professional's lines of accountability, tight deadlines, complex and conflicting objectives, and financial constraints. For people working within an organization, for example, internal evaluators, the organization also presents a significant set of environmental factors, in that its particular culture, goals, and objectives may have an impact on the way the professional judgment process operates.

Relevant professional principles and standards, such as the American Evaluation Association's "Guiding Principles for Evaluators" (1995), also form part of the judgment environment because, to some extent, they interact and condition the free exercise of judgment by professionals and replace individual judgment with collective judgment (Gibbins & Mason, 1988).

Table 12.1 Definitions of Constructs in the Model of the Professional Judgment Process

Constructs in the Model	Definitions
Values	Values are statements about what is desirable, what ought to be, in a given situation.
Beliefs	Beliefs are about what we take to be true—our assumptions about how we know what we know (our epistemologies are examples of beliefs).
Expectations	Expectations are assumptions that are typically based on what we have learned, and what we have come to accept is normal.
Shareable Knowledge	Knowledge that is typically found in textbooks or other such media; knowledge that forms the core of the formal training and education of professionals in a field.
Experience	Experience is an amalgam of our knowledge, values, beliefs, expectations, and practical know-how. For a given decision, we have a "fund" of experience that we can draw from. We can augment that fund with learning and from the consequences of the decisions we make as professionals.
Practical Know-how	Practical know-how is the knowledge that is gained through practice. It complements shareable knowledge, and is tacit, that is, is acquired from one's professional practice, and is not shareable.
Professional Judgment	Professional judgment is a process that relies on our experience, and ranges from technical judgments to deliberative judgments.
Decision	In a typical evaluation, evaluators make hundreds of decisions that collectively define the evaluation process. Decisions are choices—a choice made by an evaluator about everything from discrete methodological issues to global values-based decisions that affect the whole evaluation (and perhaps future evaluations).
Consequences	Each decision has consequences—for the evaluator and for the evaluation process. Consequences can range from discrete to global, commensurate with the scope and implications of the decision.
Decision Environment	The decision environment is the set of factors that influences the decision-making process, including the stock of knowledge that is available to the evaluator. Among the factors that could impact an evaluator decision are: professional standards; resources (including time and data); incentives (perceived consequences that induce a particular pattern of behavior) and constraints (legal, institutional, and regulatory requirements that specify the ways that evaluator decisions must fit a decision environment).

Values, Beliefs, and Expectations

Professional judgment is influenced by personal characteristics of the person exercising it. It must always be kept in mind that "judgment is a human process, with logical, psychological, social, legal, and even political overtones" (Gibbins & Mason, 1988, p. 18). Each of us has a unique set of values, beliefs, and expectations that make us who we are; and each of us subscribes to a greater or lesser degree to a set of professional norms that make us the kind of practitioner that we are. These personal factors can lead two professionals to make quite different professional judgments about the same situation (Tripp, 1993).

Among the personal characteristics that can influence one's professional judgment, expectations are among the most important. Expectations have been linked to paradigms—perceptual and theoretical structures that function as frameworks for organizing one's perspectives, even one's beliefs about what is real and what is taken to be factual. Kuhn (1962) has suggested that paradigms are formed through our education and training. Eraut (1994) has suggested that the process of learning to become a professional is akin to absorbing an ideology.

Our past experiences (including the consequences of previous decisions we have made in our practice) predispose us to understand or even expect facts, to interpret situations, and consequently to behave in certain ways rather than in others. As Abercrombie (1960) argues, "we never come to an act of perception with an entirely blank mind, but are always in a state of preparedness or expectancy, because of our past experiences" (p. 53). Thus, when we are confronted with a new situation, we automatically perceive and interpret it in whatever way makes it most consistent with our existing understanding of the world, with our existing paradigms. For the most part, we perform this act unconsciously. We are not aware of how our particular world view influences how we interpret and judge the information we receive on a daily basis in the course of our work or how it affects our subsequent behavior.

How does this relate to our professional judgment? Our expectations lead us to see things we are expecting to see, even if they are not actually there, and to not see things we are not expecting, even if they are there. Abercrombie (1960) calls our world view our "schemata" and illustrates its power over our judgment process with the following figure (Fig. 12.2).

In most cases, when we first read the phrases contained in the triangles, we do not see the extra words. As Abercrombie (1960) points out, "it's as though the phrase 'Paris in the Spring,' if seen often enough, leaves a kind of imprint on the mind's eye, into which the phrase in the triangle must be

Figure 12.2   The Three Triangles

made to fit" (p. 35). She argues that "if [one's] schemata are not sufficiently 'living and flexible,' they hinder instead of help [one] to see" (p. 29). Our tendency is to ignore or reject what does not fit our expectations. Thus, similar to the way we assume the phrases in the triangles make sense and therefore unconsciously ignore the extra words, our professional judgments are based in part on our preconceptions and thus may not be appropriate for the situation.

Expectations can also contribute to improving our judgment by allowing us to unconsciously know how best to act in a situation. When the consequences of such a decision are judged to be salutary, our expectations are reinforced.

Acquiring Professional Knowledge

Our professional training and education are key influences—they affect professional judgment in positive ways by allowing us to understand and address problems in a manner that those without the same education could not, but they also predispose us to interpret situations in particular ways. Indeed, professional education is often one of the most pervasive reasons for our acceptance of "tried and true" ways of approaching problems in professional practice. As Katz (1988) observes, "conformity and orthodoxy, playing the game according to the tenets of the group to which students wish to belong, are encouraged in . . . all professional education" (p. 552). Thus, somewhat contrary to what would appear to be common sense, professional judgment does not necessarily improve in proportion to increases in professional training and education. Similarly, professional judgment does not necessarily improve with increased professional experience, if such experience does not challenge but only reinforces already accepted ideologies. Abercrombie (1960) maintains, errors in judgment "are not easily dismissible as due to youth, ignorance, carelessness or willful stupidity, nor will the passage of time, increased experience or strengthening of moral fibre put an end to them" (p. 92).

Ayton (1998) makes the point that even experts in a profession are not immune to poor professional judgment:

> One view of human judgment is that people—including experts—not only suffer various forms of myopia but are somewhat oblivious of the fact. . . . Experts appear to have very little insight into their own judgment. . . . This oblivion in turn might plausibly be responsible for further problems, e.g. over confidence . . . attributed, at least in part, to a failure to recognize the fallibility of our own judgment. (pp. 238–239)

On the other hand, Mowen (1993) notes that our experience, if used reflectively and analytically to inform our decisions, can be an extremely positive factor contributing to good professional judgment. Indeed, he goes so far as to argue that "one cannot become a peerless decision maker without that well-worn coat of experience . . . the bumps and bruises received from making decisions and seeing their outcomes, both good or bad, are the hallmark of peerless decision makers" (p. 243).

● IMPROVING PROFESSIONAL JUDGMENT IN EVALUATION PRACTICE

Having reviewed the ways that professional judgment is woven through the fabric of evaluation practice and having shown how professional judgment plays a part in our decisions as evaluation practitioners, we can turn to discussing ways of self-consciously improving our professional judgment. Key to this process is becoming aware of one's own decision-making processes.

Guidelines for the Practitioner

Epstein (1999) suggests that a useful stance for professional practice is mindfulness. Mindfulness is the cultivation of a capacity to observe, in a nonjudgmental way, one's own physical and mental processes during tasks. In different words, it is the capacity for self-reflection that facilitates bringing to consciousness our values, assumptions, expectations, beliefs, and even what is tacit in our practice.

Although mindfulness can be linked to religious and philosophical traditions, it is a way of approaching professional practice that offers opportunities to continue to learn and improve (Epstein, 2003). A mindful practitioner is one who has cultivated the art of self-observation (cultivating the compassionate observer). Epstein (2003) characterizes mindful practice this way:

> When practicing mindfully, clinicians approach their everyday tasks with critical curiosity. They are present in the moment, seemingly undistracted, able to listen before expressing an opinion, and able to be calm even if they are doing several things at once. These qualities are considered by many to be prerequisite for compassionate care. (p. 2)

The objective of mindfulness is to see what is, rather than what one wants to see or even expects to see. Epstein suggests that there are at least three ways of nurturing mindfulness: mentorships with practitioners who are themselves well-regarded in the profession; reviewing one's own work; and meditation to cultivate a capacity to observe one's self.

In order to cultivate the capacity to make sound professional judgments it is essential to become aware of the unconscious values and other personal factors that may be influencing one's professional judgment. For only through coming to realize how much our professional judgments are influenced by these personal factors can we strive to become more self-aware and work toward extending our conscious control of them and their impacts on our judgment. As Tripp (1993) argues, "without knowing who we are and why we do things, we cannot develop professionally" (p. 54). By increasing our understanding of the way we make professional judgments, we improve our ability to reach deliberate, fully thought-out decisions rather than simply accepting as correct the first conclusion that intuitively comes to mind.

But how can we, as individuals, learn what factors are influencing our own personal professional judgment? One way is to conduct a systematic questioning of professional practice (Fish & Coles, 1998). Professionals should consistently reflect on what they have done in the course of their work and then investigate the issues that arise from this review. Reflection should involve articulating and defining the underlying principles and rationale behind our professional actions and should focus on discovering the "intuitive knowing implicit in the action" (Schön, 1988, p. 69).

Tripp (1993) suggests that this process of reflection can be accomplished by selecting and then analyzing critical incidents that have occurred during our professional practice in the past ("critical incident analysis"). A critical incident can be any incident that occurred in the course of our practice that sticks in our mind and hence, provides an opportunity to learn. What makes it critical is the reflection and analysis that we bring to it. Through the process of critical incident analysis, we can gain an increasingly better understanding of the factors which have influenced our professional judgments. As Fish and Coles (1998) point out: "Any professional practitioner setting out to offer and reflect upon an autobiographical incident from any aspect of professional practice is, we think, likely to come sooner or later to recognize in it the judgments he or she made and be brought to review them" (p. 254). For it is only in retrospect, in analyzing our past decisions, that we can see the complexities underlying what at the time may have appeared to be a straightforward, intuitive professional judgment. "By uncovering our judgments . . . and reflecting upon them," Fish and Coles (1998) maintain, "we believe that it is possible to develop our judgments because we understand more about them and about how we as individuals come to them" (p. 285).

Self-consciously challenging the routines of our practice, the "high hard ground" that Schön refers to in the quote at the outset of this chapter, is an effective way to develop a more mindful stance. In our professional practice, each of us will have developed routines for addressing situations that occur frequently. As Tripp (1993) points out, although routines

> may originally have been consciously planned and practiced, they will have become habitual, and so unconscious, as expertise is gained over time. Indeed, our routines often become such well-established habits that we often cannot say why we did one thing rather than another, but tend to put it down to some kind of mystery such as "professional intuition." (p. 17)

A key way to critically reflect on our professional practice and understand what factors influence the formation of our professional judgments is to discuss our practice with our colleagues. Colleagues, especially those who are removed from the situation at hand or under discussion, can act as "critical friends" and can help in the work of analyzing and critiquing our professional judgments with an eye to improving them. With different education, training, and experience, our professional peers often have different perspectives from us. Consequently, involving colleagues in the process of analyzing and critiquing our professional practice allows us to compare our ways of interpreting situations and choosing alternatives for action with other professionals. Moreover, the simple act of describing and summarizing an issue so that our colleagues can understand it can reveal and provide much insight into the professional judgments we have incorporated.

Implications for Training Evaluation Practitioners

There is considerable interest in the evaluation field in outlining competencies that define sound practice (King et al., 2001). Although there are different versions of what these competencies are, there is little emphasis on acquiring professional judgment skills as a competency. Efforts to establish whether practitioners themselves see judgment skills as being important indicate a broad range of views, reflecting some important differences as to what evaluation practice is and ought to be (King et al., 2001).

If we consider linkages between types of professional judgment and the range of activities that comprise evaluation practice, we can see that some kinds of professional judgment are more important for some clusters of activities than others. But for many evaluation activities, several different kinds of professional judgment can be relevant. Table 12.2 summarizes clusters of competencies that reflect the design and implementation of a typical program evaluation or a performance measurement system. These clusters are based on the outlines for the design and implementation of program evaluations and performance measurement systems included in Chapters 1 and 9, respectively. Although they are not comprehensive, that is, do not represent the full range of competencies for the profession summarized in King et al. (2001), they illustrate the ubiquity of professional judgment in our practice.

Table 12.2. suggests that for most clusters of evaluation competencies, several different types of professional judgment are in play. The notion that somehow, we could practice by exercising only technical and procedural professional judgment is akin to staying on Schön's (1987) "high hard ground."

Training Evaluators: Ways of Inculcating the Capacity to Render Sound Professional Judgments

Professional judgment depends substantially on being able to develop and practice the craft of evaluation. Schön (1987) and Tripp (1993) among others have emphasized the importance of practice as a way of cultivating sound professional judgment. Although textbook knowledge (knowing what) is also an essential part of every evaluator's toolkit, a key part of evaluation curricula are opportunities to acquire experience.

There are at least five complementary ways that evaluation curricula can be focused to provide opportunities for students to develop their judgment skills. Some activities are more discrete, that is, are relevant for developing skills that are specific. These are generally limited to a single course or even a part of a course. Others are more generic, offering opportunities to acquire experience that spans entire evaluation processes. These are typically activities that integrate coursework in real work experiences. Table 12.3

Table 12.2 Types of Professional Judgment That Are Relevant to Program Evaluation and Performance Measurement

Clusters of Program Evaluation Competencies	Types of Professional Judgment			
	Technical	Procedural	Reflective	Deliberative
Negotiating the terms of reference for the evaluation	Yes	Yes	Yes	Yes
Assessing relevant previous work	Yes	Yes	Yes	
Describing the program and its environment	Yes	Yes	Yes	
Selecting evaluation methods	Yes	Yes	Yes	
Assessing evaluability options	Yes	Yes	Yes	Yes
Conducting the evaluation	Yes	Yes	Yes	
Reporting and disseminating the results	Yes	Yes	Yes	Yes
Clusters of Performance Measurement Competencies				
Assessing organizational capacity to design, implement and sustain performance measurement systems	Yes	Yes	Yes	Yes
Identifying performance measures and the information to support them	Yes	Yes	Yes	Yes
Analysis and reporting of performance results	Yes	Yes	Yes	Yes
Adjusting the performance measurement system over time	Yes	Yes	Yes	Yes

summarizes five ways that academic programs can inculcate professional judgment capacities in their students.

The types of learning activities in Table 12.3 are typical of many programs that train evaluators, but what is important is realizing that each of these kinds of activities contributes directly to developing a set of skills that all practitioners need and will use in all their professional work. In an important way, identifying these learning activities amounts to making explicit what has largely been tacit in our profession.

Table 12.3 Learning Activities to Increase Professional Judgment Capacity in Novice Practitioners

Learning Activities	Examples
Course-based Activities	
Problem/Puzzle Solving	Develop a coding frame and test the coding categories for intercoder reliability for a sample of open-ended responses to an actual client survey that the instructor has provided
Case Studies	Make a decision for an evaluator who finds herself caught between the demands of her superior (who wants evaluation interpretations changed) and the project team who see no reason to make any changes
Simulations	Using a scenario and role playing, negotiate the terms of reference for an evaluation
Course Projects	Students are expected to design a practical, implementable evaluation for an actual client organization
Program-based Activities	
Practice	Students work as apprentice evaluators in organizations that design and conduct evaluations, for extended periods of time (at least 4 months)

EVALUATION AS A CRAFT: IMPLICATIONS ● FOR PROFESSIONALIZING EVALUATION

Evaluation is a craft. It has both a methodological aspect, where practitioners are applying tools, and an aesthetic aspect, which entails developing an appreciation for the art of design, the conduct of research, and the interpretation of results. As Berk and Rossi (1999) contend, mastering a craft involves more than learning the techniques and tools of the profession; it involves developing "intelligence, experience, perseverance, and a touch of whimsy" (p. 99), which all form part of professional judgment. Traditionally, persons learning a craft apprenticed themselves to more senior members of the trade. They learned by doing with the guidance and experience of the master craftsperson.

We have come to think that evaluation can be taught in classrooms, often in university settings or in professional development settings. Although these experiences are useful, they are no substitute for learning how evaluations are actually done. Apprenticing to a person or persons who are competent senior practitioners is an important part of becoming a practitioner of the craft. Some evaluators apprentice themselves in graduate programs, preparing master's or doctoral theses with seasoned practitioners. Others work with practitioners in work-experience settings (co-op placements, for example). Still others join a company or organization at a junior level and, with time and experience, assume the role of full members of the profession.

Apprenticeship amounts to learning by doing. It complements what can be learned in classrooms, from textbooks and other such sources. Learning by doing facilitates practitioners' developing tacit knowledge (Polanyi & Grene, 1969). Cultivating a capacity to reflect on one's practice can render at least some kinds of tacit knowledge explicit. Schön (1987) points out that an ideal way to learn a profession is to participate in practical projects wherein students design for actual situations, under the guidance of instructor/ coaches who are themselves seasoned practitioners. Students learn by doing and also have opportunities, with the guidance of coaches, to critically reflect on their practice.

In evaluation, an example of such an opportunity might be a course that is designed as a hands-on workshop to learn how to design and conduct a program evaluation. Students would work with an instructor who would arrange for client organizations who want evaluations done, to participate in the course. Students would work in teams, and teams would be matched with clients. As the course progressed, each team would be introduced to the skills that are needed to meet client and instructor expectations for that part of the evaluation process. There would be tutorials to learn skills that are needed for the teams' work, and opportunities for teams to meet as a class to share their experiences and learn from each other and the instructor. Clients would be invited into these sessions to participate as stakeholders and provide the class and the instructor with relevant and timely feedback. The teams would be expected to gather relevant lines of evidence, once their evaluation was designed, and analyze the evidence. Written reports for the clients would be the main deliverables for the teams, together with oral presentations of the key results and recommendations in class, with the clients in attendance.

● TEAMWORK AND PROFESSIONAL JUDGMENT

Evaluators and managers often work in organizational settings where teamwork is expected. Successful teamwork requires establishing norms and expectations that encourage good communication and sharing of information and a joint commitment to the task at hand. Being able to select team members and foster a work environment wherein people are willing to trust each other and be open and honest about their own views on issues is conducive to generating information that reflects "reality." Even though there will still be individual biases, the views expressed will be more valid than some preconceived view that reflects the perceptions of a dominant individual or coalition in the group. An organizational culture that emulates features of learning organizations (Garvin, 1993) will tend to produce information that is more valid as input for making decisions and evaluating policies and programs.

Managers and evaluators who have the skills and experience to be able to call on others and, in doing so, be reasonably confident that honest views about an issue are being offered, have a powerful tool to complement their own knowledge and experience and their systematic inquiries. Good professional judgment, therefore, is partly about selecting and rewarding people who themselves have demonstrated a capacity to deliver sound professional judgment.

SUMMARY ●

Program evaluation is partly about learning tools and rules to apply them. But, because most evaluation settings offer only roughly appropriate opportunities to apply tools that are often designed for social science research settings, it is essential that evaluators learn the craft of working with round pegs for square holes. Evaluators and managers have in common the fact that they are usually trained in settings that idealize the applications of the tools that they learn. Then, when they enter the world of practice, they must adapt what they have learned. What works is determined by their experiences. Experience becomes the foundation not only of when and how to apply tools but, more importantly, the essential basis for interpreting the information that is gathered in a given situation.

Evaluators have the comparative luxury of time and resources to examine a program or policy that managers usually have to judge *in situ*, as it were. That there are rarely sufficient resources to apply the tools that would yield the highest quality of data is a limitation that circumscribes what we do, but does not mean that we should stop asking whether and how programs work.

This chapter emphasizes the central role played by professional judgment in the practice of professions, including evaluation. It follows that professional programs, courses in universities, and textbooks should underscore for students the importance of developing and continuously improving their professional judgment skills, as opposed to focusing only on learning methods, facts, and exemplars. New practitioners in a profession

> begin to recognize that practice is much more messy than they were led to believe [in school], and worse, they see this as their own fault—they cannot have studied sufficiently well during their initial training. . . . [T]his is not true. The fault, if there is one, lies in the lack of support they receive in understanding and coping with the inevitably messy world of practice. (Fish & Coles, 1998, p. 13)

They continue:

> Learning to practice in a profession is an open capacity, cannot be mastered and goes on being refined for ever. Arguably there is a major onus on those who teach courses of preparation for professional practice to demonstrate this and to reveal in their practice its implications. (p. 43)

The ubiquity of judgment in evaluation practice suggests that as a profession we need to do at least three things. First, we need to fully acknowledge the importance of professional judgment and the roles it plays in the diverse ways we practice evaluation. Second, we need to understand how our professional judgments are made—the factors that condition our own judgments. Third, we need to work toward self-consciously improving the ways we incorporate into the education and training of evaluators, opportunities for current and future practitioners to improve their professional judgments. Embracing professional judgment is an important step toward a more mature and self-reflective evaluation profession.

● DISCUSSION QUESTIONS

1. Take a position for or against the following proposition and develop a strong one-page argument that supports your position. This is the proposition, "Be it resolved that experiments, where program and control groups are randomly assigned, are the only valid way to determine the effectiveness of programs."

2. What do evaluators and program managers have in common?

3. What is tacit knowledge? How does it differ from public knowledge?

4. In this chapter, we said that learning to ride a bicycle is partly tacit. For those who want to challenge this statement, try to describe learning how to ride so that a person who has never before ridden a bicycle could get on one and ride it right away.

5. What is mindfulness and how can it be used to develop sound professional judgment?

6. Why is teamwork an asset for persons who want to develop sound professional judgment?

7. What would be required to make evaluation more professional, that is, having the characteristics of a profession?

REFERENCES ●

Abercrombie, M. L. J. (1960). *The anatomy of judgment: An investigation into the processes of perception and reasoning.* New York: Basic Books.

Altschuld, J. W. (1999). The certification of evaluators: Highlights from a report submitted to the Board of Directors of the American Evaluation Association. *American Journal of Evaluation, 20*(3), 481–493.

American Evaluation Association. (1995). Guiding principles for evaluators. *New Directions for Program Evaluation, 66,* 19–26.

Ayton, P. (1998). How bad is human judgment? In G. Wright & P. Goodwin (Eds.), *Forecasting with judgement.* Chichester, UK: Wiley.

Basilevsky, A., & Hum, D. (1984). *Experimental social programs and analytic methods: An evaluation of the U.S. income maintenance projects.* Orlando, FL: Academic Press.

Berk, R. A., & Rossi, P. H. (1999). *Thinking about program evaluation* (2nd ed.). Thousand Oaks, CA: Sage.

Cronbach, L. J. (1980). *Toward reform of program evaluation.* San Francisco: Jossey-Bass.

Epstein, R. M. (1999). Mindful practice. *JAMA: Journal of the American Medical Association, 282*(9), 833–839.

Epstein, R. M. (2003). Mindful practice in action (I): Technical competence, evidence-based medicine, and relationship-centered care. *Families, Systems and Health, 21*(1), 1–9.

Eraut, M. (1994). *Developing professional knowledge and competence.* Washington, DC: Falmer Press.

Fish, D., & Coles, C. (1998). *Developing professional judgment in health care: Learning through the critical appreciation of practice.* Boston: Butterworth-Heinemann.

Ford, R., Gyarmati, D., Foley, K., Tattrie, D., & Jimenez, L. (2003). Can work incentives pay for themselves? Final report on the Self-Sufficiency Project for welfare applicants. Ottawa, ON: Social Research and Demonstration Corporation.

Garvin, D. A. (1993). Building a learning organization. *Harvard Business Review, 71*(4), 78–90.

Gibbins, M., & Mason, A. K. (1988). Professional judgment in financial reporting. Toronto, ON: Canadian Institute of Chartered Accountants (CICA-ICCA).

Gustafson, P. (2003). *How random must random assignment be in random assignment experiments?* Ottawa, ON: Social Research and Demonstration Corporation.

House, E. R., & Howe, K. R. (1999). *Values in evaluation and social research.* Thousand Oaks, CA: Sage.

Hum, D., & Simpson, W. (1993). Economic response to a Guaranteed Annual Income: Experience from Canada and the United States. *Journal of Labor Economics, 11*(1), S263-S296.

Human Resources Development Canada. (1998). *Quasi-experimental evaluation. Publication #SP-AH053E-01–98.* Ottawa, ON: Evaluation and Data Development Branch.

Katz, J. (1988). Why doctors don't disclose uncertainty. In J. Dowie & A. S. Elstein (Eds.), *Professional judgment: A reader in clinical decision making.* Cambridge, UK: Cambridge University Press.

Kelling, G. L. (1974). *The Kansas City preventive patrol experiment: A technical report.* Washington, DC: Police Foundation.

King, J. A., Stevahn, L., Ghere, G., & Minnema, J. (2001). Toward a taxonomy of essential evaluator competencies. *American Journal of Evaluation, 22*(2), 229–247.

Kuhn, T. S. (1962). *The structure of scientific revolutions.* Chicago: University of Chicago Press.

Levin, H. M., & McEwan, P. J. (Eds.). (2001). *Cost-effectiveness analysis: Methods and applications* (2nd ed.). Thousand Oaks, CA: Sage.

Mark, M. M., Henry, G. T., & Julnes, G. (2000). *Evaluation: An integrated framework for understanding, guiding, and improving policies and programs.* San Francisco: Jossey-Bass.

Modarresi, S., Newman, D. L., & Abolafia, M. Y. (2001). Academic evaluators versus practitioners: Alternative experiences of professionalism. *Evaluation and Program Planning, 24*(1), 1–11.

Mowen, J. C. (1993). *Judgment calls: High-stakes decisions in a risky world.* New York: Simon & Schuster.

Office of Management and Budget. (2004). *What constitutes strong evidence of a program's effectiveness?* Retrieved July 2004, from http://www.whitehouse.gov/omb/part/2004_program_eval.pdf

Patton, M. Q. (1997). *Utilization-focused evaluation: The new century text* (3rd ed.). Thousand Oaks, CA: Sage.

Pechman, J. A., & Timpane, P. M. (Eds.). (1975). *Work incentives and income guarantees: The New Jersey negative income tax experiment.* Washington, DC: Brookings Institution.

Polanyi, M. (1958). *Personal knowledge: Towards a post-critical philosophy.* Chicago: University of Chicago Press.

Polanyi, M., & Grene, M. G. (1969). *Knowing and being: Essays.* Chicago: University of Chicago Press.

Schön, D. A. (1987). *Educating the reflective practitioner: Toward a new design for teaching and learning in the professions.* San Francisco: Jossey-Bass.

Schön, D. A. (1988). From technical rationality to reflection-in-action. In J. Dowie & A. S. Elstein (Eds.), *Professional judgment: A reader in clinical decision making.* Cambridge, UK: Cambridge University Press.

Schwandt, T. A. (2000). Three epistemological stances for qualitative enquiry. In N. K. Denzin & Y. S. Lincoln (Eds.), *Handbook of qualitative research* (2nd ed., pp. 189–213). Thousand Oaks, CA: Sage.

Scriven, M. (1994). The final synthesis. *Evaluation Practice, 15*(3), 367–382.

Scriven, M. (In progress, unpublished). *Causation.* New Zealand: University of Auckland.

Shadish, W. R., Cook, T. D., & Campbell, D. T. (2002). *Experimental and quasi-experimental designs for generalized causal inference.* Boston: Houghton Mifflin.

Smith, M. L. (1994). Qualitative plus/versus quantitative: The last word. *New Directions for Program Evaluation, 61,* 37–44.

Tripp, D. (1993). *Critical incidents in teaching: Developing professional judgement.* London: Routledge.

U.S. Department of Education. (2002). *No Child Left Behind Act of 2001.* Pub. L. No. 107–110, 115 Stat. 1425, 2004.

Watson, K. F. (1986). Programs, experiments, and other evaluations: An interview with Donald Campbell. *Canadian Journal of Program Evaluation, 1*(1), 83–86.

Weiss, C. H. (1998). *Evaluation: Methods for studying programs and policies* (2nd ed.). Upper Saddle River, NJ: Prentice Hall.

GLOSSARY

accountability   responsibility for the expenditure, administrative and programmatic activities that occur in organizational units over which one has formal authority

activity-based accounting   accounting systems that calculate the costs of producing particular program outputs, so that it is possible to easily calculate technical efficiency

adequacy   the extent to which the program outcomes were sufficient to meet the needs of a program

after-only experimental design   a research design that does not measure program/control differences before the treatment begins

allocative efficiency   allocating resources based on whether the benefit to be obtained from one service is more or less than the benefit obtained from alternate uses of the resources

ambiguous temporal sequence   this internal validity threat arises where "cause" and the "effect" variables could plausibly be reversed so that the "effect" variable becomes the "cause" and vice versa

appropriateness   the extent to which the logic of the program is the best means to achieve the objectives of the program

attribution   the extent to which the program, and not some other factor(s) in the program environment, caused the observed outcomes

baseline measure   a measurement used as a starting-point (usually before a program is implemented), against which future results are compared

before-after design   a research design that compares one measurement or data point before the program is implemented with one measurement after implementation

behavioral goals   goals in organizations that are consistent with the assumption that individuals and coalitions will look out for their own well-being—they will tend to be self-interested rational actors

benchmark a standard or point of reference (often some standard of best practices) against which program processes or outcomes can be compared

benefit stream forecasted or actual benefits that are expected to occur or actually occur over time in a program or project

benefit-cost ratio the ratio of total discounted program benefits to total discounted program costs

bias a systematic distortion in a measurement instrument which results in data that tend to be either too high or too low in relation to the true value of a measure

case studies methods of inquiry that focus on intensive data collection and analysis that investigates only a few units of analysis

case study design a research design were the program is followed by observed results, and there are no opportunities to construct comparisons (before-after or no program group[s]) to assess the incremental outcomes of the program

causal chain a set of connected causal relationships

causal linkages (same as **cause and effect linkages**) intended causal relationships between the constructs in a program logic model

causal modeling a multivariate statistical procedure that calculates the significance and strength of the relationships among a set of independent variables, as well as the relationships between each independent variable and a dependent variable

causal relationship one variable is said to cause another where: the causal variable occurs before the effect variable; the cause and effect variables co-vary, that is, as one changes, the other one also changes (either positively or negatively) and there are no other variables that could plausibly account for the co-variation between the cause and the effect variables

causally connected two variables are said to be causally connected when one is correlated with the other, the first occurs before the second, and there is no other plausible rival factors that could account for the observed relationship

ceteris paribus all other things being equal, that is, all influences are held constant except the one that is of immediate interest

closed-ended questions questions that are structured so that the categories of possible responses are constructed by the evaluator before the data are collected

comparative time-series design   a research design that relies on data collected at multiple points in time (before and after program implementation) for both the program group and a control group

comparison group   a group of units of analysis (usually people) who are not exposed to the program and who are compared to the program group

compensatory equalization of treatments   a construct validity threat where the program is offered to the control group as well as the program group, resulting in an inability to sort out the program effects

compensatory rivalry   a construct validity threat where the control group is exposed to something better than the status quo situation, which biases the comparison of program and control group results

conceptual framework   a set of constructs that provides us with definitions and categories that permit us to structure our thinking about social processes

concurrent validity   validity related to the strength of the correlations between or among measures of a construct

confidence interval   when sample descriptive statistics are calculated (a mean, for example), and then generalized to the population from which the sample was randomly drawn, the interval of possible values of the population mean, centered around the sample mean, is the confidence interval

confirmatory factor analysis   the use of factor analysis (a multivariate statistical procedure for data analysis) to confirm the underlying dimension(s) in an empirical assessment of the internal structure of a measure

construct validity   the extent to which the variables used to measure program constructs convincingly represent the constructs in the program logic model

constructs   the words or phrases in logic models that we use to describe programs, including the cause and effect linkages in the program

constructivist   a philosophical view of the world that assumes that people's perceptions are relative and that reality is socially constructed; there is no foundational reality

content validity   the extent to which a measure "captures" the intended content of a construct

control group   a group of units of analysis (usually people) who are not exposed to the experiment or program and who are compared to the program group

convergent validity   the extent to which measures of two or more constructs that are theoretically related correlate or co-vary with each other

core technologies means-ends relationships that define the competencies that an organization has, which are believed by organizational participants to be workable in accomplishing program/organizational objectives

correlation the extent to which the variance of one variable co-varies with the variance of another variable—correlations can be either positive or negative and can vary in strength between −1 (perfect negative correlation) to +1 (perfect positive correlation)

correlational validity the extent to which two measures of a given construct correlate with each other

cost-benefit analysis an evaluation of the costs and the benefits of a policy, program, or project wherein all the current and future costs and benefits are converted to current dollars

cost-effectiveness   the ratio of program inputs (expressed in monetary units) to program outcomes

cost-effectiveness analysis   a comparison of the costs and outcomes of policy, program, or project alternatives such that ratios of costs per unit of outcome are calculated

cost-utility analysis   a comparison of costs and estimated utilities of program outcomes that weights and combines outcomes so that alternatives can be compared

counterfactual condition   offering evidence of what would have happened without the program

co-variation   as the values of one variable change (either increasing or decreasing), the values of the other variable also change—co-variation can be positive or negative

criterion sampling   a sampling procedure whereby all cases are chosen that meet some criterion; useful for quality assurance

diffusion of treatments   a construct validity threat where interactions between the program group and the control group offer ways for the control group to learn about the intended treatment, weakening the intended differences between the two groups

direct costs   costs associated with the actual operation of a program, including operating costs and support costs

discount rate the rate of interest used in discounting costs and benefits, that is, converting all costs and benefits over the life of the policy, program, or project into net present values

discounting the process of determining the net present value of a dollar amount of costs or benefits

discriminant validity the extent to which the measures of two or more constructs that are not theoretically related do not correlate with each other

disproportionate stratified sample similar to stratified sample (see definition below) except that one or more strata are randomly sampled so that the number of cases selected is greater (or less) than the fraction/proportion that that stratum is of the whole population

economic costs costs of a project or program that are estimated using the opportunity costs (value of the next highest priced use) of the human resources, capital, technology, and other inputs

economic efficiency the net social value of a project or program, estimated by subtracting the discounted social costs from the discounted social benefits

economy attaining the right program inputs at the lowest possible cost

effectiveness the extent to which the observed outcomes are consistent with the intended objectives

efficiency attaining the most program outputs possible for each program input (usually expressed in monetary units)

empowerment evaluation the use of evaluation concepts, techniques, and findings to help program managers and staff evaluate their own programs and thus improve practice and foster self-determination in organizations

environmental factors factors in the surroundings of a program that may have an effect on it and on the intended outcomes

environmental scan an analysis of trends and key factors in an organization's environment that may have an impact on it

epistemological beliefs beliefs about how we can know ourselves and the physical and social world around us

evaluation assessment a systematic study of the options, including their strengths and weaknesses, when a program evaluation is being planned

evaluation study the process of designing, conducting, and reporting a program evaluation

evidence-based decision making a philosophy of management that emphasizes the importance of using defensible evidence as a basis for making decisions—sometimes associated with performance management

***ex ante* analyses** analyses (usually cost-benefit, but can also be cost-effectiveness or cost-utility analysis) that are done before a program, policy, or project is implemented

***ex ante* evaluation** an evaluation that is conducted before a program is implemented

***ex post* analyses** analyses that are done after a policy, program, or project is implemented

***ex post* evaluation** an evaluation that is conducted after a program has been implemented

experimental design a research design involving one or more treatment (program) and control groups, where program and control participants are randomly assigned to the groups, ensuring that the groups are equal except for the program itself

external validity the extent to which the results of an evaluation can be generalized to other times, other people, other treatments, and other places

face validity where an evaluator or experts judge that a measurement instrument appears to be adequately measuring what it is intended to measure

flow chart a model that outlines the steps in a process, including flows of clients and decision points along the way

focus group a group of persons (usually a maximum of 12) selected for their relevance for a particular evaluation question—usually focus groups are facilitated by one or more persons who guide the discussion and record the proceedings

formative evaluation an evaluation designed to provide feedback and advice for improving a program

gestalt an organized whole that is perceived to be more than the sum of its parts—a pattern that organizes, interprets, and makes data meaningful

goal a broad statement of intended outcomes for a program, line of business or an organization—goals are typically intended to guide the formation of (more specific) objectives that can be linked to goals

Hawthorne effect a construct validity threat where there are unintended research results caused by the subjects' knowing they are participants in a research process and behaving differently than they would if there was no experiment being conducted

high-probability program technologies program technologies (means-ends relationships) that are highly likely, if fully implemented, to achieve the intended program outcomes

history an internal validity threat where changes in the program environment coincide with or mask program effects, biasing the results of the evaluation

holding constant the process of using either research design or statistics to isolate one intended cause and effect linkage so that it can be tested with evidence

holistic approach seeking patterns that provide an overall understanding of the evaluation data, including and integrating the perspectives of different stakeholders

implementation objectives statements of what needs to happen to get a program producing outputs—they focus on program implementation and not on program outcomes

implicit design a posttest only design with no control group—the evaluation occurs after the program is implemented and there are no non-program comparison groups

incommensurability the hypothesis that theories (physical or social theories) that focus on the same phenomena consist in part of key definitions and assumptions that make it impossible to fully translate the constructs (language) used in one theory into language used in another one

incremental effects outcome changes that result from a program

index a measure based on combining the data (either weighted or unweighted) from two or more other measures, usually from a survey

indirect costs costs caused by a policy, program, or project that occur in the environment, and are not intended

individually necessary and jointly sufficient conditions when we speak about conditions for determining whether a relationship between two variables is causal, we specify three criteria: temporal asymmetry, co-variation, and no plausible rival hypotheses—these are individually necessary for causality and together are jointly sufficient to determine whether cause and effect exists

inductive approach a process that begins with data, and constructs patterns that can be generalized

instrumentation   an internal validity threat where changes in the measurement instrument(s) used in the evaluation coincide with the implementation of the program, making it very difficult to distinguish program effects from effects due to changes in the measurement processes

intangible costs   costs that cannot easily be expressed in dollar terms

internal rate of return   the discount rate that balances total costs and benefits, producing a net present benefit of 0 or a cost-benefit ratio of 1

internal structure validity   validity related to the coherence of a pool of items that are collectively intended to be a measure of a construct

internal validity   the extent to which there are no plausible rival hypotheses that could explain the linkage between the program and the observed outcomes—an internally valid research design eliminates all plausible rival hypotheses, allowing a "clean" test of the linkage

interval level of measurement   a level of measurement where there is a unit of measurement, that is, values of the variable are all equal intervals but there is no natural zero value in the scale

learning organization   an organization that is characterized by double-loop learning, that is, acquiring and using information to correct its performance in relation to current objectives as well as assessing and even changing its objectives

level of confidence   in generalizing to a population from the description of a sample (the mean, for example), how much error are we willing to tolerate in estimating a range of values that will include the population mean—the higher the level of confidence we pick (e.g., the 99% level instead of the 95% level), the greater the likelihood that our range of values (our confidence interval) will, in fact, capture the true population mean

levels of analysis problem   a situation in performance measurement where performance data for one level in an organization (the program level, for example) is used to (invalidly) infer performance at another level (individuals working within the programs, for example)

levels of measurement   a hierarchy of measurement procedures that begins with classification (nominal measurement), proceeds through ranking (ordinal measurement), then to interval (counting the amounts of a characteristic), and ends with ratio measures (interval measures that have a natural zero point)

Likert-scale items   statements worded so that survey respondents can respond by agreeing or disagreeing with the statement, usually on a five-point scale from "strongly disagree" to "strongly agree"

linking constructs transition factors in some program logics that are necessary to connect the program outputs to the intended program outcomes

low-probability program technologies program technologies (intended cause and effect relationships) for which there is limited evidence that they will achieve the intended program outcomes

marginal cost the cost of producing one additional unit of output

marginal value the value of one more unit of output produced

maturation an internal validity threat where natural changes over time in the subjects being studied coincide with predicted program effects

means-ends relationship a causal relationship between or among factors such that one is affected by the other(s)—one is said to cause the other

measurement the procedures that we use to translate a construct into observable data

measurement instrument the instrument that implements the procedures we use to translate a construct into observable data

medium probability program technologies program technologies for which there is considerable evidence that implementing the program as intended will produce the desired outcomes

meta-analysis the same as meta-evaluation; a synthesis of existing program evaluation studies in a given area, designed to summarize current knowledge about a particular type of program

meta-evaluation a synthesis of existing program evaluation studies in a given area, designed to summarize current knowledge about a particular type of program

mortality an internal validity threat where the withdrawal of subjects during the evaluation process interferes with before-after comparisons

multiple-causation problem the problem of attributing the observed costs and benefits to a program, as opposed to other factors (rival hypotheses)

multivariate statistical analysis statistical methods that allow for the simultaneous assessment of the influence of two or more independent variables on one or more dependent variables

naturalistic approach involves using research designs (usually qualitative) that do not attempt to manipulate or control the program setting

needs assessment a study that measures the nature and extent of the need for a program, either before a new program is developed or during its lifetime

net present benefit the total value of program benefits expressed in current dollars minus the total value of program costs expressed in current dollars

net present value the present monetary value of the benefits less the present value of the costs.

net social value The economic value of a project or program once net present (discounted) costs have been subtracted from net present (discounted) benefits

new public management An approach to public sector reform that emphasizes business-like practices for organizations, including performance measurement and managing for results

nominal the most basic level of measurement, where the variable consists of two or more mutually exclusive categories

nominal prices prices (either benefits or costs) that have *not* been adjusted to remove the effect of inflation over time

objective a statement of intended outcomes that is focused and time-specific, that is, achievable in a specified time frame

objectivity a two-stage process involving scrutable methods and replication of findings by independent testing of hypotheses

observables when we translate constructs into (measurable) variables, these variables are the observables in the evaluation

open systems approach conceptualizing programs as open systems, that is, sets of means-ends linkages that affect and are affected by their environment

open-ended questions questions where the respondent can answer in his or her own words, and then categories are created by the analyst to classify responses after all the data are collected

open-system a bounded structure of means-ends relationships that affects and is affected by its environment

open-systems metaphor a metaphor from biology or from engineering that offers a way of describing programs as open systems

operating costs costs associated with items that contribute to the operation of a program, such as salaries and supplies

operationalized when we measure constructs, we sometimes say that the constructs have been operationalized

opportunistic sampling sampling strategy where participants are selected based on their connection to emerging research questions and by following leads

opportunity costs the cost that is equivalent to the next-best economic activity that would be forgone if a project or program proceeds

ordinal level of measurement a level of measurement where the variable is a set of categories that are ranked on some underlying dimension

organizational logic models logic models for whole organizations that link organizational goals (or business line goals) to objectives and hence, to strategies and performance measures

paradigm a particular way of seeing and interpreting the world, akin to a belief system

patched-up research designs in program evaluations where several research designs have been combined to offset the weaknesses in any one of them, we refer to this as a patched-up research design

pecuniary benefits program benefits that are offset by other costs so that, from the point of view of society, there is no overall benefit

pecuniary costs program costs that are offset by other benefits so that, from society's perspective, there is no overall cost

performance management organizational management that relies on evidence about policy and program accomplishments to connect strategic priorities to outcomes and make decisions about current and future directions

performance management cycle a normative model of organizational planning and actions that emphasizes the importance of stating clear goals and objectives, translating these into policies and programs, implementing and then assessing and reporting outcomes so that goals and objectives can be appropriately modified

performance measurement the process of designing and implementing quantitative and qualitative measures of program results, including outputs and outcomes

performance measures quantitative and qualitative measures of program or organizational results, including outputs and outcomes

plausible rival hypotheses variables that are shown either by evidence or by judgment to influence the relationship between a program and its intended outcome(s)

politics the authoritative (either formally or informally) allocation of values within organizations

population a group of people that may or may not be from the same geographic area, who receive services from public sector or nonprofit organizations

positivist a philosophical view of the world that assumes that our perceptions are factual and the process of testing hypotheses involves comparing predictions to the facts

positivistic social science social science that is premised on the belief that it is possible to develop and test theories and hypotheses about human behavior using methodologies and factual data that assume that there is a knowable social reality that is independent of our subjective perceptions

postpositivistic social science social science that is based in part on the epistemological belief that there is a reality independent of our senses, but we have no way of knowing for sure what that reality is

posttest only design where measurements occur only after being exposed to the program

predictive validity the extent to which a measure of one construct can be used to predict the measures of other constructs in the future

pretest a test of a measurement instrument prior to its actual use, designed to identify and correct problems with the instrument

pretest posttest design where measurements occur before and after being exposed to the program and the two sets of results are compared

primary data data gathered by the evaluator specifically for a current needs assessment or evaluation

professional judgment combining experience, which is influenced by beliefs, values, and expectations, with evidence to construct findings, conclusions, and recommendations in program evaluations

program a group of related, purposive activities that is intended to achieve one or several related objectives

program activities the work done in a program that produces the program outputs

program components major clusters of activities in a program that are intended to drive the process of producing outcomes

program constructs words or phrases that describe key features of a program

program effectiveness the extent to which the program achieves its intended outcomes

program environment the surroundings and conditions within which the program is situated

program evaluation a systematic process for gathering and interpreting information intended to answer questions about a program

program implementation converting program inputs into the activities that are needed to produce outputs

program inputs the resources consumed by program activities

program logic model a way of representing a program as an open system that categorizes program activities and outlines the intended flow of activities from outputs to outcomes

program logics models of programs that categorize program activities and link activities to results (outputs and outcomes)

program objectives statements of intended outcomes for programs that ideally: specify the target group; specify the magnitude and the direction of expected change; specify the time frame for achieving the result; and are measurable

program outcomes (intended) the results occurring in the environment of a program that it is designed to achieve

program outcomes (observed) what a program appears, through a process of measurement, to have achieved

program outputs the work produced by program activities

program processes the activities in a program that produce outputs

program rationale the ways (if any) that a program fits into current and emerging priorities of the government or agency that has sponsored/funded it

program technologies means-ends relationships that are used in programs to achieve the program objectives

proportionate stratified sample similar to a stratified sample (see the definition below), but in each stratum, the number of cases selected is proportional to the fraction/percentage that that stratum is of the whole population

proxy measurement a measure that substitutes for another, using measures of outputs to measure outcomes

qualitative evaluation methods evaluation methods that rely on narrative, non-numerical, data

qualitative program evaluations evaluations that rely on words (instead of numbers) as the principal source of data

quality-adjusted life-years a method of estimating utility that assigns a preference weight to each health state, determines the time spent in each state, and estimates life expectancy as the sum of the products of each preference weight and the time spent in each state

quantitative evaluation methods evaluation methods that rely on numerical data

quasi-experimental research designs that do not involve random assignment to program and control groups, but do include comparisons that make it easier to sort out the cause and effect linkages that are being tested

random sample a sample that is selected at random from the population, where each member of the population has an equal or known chance of being selected, which enables the research results to be generalized to the whole population

randomized experiments research designs that involve randomly assigning units of analysis (usually people) to program and control groups

ratio level of measurement a level of measurement where there is a unit of measurement, that is, values of the variable are all equal intervals *and* there is a natural zero value in the scale

real benefits program benefits that represent net gains to society

real costs program costs that represent net losses to society

real prices prices that have been adjusted to remove the effect of inflation over time

recursive causal models causal models that specify one-way causal relationships among the variables in the model

relevance the extent to which the objectives of the program are connected to the assessed needs

reliability the extent to which a measurement instrument produces consistent results over repeated applications

representative sample   when the characteristics of a sample (demographic characteristics, for example) match those same characteristics for the population, we say that the sample is representative

research design   the overall method and procedures that specify the comparisons that will be made in the evaluation

resentful demoralization   a construct validity threat where the control group reacts negatively as a result of being the control group, which unintentionally biases the results of the evaluation

response process validity   the extent to which respondents to an instrument that is being validated demonstrate engagement and sincerity in the way that they participated

results-based management   a philosophy of management that emphasizes the importance of program or organizational results in managing the organization, its programs, and its people

rival hypotheses   factors in the environment of a program that operate on both the program and its intended outcomes in such a way that their effects could be mistaken for the outcomes that the program itself produces

sampling   process of selecting cases or units of analysis from a population so that we can generalize the findings from the sample to the population

sampling error   estimated range of population percentages that could be true, given a particular sample percentage—the greater the sample size, the smaller the sampling error

sampling strategy/procedure   the process through which a sample is selected

scientific standards/methods   research methods that lend themselves to repeatability and, hence, claims of objectivity

scrutability   characteristics of methods and procedures in research that make them transparent and replicable

secondary data   data that have been previously gathered for purposes other than a current needs assessment or evaluation

selection   an internal validity threat where differences between the program group and the control group before the program is implemented could account for observed differences in outcomes between the program and control groups

selection-based interactions   where selection interacts with other threats to internal validity to bias the results of the evaluation

sensemaking our innate capacity to construct patterns from sensory information

sensitivity analysis a process of calculating the benefits and costs using a range of possible discount rates to see what the maximum and minimum net present benefit is for a project, program, or policy

single time-series design a pretest posttest design, with no control group, where there are multiple observations before and after the program is implemented

skip factor a fixed number that defines how many cases need to be counted from a population list of all cases before the next case is drawn— skip factors are used in systematic sampling

snowball sampling sampling strategy where additional participants are identified based on information provided by previous participants

social scientific approach an approach to program evaluation that relies on the tools and methods of social science, and also relies on the belief that it is possible to construct and test hypotheses and theories about a social reality that exists independent of our perceptions of it

statistical conclusions validity the extent to which we can be confident that we have met the statistical requirements needed to calculate the existence and strength of the co-variation between the independent (cause) and dependent (effect) variables

statistical regression an internal validity threat where subjects' posttest scores tend to regress towards the average score for all persons in the population, usually because the pretest scores were extreme relative to others in the population

strategic goals high-level organizational goals that are intended to be a basis for the more specific objectives of programs

stratified sample a probabilistic (having an element of randomness in the selection process) sample that divides a population into groups or strata, and samples randomly from each one

strategy summary of activities that are intended to contribute to the achievement of an objective

summative evaluation an evaluation designed to provide feedback and advice about whether or not a program should be continued, expanded, or contracted

support costs costs that accrue to a program or project due to the support services and facilities that are used (examples include overhead costs, space rental, depreciation of equipment)

survey a measuring instrument where information is gathered from units of analysis (usually people) generally through the use of a questionnaire

systematic sample a sample drawn where the ratio of the population size compared to the sample size is used to calculate a skip factor (defined above as intervals from which cases are sampled)—the first case in the sample is randomly selected in the first interval and from that point onwards, each additional case is selected by counting a fixed number (the skip factor) and then selecting that case

tacit knowledge the capacity we have as human beings to integrate "facts"—data and perceptions—into patterns

tangible costs costs that can be easily expressed in dollar terms

technical efficiency attaining the most program outputs possible for each program input (usually expressed in monetary units)

temporal asymmetry where the independent variable precedes the dependent variable

testing a threat to internal validity where taking a pretest familiarizes the subjects with the measurement instrument and unintentionally biases their responses to the posttest

thematic analysis the process of categorizing ideas (words, phrases, sentences, paragraphs) in narrative data

treatment groups persons (usually) who are provided with a program or some other intervention that is being evaluated

triangulation the process of collecting data to answer an evaluation question from a variety of sources and/or using a variety of measurement procedures

unit of analysis the cases (often people) that are the main focus of an evaluation—we measure (observe) characteristics of the units of analysis and these observations become the data we analyze in the evaluation

units of effectiveness numerical measures of outcomes that can be counted (added and subtracted) and treated as interval measures

utilization the extent to which the program evaluation process and results (findings, conclusions, and recommendations) are deemed by stakeholders to be useful to them

validity   the extent to which a measuring instrument measures what it is intended to measure

value-for-money   an important goal for public officials and other stakeholders who are concerned with whether taxpayers and citizens are receiving efficient and effective programs and services for their tax dollars

value-for-money auditing   auditing that makes broad comparisons of the inputs (usually expressed in monetary units) to the results of a program or policy, with the purposes of judging the economy, efficiency, and effectiveness of a program

variable (dependent)   a variable that we expect will be affected by one or more independent variables—in most evaluations, the observed outcomes are dependent variables

variable (independent)   an observable characteristic of our units of analysis that we expect to cause some other variable—in a research design where we have a program and a control group, the presence or absence of the program for each unit of analysis becomes an independent variable

variables   in program evaluations and performance measurement systems, variables are the results of our efforts to measure constructs—they are the observables that are analyzed and reported across the units of analysis in our evaluation

verstehen   understanding based on a process of learning the subjective realities of people

INDEX

ABOUT THE AUTHORS

Laura Hawthorn holds a Master of Arts degree in Canadian history from Queen's University in Ontario, Canada and a Master of Public Administration degree from the University of Victoria. After completing her MPA, she worked for several years as Manager of Assessment Policy with the British Columbia Workers' Compensation Board. She is currently the proprietor of her own business in Vancouver, TFL Equine Enterprises, where she is following a lifelong dream of working with horses.

James C. McDavid (PhD, Indiana, 1975) is a professor of Public Administration at the University of Victoria in British Columbia, Canada. He is a specialist in program evaluation, performance measurement, and the production of local government services. He has conducted extensive research and evaluations focusing on federal, state, provincial, and local governments in the United States and Canada. His published research includes comparisons of the productivity and efficiency of local government service production across Canada, which has become a benchmark for comparisons of alternative production arrangements for local government services. He is currently a member of the editorial board of the *Canadian Journal of Program Evaluation.*

Dr. McDavid has organized and presented workshops for public sector managers on service production, program evaluation, performance measurement, contract management, and other related topics. In 2006, he will teach an online seminar in program evaluation and performance measurement, as part of the online MPA Program in the School of Public Administration.

In 1993, Dr. McDavid won the prestigious University of Victoria Alumni Association Teaching Award. In 1996, he won the J. E. Hodgetts Award for the best English-language article published in *Canadian Public Administration.* From 1990 to 1996, he was Dean of the Faculty of Human and Social Development at the University of Victoria. In 2004, he was named a Distinguished University Professor at the University of Victoria and was also Acting Director of the School of Public Administration during that year. In 2005, he was nominated by the Learning and Teaching Center at the University of Victoria for a national 3M Teaching Award.